Nikola Tesla: Colorado Springs Notes, 1899-1900

CONTENTS

INTRODUCTION

In 1898 Tesla's creativity in the field of high frequencies was at its peak. From his initial ideas in 1890 and his first, pioneering steps, he had worked with such intensity that many of the inventions and discoveries which he had given the world by this time have remained unsurpassed to this day. Even the loss of his laboratory on Fifth Avenue in 1895, a severe blow for him, did not hold him back for long. He soon resumed his experiments in a new laboratory, on Houston Street, continuing to make new discoveries and inventions applying them with unflagging energy.

Tesla's polyphase system essentially solved the problem of generating, transmitting and utilization of electrical power. When he started working on high frequencies, he almost immediately began to perceive their vast possibilities for wireless transmission of "intelligible signals and perhaps power". He worked on the practical development of his first ideas of 1891—1893 at such a rate that by 1897 he had already patented a system for wireless transmission of power and an apparatus utilizing this system. Shortly before this, during the ceremonial opening of the hydroelectric power plant on Niagara, at a time when the world was only just coming around to Tesla's polyphase system which for the first time in history enabled the transfer of electrical power over distance, he said: "In fact, progress in this field has given me fresh hope that I shall see the fulfillment of one of my fondest dreams; namely, the transmission of power from station to station without the employment of any connecting wires." [16]

Always true to the principle that ideas must be experimentally verified, Tesla set about building powerful high-frequency generators and making experiments in wireless power transmission. The Nikola Tesla Museum in Belgrade posesses a Tesla's own slide which confirms that the experiment described in the patent "System of transmission of electrical energy" [13] was in fact carried out before the Examiner-in-Chief of the U.S. Patent Office. For experimental verification of his method of wireless power transmission "by conduction through the intervening natural medium", on the global scale Tesla needed still higher voltages and more room (in the Houston Street laboratory he generated volages of 2 to 4 MV using a high-frequency transformer with a coil diameter of 244 cm), so towards the end of 1898 he began looking for a site for a new laboratory. Mid-1899 he finally decided on Colorado Springs, a plateau about 2000 m above sea level, where he erected a shed large enough to house a high-frequency transformer with a coil diameter of 15 meters!

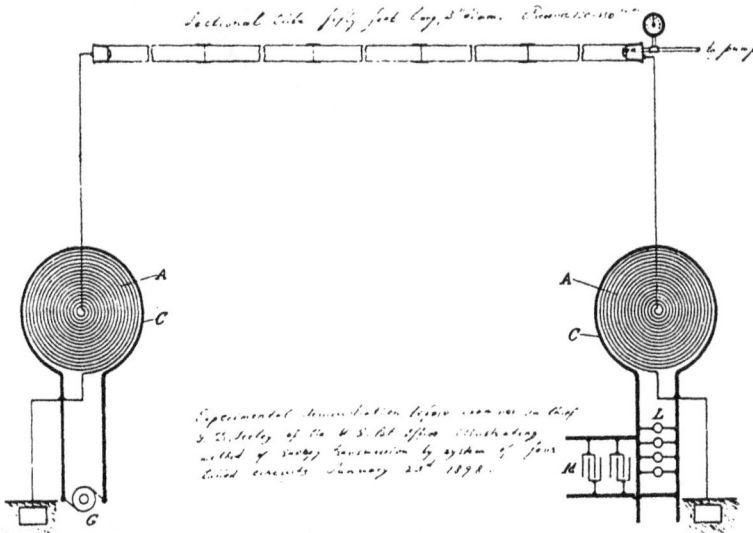

Fig. 1c. Diagram of an apparatus demonstrating transmission of electric power through rarified gas (Tesla's own slide now at the Nikola Tesla Museum, Belgrade)

Tesla's arrival in Colorado Springs was reported in the press. According to the Philadelphia "Engineering Mechanics" Tesla arrived on the 18th of May 1899 (according to[68] he left New York on 11th May 1899), with the intention of carrying out intensive research in wireless telegraphy and properties of the upper atmosphere. In his article "The transmission of electric energy without wires" (1904[1]) Tesla writes that he came to Colorado Springs with the following goals:

1. To develop a transmitter of great power.

2. To perfect means for individualizing and isolating the energy transmitted.

3. To ascertain the laws of propagation of currents through the earth and the atmosphere.

Tesla had some ten years of experience with high frequency AC behind him by the time he moved to Colorado Springs. In 1889, on his return from Pittsburg where he had been working as a consultant to Westinghouse on the development of his polyphase system, he began work on the construction of an alternator for generating currents at much higher frequencies than those used in ordinary power distribution. In 1890 he filed applications for two patents[2] for alternators working at over 10 kHz. One of these patents was in conjunction with a method for achieving quiet operation of arc lamps, but this was in fact a first step towards a new application of alternating currents, which soon became known as "Tesla currents". Tesla's alternators were an important milestone in electrical engineering and were the prototypes for alternators which were used some quarter-century later for driving high-power radio transmitters, and later on also for inductive heating.[24]

Soon after he had started his research in high frequencies Tesla discovered there specific physiological action and suggested the possibility of medical applications. He did a lot of work on the utilization of high frequency AC for electric lighting by means of rarefied gas tubes of various shapes and types. During 1891 he publicized his results in journals[3], patent applications[15] and in his famous lecture to the AIEE at Columbia College[4]. This lecture, before a gathering of eminent electrical engineers, brought Tesla widespread recognition and soon made him world-famous. This success was due in good measure to his convincing experiments too, which included a demonstration of rarefied gas luminescing in a tube not connected by wires to the source of power. This was the first experiment demonstrating wireless power transmission, and marked the birth of an idea to which Tesla was subsequently to devote a great part of his life. The necessary powerful electric field was created between the plates of a condenser connected across the secondary of a high-frequency transformer, whose was connected via a series condenser to a high-frequency alternator. The system worked best when the primary and secondary circuits were in resonance. Tesla also made use of the resonant transformer with his spark oscillator, enabling easy and efficient generation of high-frequency AC from a DC or low frequency source. This oscillator was to play a key role in the development of HF engineering. Only a few years later it was to be found among the apparatus of practically every physics laboratory, under the name of the Tesla coil[20].

The first record of Tesla's high-frequency coupled oscillatory circuit with an air--cored transformer is to be found in Patent No. 454622 of 23 June 1891 (application filed 25 April 1891) under the title "System of electric lighting". The oscillator converts low--frequency currents into "current of very high frequency and very high potential", which then supplies single-terminal lamps (see Fig. 2c). Induction coil PS produces a high secon-

No. 454,622.   Patented June 23, 1891.

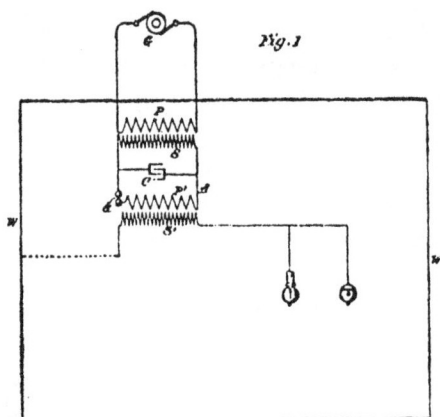

Fig. 2c. System of electric lighting

dary voltage which charges condenser C until a spark occurs across air gap a. The discharge current flows through the air gap and the primary of the high-frequency induction coil P'. The discharge of the condenser in this case differs from the discharge through coil with ohmic resistance studied by Henry[22], already known by that time. In Tesla's oscillator the energy of the high-frequency oscillations in the primary circuit is gradually transferred to the secondary circuit. The secondary circuit contains the distributed capacity of the secondary winding and the wiring and the capacity of the load, and is thus also a resonant circuit. After energizing of the secondary circuit, the remaining energy is returned to the primary, then back to the secondary, and so on until losses reduce it sufficiently to interrupt spark across a in the primary circuit. Then condenser C begins to recharge from source G via induction coil (transformer) PS, Oberbeck[29] published a theoretical analysis of Tesla's oscillator in 1895.

Tesla presented much new information about his discharge oscillators and his further research on high frequency currents in the lecture he gave to the IEE in London, February 1892, which he subsequently repeated in London and then in Paris[5]. He described at length the construction of a type of air-cored HF transformer and drew attention to the fact that the secondary voltage cannot even approximately be estimated from the primary/secondary turns ratio. Tesla also did a lot of work on improvements of the spark gap and described several designs, some of which were subsequently attributed to other authors[24]. In describing the apparatus with which he illustrated this lecture he explained several ways for interrupting arcs with the aid of a powerful magnetic field; using compressed air; multiple air gaps in series; single or multiple air gaps with rotating surfaces.

He describes how the capacity in the primary and secondary circuits of the HF transformer should be adjusted to get the maximum performance, stating that so far insufficient attention had been paid to this factor. He experimentally established that the secondary voltage could be increased by adding capacity to "compensate" the inductance of the secondary (resonant transformer).

He demonstrated several single-pole lamps which were connected to the secondary, describing the famous brush-discharge tube and expressing the opinion that it might find application in telegraphy. He noted that HF current readily passes through slightly rarefied gas and suggested that this might be used for driving motors and lamps at considerable distance from the source, the high-frequency resonant transformer being an important component of such a system.

The drawing shown in Fig. 3c dates from early on during Tesla's work with high frequencies, 1891—1892. It is taken from Tesla's original slide found in the archives of

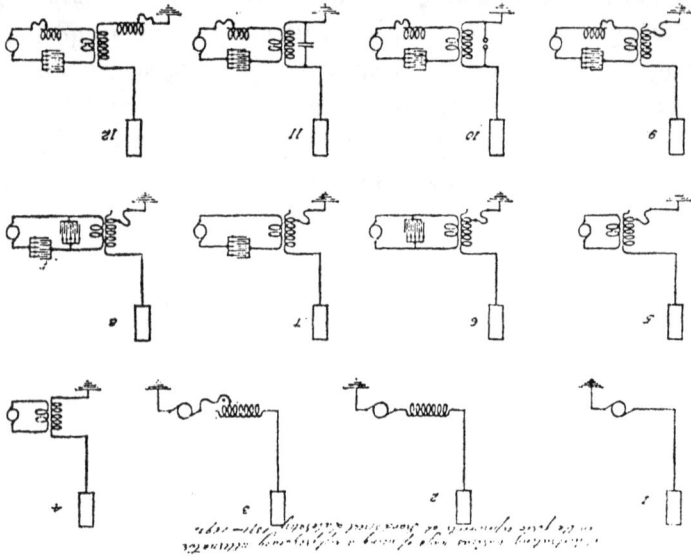

Fig. 3c. Various connections of HF transformer used by Tesla in 1891—1892
(Tesla's own slide now at the Nikola Tesla Museum, Belgrade)

the Nikola Tesla Museum in Belgrade. According to Tesla's caption these diagrams are "Illustrating various ways of using highfrequency alternator in the first experiment at Grand Street Laboratory 1891—1892". It seems that Tesla made these to prove his priority in a patent suit[35]. Only some of these diagrams have been published in [4, 6, 13], so that this is an important document throwing new light on an exceptionally fertile but relatively little known period of Tesla's work. It is, for example, clear from these diagrams that he introduced an HF transformer in the open antenna circuit. Circuits like that in Fig. 3c—4 are to be found later in two patents filed in 1897[13,14] on his apparatus and system for wireless transmission of power (these patents refer to Tesla's disruptive discharge oscillator as an alternative to the high-frequency alternator).

In February 1893 Tesla held a third lecture on high-frequency currents before the Franklin Institute in Philadelphia[6], and repeated it in March before the National Electric Light Association in St. Louis. The most significant part of this lecture is that which refers to a system for "transmitting intelligence or perhaps power, to any distance through the earth or environing medium". What Tesla described here is often taken to be the foundation of radio engineering, since it embodies principles ideas of fundamental importance, viz.: the principle of adjusting for resonance to get maximum sensitivity and selective reception, inductive link between the driver and the tank circuit, an antenna circuit in which the antenna appears as a capacitive load[71]. He also correctly noted the importance of the choice of the HF frequency and the advantages of a continuous carrier for transmitting signals over great distances[12].

Between 1893 and 1898 Tesla applied for and was granted seven American patents on his HF oscillator as a whole[25], one on his HF transformer[26], and eight on various types of electric circuit controller[27]. In a later article[28] Tesla reviews his work on HF oscillators and reports that over a period of eight years from 1891 on he made no less than fifty types of oscillator powered either by DC or low-frequency AC.

Along with his work on the improvement of his HF oscillators Tesla was continuously exploring applications of the currents they produced. His work on the improvement of X-ray generating apparatus is well known — he reported it in a series of articles in 1896 and 1897[7] and in a lecture to the New York Academy of Sciences[17]. In a lecture before the American Electro-Therapeutic Association in Buffalo September 1898[18] he described applications of the HF oscillator for therapeutic and other purposes. The same year he took out his famous patent "Method of and apparatus for controlling mechanism of moving vessels or vehicles"[59], which embodies the basic principles of telemechanics a field which only began to develop several decades after Tesla's invention.

On 2nd September 1897 Tesla filed patent application No. 650343, subsequently granted* as patent No. 645576 of 20th March 1900[13] and patent No. 649621 of 15th May 1900[14]. Unlike other radio experimenters of the time who worked either with damped oscillations at very high frequencies[43], Tesla investigated undamped oscillations in the low HF range. While others principally developed Hertz's apparatus with a spark-gap in the tank circuit (Lodge, Righi, Marconi, and others) and improved the receiver by

---

* The second of the two patents by which Tesla protected his apparatus for wireless power transmission, known as the "system of four tuned circuits", is particularly important in the history of radio. It was a subject of a long law suit between the Marconi Wireless Telegraph Company of America and the United States of America alleged to had used wireless devices that infringed on Marconi's patent No. 763772 of 28th June 1904. After 27 years the U.S. Supreme Court in 1943 invalidated the fundamental radio patent of Marconi as containing nothing which was not already contained in patents granted to Lodge, Tesla and Stone[65].

introducing a sensitive coherer (Branly, Lodge, Popov, Marconi, and others), he set about implementing his ideas of 1892—1893. How far he had got in verifying his ideas for wireless power transmission before coming to Colorado Springs may be seen from patent No. 645576 and the diagram in Fig. 1c.

Tesla based his hopes for wireless power transmission on the global scale on the principle that a gas at low pressure is an excellent conductor for high frequency currents. Since the limiting pressure at which the gas becomes a good conductor is higher the higher the voltage, he maintained that it would not be necessary to elevate a metal conductor to an altitude of some 15 miles above sea level, but that layers of the atmosphere which could be good conductors could be reached by a conductor (in fact an aerial) at much lower altitudes. "Expressed briefly, (cit. patent 645576) my present invention, based upon these discoveries, consists then in producing at one point an electrical pressure of such character and magnitude as to cause thereby a current to traverse elevated strata of the air between the point of generation and a distant point at which the energy is to be received and utilized". Figure 1c proves that Tesla did actually carry out an experimental demonstration of power transmission through rarefied gas before an official of the Patent Office. From the patent it may be seen that the pressure in the tube was between 120 and 150 mm Hg. At this pressure, and with the circuits tuned to resonance, efficient power transfer was acheived with a voltage of 2—4 million volts on the transmitter aerial. In the application Tesla also claims patent rights to another, similar method of transmission, also using the Earth as one conductor, and rendered conductive high layers of the atmosphere as the other*.

Tesla spent about eight months in Colorado Springs. Something of his work and results from this period can be gleaned from articles in "American Inventor" and "Western Electrican". For instance, it is stated that Tesla intended to carry out wireless transmission of signals to Paris in 1900. An article of November 1899 reports that he was making rapid progress with his system for wireless transmission of signals and that there was no way of interfering with messages sent by it. Tesla returned to New York on the 11th of January 1900[68].

The diary which Tesla kept at that time gives a detailed day-by-day description of his research in the period from 1st June 1899 to 7th January 1900. Unlike many other records in the archives of the Nikola Tesla Museum in Belgrade, the Colorado Springs diary is continuous and orderly. Since it was not intended for publication, Tesla probably kept it as a way of recording his research results. It could perhaps also have been a safety measure in case the laboratory should get destroyed, an eventuality by no means unlikely considering the dangerous experiments he was performing with powerful discharges. Some days he made no entries, but usually explained why at the beginning of the month.

---

* In the late eighties of the last century very little was known about the radiation and propagation of electromagnetic waves. Following the publication of Hertz's research[23] in 1888, which provided confirmation of Maxwell's dynamic theory of the electromagnetic field published in 1865[60], scientists became more and more convinced that electromagnetic waves behaved like light waves, propagating in straight lines. This led to pessimistic conclusions about the possible range of radio stations, which were soon refuted by experiments using the aerial-earth system designed by Tesla in 1893[6]. Tesla did not go along with the general opinion that without wires "electrical vibrations" could only propagate in straight lines, being convinced that the globe was a good conductor through which electric power could be transmitted. He also suggested that the "upper strata of the air are conducting" (1893), and "that air strata at very moderate altitudes, which are easily accessible, offer, to all experimental evidence, a perfect conducting path" (1900)[41]. It is interesting to note that this mode of propagation of radio waves was initially considered as something different from other modes[61] then to be forgotten until recent years. In the 1950's Schumann, Bremmer, Budden, Wait, Galejs and other authors[34], working on the propagation of very low (3 to 30 kHz) and extremely low (1 to 3000 Hz) electromagnetic waves, founded their treatment on essentialy the same principles as Tesla.

According to his notes, Tesla devoted the greatest proportion of his time (about 56%) to the transmitter, i.e. the high-power HF generator, about 21% to developing receivers for small signals, about 16% to measuring the capacity of the vertical antenna, and about 6% to miscellaneous other research. He developed a large HF oscillator with three oscillatory circuits with which he generated voltages of the order of 10 million volts. He tried out various modifications of the reciever with one or two coherers and special preexcitation circuits. He made measurements of the electromagnetic radiations generated by natural electrical discharges, developed radio measurement methods, and worked on the design of modulators, shunt-fed antennas, etc.

The last few days covered by the diary Tesla devoted to photographing the laboratory inside and out. He describes 63 photographs in all, most of them showing the large oscillator in action with masses of streamers emerging from the outer windings of the secondary and the "extra coil". He probably derived special satisfaction from observing his artificial lightning, now a hundred times longer than the small sparks produced by his first oscillator in the Grand Street Laboratory in New York. By then many leading scientists had been experimenting with "Tesla" currents but Tesla himself was still in the vanguard with new and unexpected results. When he finally finished his work in Colorado Springs he published some photographs of the oscillator in a blaze of streamers causing as much astonishment as had those from his famous lectures in the USA, England and France in 1891—1893. The famous German scientist Slaby wrote that the apparatuses of other radio experimenters were mere toys in comparison with Tesla's in Colorado Springs.

The descriptions of the photographs in the diary also include detailed explanations of the circuitry and the operating conditions of the oscillator. The photographs themselves give an impressive picture of the scale of these experiments. Tesla maintains that bright patches on some of the photographs were a consequence of artificially generated fireballs. He also put forward a theory to explain this, still today somewhat enigmatic phenomenon. Research on fireballs was not envisaged in his Colorado Springs work plan, but belonged to the special experiments which, in his own words, "were of an interest, purely scientific, at that time"[68], which he carried out when he could spare the time.

Tesla used some parts of the diary in drawing up the patent applications which he filed between 1899 and 1902. Keeping such notes of his work was more a less a constant practice; they provided him with an aide-mémoire when preparing to publish his discoveries.

The diary includes some descriptions of nature, mostly the surroundings of the laboratory and some meteorological phenomena, but only with the intention of bringing out certain facts of relevance to his current or planned research.

Immediately after he finished work at Colorado Springs Tesla wrote a long article entitled "The problem of increasing human energy" in which he often mentions his results from Colorado Springs[41]. In 1902 he described how he worked on this article[68]: "The Century" began to press me very hard for completing the article which I have promised to them, and the text of this article required all my energies. I knew that the article would pass into history as I brought, for the first time, results before the world which were far beyond anything that was attempted before, either by myself or others".

The article really did create a sensation, and was reprinted and cited many times. The style he uses in describing Colorado Springs research differs greatly from that of the diary.

Tesla wrote about his Colorado Springs work again in 1904[1]. Some interesting data is to be found in his replies before the United States Patent Office in 1902, in connection with a patent rights dispute between Tesla and Fessenden[68]. This document includes statements by Tesla's assistant Fritz Lowenstein and secretary George Scherff. Tesla took Lowenstein on in New York in April 1899. At the end of May that year he summoned him to Colorado Springs, where Lowenstein remained until the end of September, when family matters obliged him to return to Germany. Tesla was satisfied with him as an assistant and asked him to return later, which he did, again becoming Tesla's assistant in February 1902.

Tesla did not break off his research in the field of radio after visiting Colorado Springs. Upon returning to New York on the 11th of January 1900[68] he took energetic steps to get backing for the implementation of a system of "World Telegraphy". He erected a building and an antenna on Long Island, and started fitting out a new laboratory. From his subsequent notes we learn that he intended to verify his ideas about resonance of the Earth's globe, referred to in a patent of 1900[42]. The experiments he wanted to perform were not in fact carried out until the sixties of this century, when it was found that the Earth resonates at 8, 14 and 20 Hz.[34] Tesla predicted that the resonances would be at 6, 18 and 30 Hz. His preoccupation with this great idea slowed down the construction of his overseas radio station, and when radio transmission across the Atlantic was finally acheived with a simpler apparatus, he had to admit that his plans included not only the transmission of signals over large distances but also an attempt to transmit power without wires. Commenting on Tesla's undertaking, one of the world's leading experts in this field, Wait[21], has written: ... "From an historical standpoint, it is significant that the genius Nikola Tesla envisaged a world wide communication system using a huge spark gap transmitter located in Colorado Springs in 1899. A few years later he built a large facility in Long Island that he hoped would transmit signals to the Cornish coast of England. In addition, he proposed to use a modified version of the system to distribute power to all points of the globe. Unfortunately, his sponsor, J. Pierpont Morgan, terminated his support at about this time. A factor here was Marconi's successful demonstration in 1901 of transatlantic signal transmission using much simpler and far cheaper instrumentation. Nevertheless, many of Tesla's early experiments have an intriguing similarity with later developments in ELF communications.

Tesla proposed that the earth itself could be set into a resonant mode at frequencies of the order of 10 Hz. He suggested that energy was reflected at the antipode of his Colorado Springs transmitter in such a manner that standing wave were set up."

In a letter to Morgan[69] early in 1902 Tesla explained his research, in which he envisaged three "distinct steps to be made: 1) the transmission of minute amounts of energy and the production of feeble effects, barely perceptible by sensitive devices; 2) the transmission of notable amounts of energy dispensing with the necessity of sensitive devices and enabling the positive operation of any kind of apparatus requiring a small amount of power; and 3) the transmission of power in amounts of industrial significance. With the completion of my present undertaking the first step will be made". For the experiments with transmission of large power he envisaged the construction of a plant at Niagara to generate about 100 million volts.[1]

However, Tesla did not succeed in getting the necessary financial backing, and after three years of abortive effort to finish his Long Island station he gave up his plans and turned to other fields of research. He wrote several times about his great idea for wireless transmission of power, and remained convinced to his death that it would one day become reality. Today, when we have proof of the Earth's resonant modes (Schumann's resonances[34]), and it is known that certain waves can propagate with very little attenuation, so little that standing waves can be set up in the Earth-ionosphere system, we can judge how right Tesla was when he said that the mechanism of electromagnetic wave propagation in "his system" was not the same as in Hertz's system with collimated radiation. Naturally, Tesla could not have known that the phenomena he was talking about would only become pronounced at very low frequencies, because it seems he was never able to carry out the experiments which he had so brilliantly planned, as early as 1893.[6] It is gratifying that after so many years Tesla's name is rightfully reappearing in papers dealing with the propagation of radio waves and the resonance of the Earth[21, 54, 62, 72]. In a recent book of a well known scientist (Jackson[54]) it is stated that "this remarkable genius clearly outlines the idea of the earth as a resonating circuit (he did not know of the ionosphere), estimates the lowest resonant frequency as 6 Hz (close to 6.6 Hz for a perfectly conducting sphere), and describes generation and detection of these waves. I thank V.L. Fitch for this fascinating piece of history". We believe that further studies of Tesla's writings will reveal some interesting details of his ideas in this field.

The publication of the Colorado Springs diary, a unique record of the work of a genius, means an enrichment of the scientific literature, not only in that throws light on a particularly interesting period of Tesla's creativity, but also as a source for the study of his work as a whole, and particularly of his part in the development of radio. It also facilitates the identification of many documents now at the Nikola Tesla Museum in Belgrade which lack date or description.

The preparation of this manuscript for publication required considerable time and labor in order to present its content in a form not deviating essentially from the original but more accessible to study. No alterations have been made even where the original contains certain minor errors, sometimes also in the use of power and energy units; some more important calculation errors which influence the conclusions drawn are also reproduced but are noted. A section at the end of the book contains commentaries on the Diary with explanatory notes, and a survey of his earlier work and that of other researchers. For these commentaries reference was made to the large body of literature and documents in the archives of the Nikola Tesla Museum in Belgrade.

Aleksandar Marinčić

Nikola Tesla
Colorado Springs Notes
1899-1900

Colorado Springs Notes

June 1—30, 1899

To this to be added two applications filed with Curtis and some other patent matters chiefly foreign.

Colorado Springs   June 4. 1899.

Telephony without wires.

General observations.

Senders { one impulse / several impulses } for one telephone impulse.   Receiver { several impulses / one impulse } for one telephone impulse.

Form of Energy Sender {
Static action
Current "
Rays { Light / heat / Roentgen etc.
Magnetic act.
Sound "
}

Form of energy affecting receiver {
Static
too current attraction, repulsion
Magnetic
Kathode rays
heat "
light "
Kathode mechanical action
deflection by mech. displacement.
}

Instruments " to be used {
Static machines
Induction coil
Oscillator - { single terminal / two terminals
Batteries,
Dynamos {high frequency} {alt. / direc
Condenser {Rheostatic mech. / Oscillator
}

Receiving Instruments suited for the various transmitting instruments } to be worked out.

Arrangement of circuit etc. to be worked out.

The following seems to be the best plan for constructing small batteries of very high e.m.f. required for exciting vacuum tube to be used as receiver in telegraphy: As

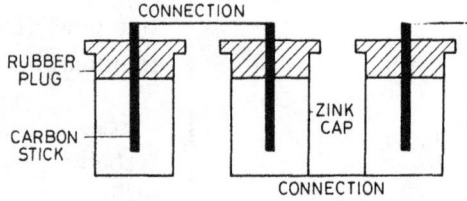

the current for exciting the tube need be only very small the battery can furnish a minute current. From previous experiments about $\dfrac{1}{20,000}$ amp. is amply sufficient. Approximate dimensions of box 1/4 cu. foot. The price will not be prohibitory. Tin caps, plugs and carbons will be readily obtainable.

The connection of the receiver is to be as in experiments in New York: If necessary the resistance R_1 will be used to strain the tube exactly just to point of breaking down. It is very important as in all sensitive devices so far used that the dielectric is strained

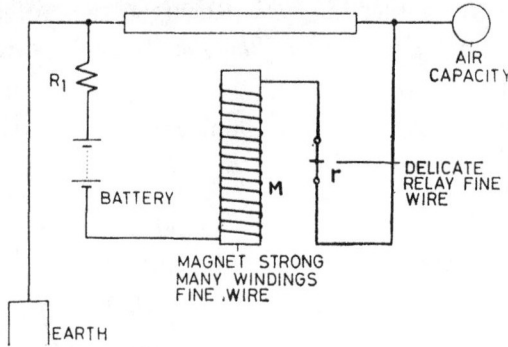

exactly to the breaking point. The magnet M is to have a resistance nearly equal to the internal resistance of the battery, so as to get best output. The relay will suit as it is with 1000 ohms resistance. The magnet must be strong to blow out tube when lighted. This device should by *very sensitive* and should break down by very minute currents propagated through the earth from a similarly connected oscillator.

Colorado Springs

June 2, 1899

- **Signalling Valuable Uses of Condenser Methods**
  - Telegraphy
    - Ships
    - Land
    - Cables (this very much so, but powerful apparatus required).
  - Protection
    - Ships } Collisions
    - Icebergs } Collisions
- **Research**
  - Animal, Plant
    - Electricity
  - Terrestrial Disturbances
    - Magnetism
    - Static } Electricity
    - Atmosph. } Electricity
    - Earth currents
    - Solar influence etc.
  - Locating on deposits
    - Magnetic
    - Non-magnetic by reflection
  - In connection with Roentgen rays, other rays and *dark rays of the sun.* **Most important.**
- **Measurements etc.**
  - Resist., current., e.m.f. etc.
  - Intensities, light, heat etc.
  - Meters
    - power
    - current
    - integrators of all kinds

Colorado Springs

June 3, 1899

Various modifications of a principle consisting in accumulating energy of feeble impulses received from a distance and utilizing magnified effect for operating a receiving device. Several modes of carrying out the same, generally considered:

Resonance —

- Condenser
  - Commutating currents and charging
  - Directing currents { by batteries / gas valves } and charging
  - Using direct currents of high tension
- Magnet
  - Commutating currents for relay
  - Directing currents through relay { batteries / valves }
- Dynamo principle
  - Commutating currents } field
  - Directing { batteries / valves } } direct
  - Using direct currents } current
  - Low frequency currents in field
    - High frequency currents in armature
      - heating device, electromagnet, condenser, etc.
    - Direct currents in armature
      - relay, field magnet, dynamo, condenser, etc.

Colorado Springs

June 4, 1899

Telephony without wires.
General observations:

Senders	one impulse several impulses	for one telephone impulse.
Receiver	several impulses one impulse	for one telephone impulse

Form of energy sender	Static action			Static Two currents attraction, repulsion Magnetic Cathode rays Heat rays Light rays Cathode mechanical action Deflection by mech. displacement
	Current action			
	Rays	light heat Roentgen etc.	Form of energy affecting receiver	
	Magnetic action			
	Sound action			

Instruments to be used	Static machine		Receiving instruments suited for the various transmitting instruments	to be worked out
	Induction coil			
	Oscillator	single terminal two terminals		
	Batteries,			
	Dynamos (high frequency)	alt. direct		
	Condenser	Rheostatic Mach. Oscillator		

Arrangements of circuit etc. to be worked out.

Colorado Springs

June 5, 1899

Induction method; results with apparatus to be used calculated from

$$M = \frac{p^2 \, s^4 \, W_p \, V_t^2}{32^2 \, D^6 \, S^2}$$ (This formula is very problematical)

M = Power in secondary or receiving circuit
p = $2\pi n$, estimated $40{,}000 = 4 \times 10^4$
s = length of side of square circuit $= 1200 = 12 \times 10^2$ cm.
W_p = power spent in primary $= 4 \times 10^{10}$ ergs assumed.
V_t = total volume of wire in both circuits $= 25 \times 10^3$ cu.cm.
D = distance from center to center of circuits (horizontal)
S = specific resistance of wires $= 1.7 \times 10^3$

Taking M to be the minimum 0.3 ergs to affect relay, it is found that with above circuits and under such conditions about 1 mile communications should be possible. With circuits 1000 meters square, about 30 miles. From this, the inferiority of the induction method would appear immense as compared with disturbed charge of ground and air method.

Colorado Springs

June 6, 1899

Arrangements with single terminal tube for production of powerful rays.

There being practically no limit to the power of an oscillator, it is now the problem to work out a tube so that it can stand any desired pressure. The tubes worked with in New York were made either with aluminium caps or without same, but in both cases a limit was found so that but a small fraction of the obtainable e.m.f. was available. If of glass, the bottom would break through owing to streamers, and if an aluminium cap were employed there would be sparking to the cap. Immersion in oil is inconvenient, likewise other expedients of this kind. The best results will probably be obtained in the end by static screening of the vulnerable parts of the tube. This idea was experimented upon in a number of ways. It is now proposed to test the arrangements indicated below:

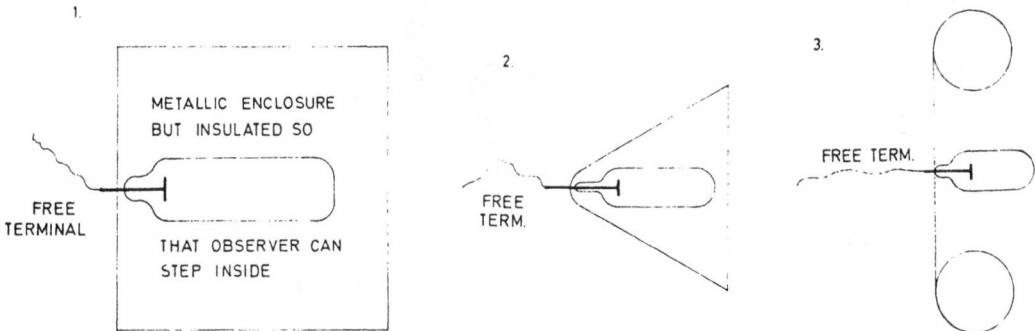

In each case there would be an insulated body of capacity so arranged that the streamers can not manifest themselves. The capacity would be such as to bring about maximum rise of e.m.f. on the free terminal.

Colorado Springs

June 7, 1899

Approximate estimate of a primary turn to be used in experimental station.

$$L_s = \pi \left[4\,A \left(\log_e \frac{8\,A}{a} - 2 \right) + 2\,a \left(\log_e \frac{8\,A}{a} - \frac{5}{4} \right) - \frac{a^2}{16\,A} \left(2 \log_e \frac{8\,A}{a} + 19 \right) \right]$$

Here A radius of circle $= 25$ feet $= 300$ inch $= 300 \times 2.54 = 762$ cm.

a ,, ,, cable $= \dfrac{13''}{32} = \dfrac{13}{32} \times 2.54 = 1.03$ cm.

$\dfrac{8\,A}{a} = 5919 \qquad \log_e \dfrac{8\,A}{a} = 3.772248 \times 2.3 = 8.6762$

$4A = 3048 \qquad 2a = 2.06 \qquad a^2 = 1.061; \qquad 16A = 12,192$

$$L_s = \pi \left[3048 \times (8.6762 - 2) + 2.06 \times (8.6762 - 1.25) - \frac{1.061}{12,192} \times (17.3524 + 19) \right]$$

last term being negligible, we have

$L_s = 3.1416 \times (3048 \times 6.6762 + 2.06 \times 7.4262) =$

$\quad = 3.1416 \times (20,349.06 + 15.3) = 3.1416 \times 20,364.36 =$

$L_s = 63,976.67$ cm.

or approx: **63,900 cm.**

Two turns in series should be approx. 255,600 cm.

Approximate estimate of inductance of primary loop used in experimental oscillator on vertical frame in New York.

Diameter of loop $= 8$ feet $= 244$ cm.

This gives $A = 122$ cm $\qquad$ diam. of cable $= \dfrac{13''}{16}$

$a = \dfrac{13''}{32} \times 2.54 = 1$ cm. nearly

$\dfrac{8\,A}{a} = 976 \qquad \log_e \dfrac{8\,A}{a} = 2.98945 \times 2.3 = 6.875735$

$a^2 = 1 \qquad 4A = 488 \qquad 16A = 1952$

$$L = \pi \left[4A \left(\log_e \frac{8A}{a} - 2 \right) + 2a \left(\log_e \frac{8A}{a} - \frac{5}{4} \right) - \frac{a^2}{16A} \left(2 \log_e \frac{8A}{a} + 19 \right) \right]$$

$$= \pi \left(488 \times 4.875735 + 2 \times 5.625735 - \frac{1}{1952} \times 32.75 \right) =$$

$$= \pi (2379.3600 + 11.2515 - \text{small fraction})$$

$$= \pi \times 2390.6115 = 3.1416 \times 2390.6115 \qquad \textbf{L} = \textbf{7210.345 cm.}$$

this is a trifle more with ends close enough **8,000 cm.**

Colorado Springs

June 8, 1899

Method and apparatus for determining self-inductances, also capacities,
particularly suitable for determining small inductances.

Since resistance can be neglected when the frequency of the currents is high the
inductances can be easily compared in the following way:

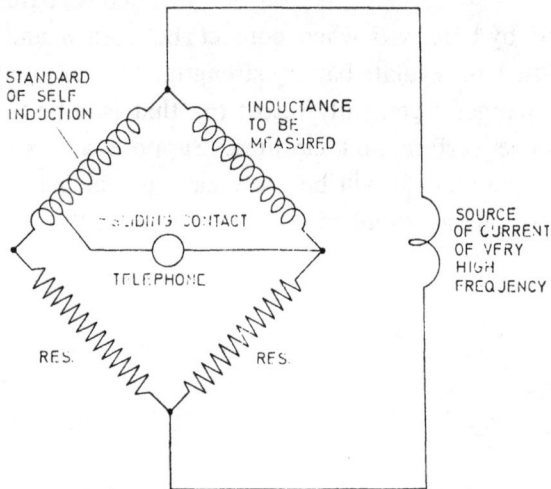

A standard of self-induction is provided with a sliding contact so that any number of
turns can be inserted. Two resistances, suitable for the source of high frequency current
and the inductance to be measured, are connected in the manner of a bridge. The two
opposite points, one movable, are connected through a telephone. When no sound is
heard then we have the two inductances — that is, the one to be measured, and that part
of the standard which corresponds to equilibrium or silence in the telephone-practically
equal if the quantities are suitably chosen. By determining inductances capacities may
from these be easily measured. It is possible that the high frequency source might be dispensed with and a *very sudden* discharge of a condenser passed through instead. The
auxiliary resistance should be so determined that the resistances in the two parts through
which the current divides are equal or nearly so.

Consider the practicability of using a column of air or other gas as detector of disturbances from a distance. This would be on the principle of the Ries thermometer as experimented with in New York. The arrangement of apparatus is illustrated in the diagram below. There is a reservoir V, preferably of polished surface, made in the manner of mirrors to reflect rays to center. In this reservoir is placed a resistance r of *minute mass*.

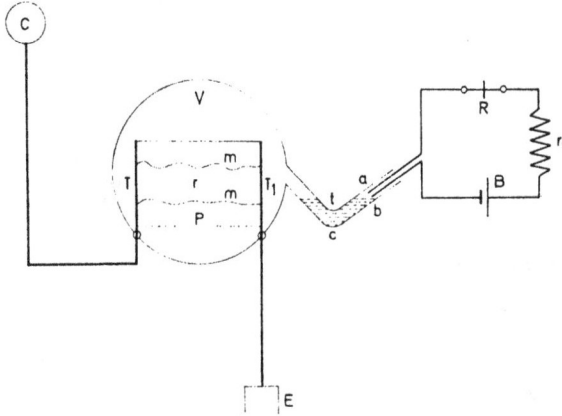

This resistance may be conveniently obtained by connecting with pencil marks $m\,m$ two terminals T and T_1, holding a glass plate P. The *mass must* be minute so that the smallest amount of current would raise the temperature of the marks or conductors and thus heat the air in the reservoir which, expanding, would drive a minute column of liquid c contained in tube t towards contacts $a\,b$. The liquid should be very light and need not be highly conducting, barely enough to allow the relay R to be worked by battery B when contact between a and b is established. The resistance r_1 may be used to regulate battery strength. The terminals $T\,T_1$ I would preferably connect in the manner I generally resort to, that is, one to the ground and the other to a body of some surface and elevated. Suppose air is used, we would want 0.1696 Ca per °C per gramme. It will be now easy to calculate how much the air can be expanded per erg of energy supplied.

(to work out)

Suppose a *very fine* mercury column were prepared of resistance R, length L and connected to a ground and an elevated insulated conductor of capacity in the manner illustrated in diagram.

Then if a current I be passed through it the energy lost in the column and converted into heat will be RI^2 watts. The current is, of course, minute and we could scarcely calculate on more than 1 erg in telegraphy being transmitted to a great distance from the transmitting station; the question is what can be done with that little amount of energy.

If the mercury be raised to a temperature t degrees above normal it will expand for each degree 0.00018 of its length hence its length will be $L+Lt\times0.00018 = L(1+0.00018\,t)$. This value is a little greater than would actually be found in a glass tube. Suppose the tube were $\dfrac{1}{10}$ mm. diam. and 10 meters long; its resistance would be approx.

1000 ohm. Then $RI^2=1000\ I^2=\dfrac{1}{10^7}$, taking one erg as energy supplied, gives $I^2=\dfrac{1}{10^{10}}$

$I=\dfrac{1}{10^5}$ amp. The column in the tube would expand for one degree $0.00018\times10=0.0018$ meters or 0.18 cm or 1.8 mm.

The volume of the column would be $0.01\times10,000=100$ cu.mm. or 0.1 cu.cm. Now this would weigh $0.1\times13.6=1.36$ gramme. The mass of this would be $\dfrac{0.00136}{9.81}$. Now to raise water $1°$C we want 41,600,000 ergs per gramme. Specific heat of mercury being 0.0319 we would want $41,600,000\times$ $\times\dfrac{320}{10,000}=41,600\times32=$ about 1,330,000 ergs.

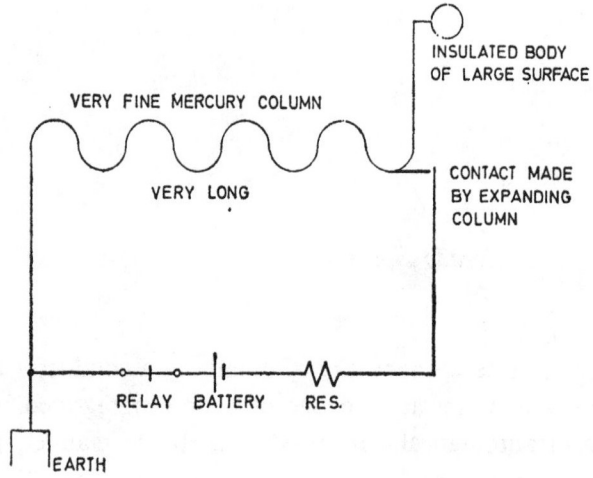

This shows that on the above assumptions, indication of disturbances by mercury column would be hardly practicable unless the column could be made *much thinner*.

Colorado Springs

June 11, 1899

The following method and apparatus for detecting feeble disturbances transmitted through a medium seem to be particularly adapted for telegraphy. The idea was followed up in New York but results were not satisfactory. Now the experiments are to be resumed with apparatus as illustrated below.

The general idea is to provide a path for the passage of a current such that it will diminish in resistance when the current passes and also such that it will be of as minute a mass as possible. The specific heat of the material forming the path for the current should also be as small as possible. The best way I have so far found is to make a mark of the required thickness with a carbonstick so as to connect two terminals through a conductor of high resistance so deposited. This conductor I preferably connect with one end to the earth and with the other to an elevated object of a large surface. The conductor is further connected in circuit with relays and batteries in any way suitable, as for instance in the arrangement here shown. Now when a feeble impulse passes it reduces the resi-

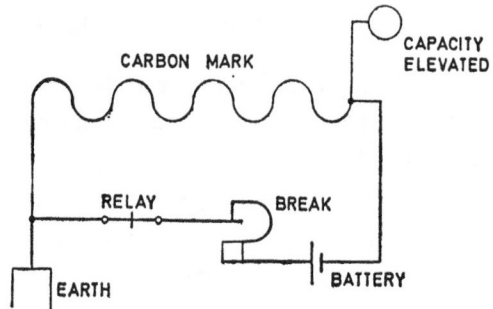

stance of the carbon and more of the battery current can pass through and so on until the relay is brought into action. The relay then, in any way suitable, breaks the current of the battery and a normal regime is established. The relay itself may be utilized to break the current or an auxiliary magnet may be employed as illustrated. The carbon mark may be connected in the manner *of a bridge* to increase sensitiveness.

This to be followed up.

Colorado Springs

June 12, 1899

A convenient way of obtaining a conductor (rather a poor one) of small mass, such as will be instantly evaporated or disintegrated by a battery current, and one which is also automatically renewed in a simple manner, is the following:

Two terminals are fastened to an insulating plate, preferably of glass, and provision is made once for a film of poorly conducting substance to be deposited on the plate thus bridging the terminals and establishing sufficient contact between them to allow a current to pass.

The best manner to carry this idea out seems to be the following:

In a small bottle, having a stopper with two terminals, is placed a quantity of iodine and the bottle is by any suitable means kept at a temperature such that the haloid is deposited in an exceedingly fine film causing a leak of the battery current through a relay. A stronger current may then be passed by establishing a suitable connection with the relay and the film of iodine may thus be destroyed and the terminals again insulated, this process being repeated in as rapid succession as may be desired. This film may be used in the detection of feeble impulses as in telegraphy through media, in which case it is connected to ground and capacity.

Colorado Springs

June 13, 1899

Arrangements of transmitting apparatus for telephony at a distance without wires.

The most difficult part in the practical solution of a problem of this kind of telephony is to control a powerful apparatus by feeble impulses such as are produceable by the human voice.

One of the best ways is to use carbon contacts as in the microphone, but when powerful currents either of great volume or high e.m.f. are used, as they must be in such cases, the problem offers great difficulties.

A solution which I have before described is offered in the following scheme, illustrated diagramatically below.

S is a source of preferably direct current as a powerful battery or dynamo, C a condenser which is connected with a primary p and break d as usual in an oscillator. The break d is such that at the number of breaks resonance is obtained.

A secondary s is provided which is connected to the ground and an insulated body of capacity and elevated as shown, and normally the adjustment is such that the secondary with its capacity and self-induction is in resonance with the primary p. From the latter a shunt is made by two contacts c c preferably of carbon. Normally these carbons touch loosely but by speaking upon funnel f they are harmonically pressed together and the primary current is diverted thus destroying the resonance and greatly diminishing effect in the secondary rhythmically with the undulations of the voice. In this manner very minute variations in the contact resistance are made to produce great variations in the intensity of the waves sent out. The breaks at d must be much above undulations of the voice.

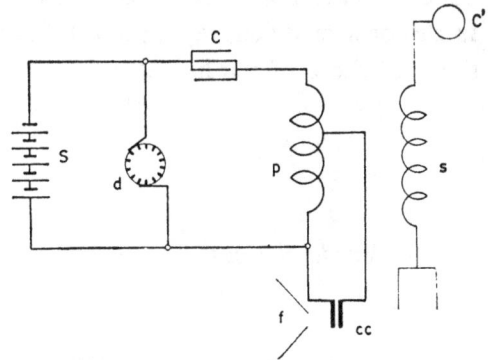

Colorado Springs

June 14, 1899

The following arrangement, considered before in a general way, seems to be particularly suitable for telephony at a distance without wires and for such purposes where it is necessary to effect control of a powerful apparatus by feeble impulses such as those produced by a human voice.

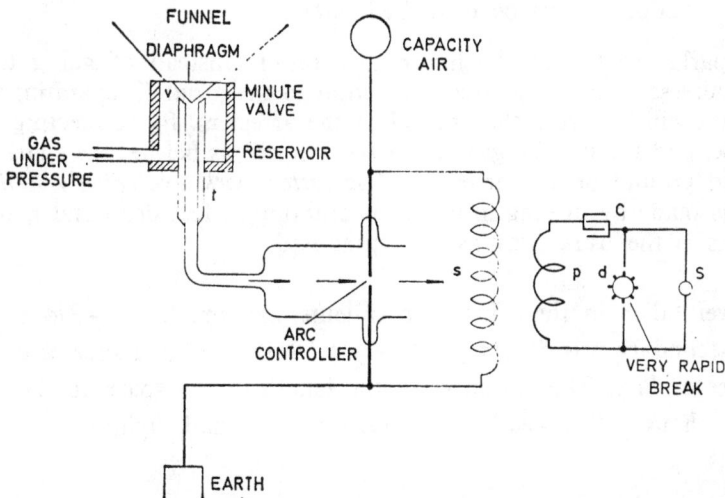

The idea is to use an ordinary oscillator, preferably one operated from a source of direct currents with a break (mercury or simply an air gap) which is much higher in frequency than the vibrations of the voice. At any rate, there will be an arc, whether in the primary of the secondary which will be blown out, or the resistance of which will be enormously increased, rhythmically with the vibrations impressed by voice or otherwise, as the case may be. The control of the arc is effected by a jet of air or other gas issuing under pressure from an orifice the opening of which is controlled in some convenient way by the vibrations. An arrangement of such apparatus is illustrated in the diagram below, the arc controlled thus being in the secondary. The source of direct currents S charges a condenser C and the discharges of the same (a very great number) through a break d and primary p energize a secondary s with the usual connection in telegraphy as I have introduced. The air or gas under pressure is controlled by a diaphragm and valve v. The outlet pipe t can be screwed up as close to the diaphragm as is necessary for the best result. In this or a modified way a powerful apparatus may be controlled by very feeble undulations, as those of the human voice.

Colorado Springs

June 15, 1899

First experiments in the station were made today. The e.m.f. of the supply transformer was 200 volts only. The break on the disk, which was driven by a Crocker Wheeler motor, varied from 800—1200 per second. ω was found to be 800 approximately.

Under these conditions the secondary from the New York high tension transformer could only charge from 3—4 jars and it was impossible to obtain more than a harmonic of the vibration of the secondary system of the oscillator, which required many more jars.

The secondary was wound on a conical framework, there being 14 turns of an average length of 130 feet each, that is approximately.

The primary was formed by one turn of cable, used in New York laboratory for the same purpose, consisting of 37 wires No. 9 covered with rubber and breading. The details of construction are to be described later.

Note: Sparks went over the lightning arresters instead of going to the ground This made it necessary to change the connection to the ground, separating that of the secondary of the oscillator from the ground of the arresters. By connecting the secondary to a water pipe, and leaving the ground of the arrester as before, the sparks ceased. This indicates a bad ground on the arresters. The *latter work exceedingly well.* The ground connection was made by driving in a gas pipe about 12 feet deep and gammoning coke around it. This is the usual way as here practised.

The power taken in these first experiments was small, 1/2—3/4 H. P. only. The spark on the secondary was 5″ long but *very thick and noisy;* indicates considerable capacity in the secondary. The variation of the length of the spark in the break did not produce much change. The weather was very stormy, hail, lightning.

Experiments were continued today. A new ground connection was made by digging a hole 12 feet deep and placing a plate of copper $20'' \times 20''$ on the bottom and spreading coke over it again, as customary. Water was kept constantly flowing upon the ground to moisten it and improve the connection but in spite of this the connection was still bad and to a remarkable degree. It is plain that the rocky formation and dryness is responsible and I think that the many cases of damage done by lightning here are partially to be attributed to poor earth connections. By keeping the water constantly running the resistance was finally reduced to 14 ohms between the earth plate and the water main. By connecting the earth plate and water main again, the lower end of the secondary being connected to the latter, sparks would again fly over the arresters. When the water main was disconnected they again ceased.

The action of the waves spreading through the ground was tested by a form of sensitive device later to be described and it was found that there was a strong vibration passing through the ground in and around the laboratory. The device was purposely unsensitive, to get an idea by comparison with former experiences in this direction. It did not respond when placed close to the oscillator, but unconnected to ground or capacity, but responded 200 feet from the shop when connected to the ground with one terminal. It responded also all along a water main, as far as it reached, although it was connected to the ground fairly well. The action on the device was still strong when there were *no sparks* from the secondary terminal. This is a good indication for the investigation of waves, stationary in the ground. It was concluded the earth resistance was still too great. Possibly the ground affects the primary and the secondary, more than assumed, by the formation of induced currents.

To be investigated.

Measurements of resistance between ground wire and water main showed the surprising fact that it was 2960 ohms, and even after half an hour watering it still was 2400 ohms, but then by continued watering it began to fall rapidly. Evidently the soil lets the water run through easily and being extremely dry as a rule it is very difficult to make a good ground connection. This may prove troublesome. The water will have to be kept flowing continuously. The high resistance explains the difficulty, from a few days before, of getting the proper vibration of the secondary. The first good ground was evidently at the point where the water main feeding the laboratory connected to the big main underground and this was several hundreds of feet away. This introduced additional length in the secondary wire which became thus too long for the quarter of the wave as calculated. The nearest connection to earth was as measured about 260 feet away and even this one was doubtful.

Measurement of inductances primary, secondary and *mutual induction.*

Readings for two primary turns in series showed:

$$I=34, \qquad E=7, \qquad R=0.015, \qquad \omega=716, \qquad I=\frac{E}{\sqrt{R^2+\omega^2 L^2}}$$

Neglecting R^2 we have $\omega L_p = 0.206$ and $L_p = \mathbf{287{,}000}$ **cm** approx.

For secondary 14 turns on conical frame average length of turn 130 feet.

$E = 57.7$ $\quad \dfrac{E}{I} = 4.57 \quad \omega^2 L^2 = 16.49 \quad L_s = \dfrac{4}{716} = 0.0056$ henry approx.

$I = 12.65$

$\omega = 716$ $\qquad \left(\dfrac{E}{I}\right)^2 = 20.98 \qquad \omega L = 4$ approx. or $L_s = \mathbf{5{,}600{,}000}$ **cm.** approx.

$R = 2.12 \qquad R^2 = 4.49$

Coefficient of mutual induction, 2 primary turns in series:

$$M = \frac{E_p}{\omega I_s} = \frac{6 \times 10^9}{716 \times 10.7} = \mathbf{783{,}300}\ \text{cm.}$$

$E_p = 6$
$I_s = 10.7$
$\omega = 716$

This will reduce L. Reduction estimated from

$$L - \frac{M^2}{N} = L\left(1 - \frac{M^2}{NL}\right) = L \times 0.64$$

Colorado Springs

June 18, 1899

Experiments were continued with the oscillator showing that proper vibration does not take place, evidently owing to some cause which is still to be explained. To see whether the trouble is due to poor induction from the primary, a coil-wound on a drum of about 30″ diam, 10″ long, 500 turns approx. of No.26 wire, used in some experiments in New York — was connected to the free end of the secondary and with this coil a great rise was obtained, streamers about 12″ long being obtained on the last free turn even with a small excitation of secondary. The trouble seems to be due to internal capacity. The total length of a quarter wave with coil was about 2400 feet, which agrees fairly with the calculation from the vibration of the primary circuit. The experiments with the coil show strikingly the advantage of an *extra coil*, as I call it, already noticed in experiments in New York; that is, a coil practically not inductively connected but merely used to raise the impressed electromotive force.

Measurements of inductance of the secondary as used: 12 turns on tapering frame 1 1/4″ apart from center to center showed:

Current through secondary	E.m.f. on terminals	ω
10.9	74 V	710

from this L'_s was found $= \mathbf{9{,}500{,}000}$ **cm.**

Readings for mutual induction:

	E.m.f. on primary (one turn)	ω
10.9	4.75	710

gave $M' = \dfrac{E}{I\omega} = 0.00062$ H $\quad$ or $\quad$ **620,000 cm.**

Compared with the first winding (14 turns far apart) the second winding was better because of both higher self-induction and greater mutual induction coefficient.

Measurement of capacity of condenser in sections:

The condenser was compared today with 1/2 mfd. standard by wirebridge and telephone receiver, according to the Maxwell method. There are 80 sections in the condenser, 40 on each side, which can be connected by plugs as desired.
They are: $1+2_1+2_2+5+10+10+20+30=80$
The measurements made by Mr. L. today gave 0.153 mfd per unit.

This to be verified.

Colorado Springs

June 19, 1899

Sensitive automatic device for receiving circuits in telegraphy through the natural media, purposes of tuning, etc. The device in simple form is illustrated in the diagram below.

In a small glass tube t are fixed two thin wires $w\,w$ of soft iron or steel carrying contact points of platinum $c\,c$ on the top. A spool S wound with wire surrounds tube t. The contact points are shaped so that the wires can deviate considerably without the separation becoming too great. When the current passes through coil S the wires $w\,w$ are separated and the distance between the contact points, $c\,c$ increased. The tube is moderately exhausted. The dielectric between the points is strained, as in sensitive powders, very nearly to the point of breaking down by means of a battery and when the disturbance reaches the circuit the dielectric gives way under the increased strain and the battery current passes through coil S separating the terminals and now breaking the battery current. It is supposed in this instance that the contact points $c\,c$ and coil S are connected in series with the battery, but the connection may be made in many other ways for the purpose of securing the same result — that is of automatically interrupting the current after the signal has been received. The contact points must be very close together and pointed. Stops $p\,p\,p$ are provided to limit the movement of the wires $w\,w$ and prevent vibration upon each action. An additional coil may be placed upon S for the purpose of adjusting the wires so the points will be at the required minute distance from each other, which is easily effected by graduating the strength of the current passing through the additional coil and an independent relay may be connected in the circuits in any convenient way for registering the signals. The degree of vacuum may also be made adjustable. In the first device coil S had 24 layers, 94 turns per layer=2256 turns, No. 21 wire, res. 14.7 ohms.

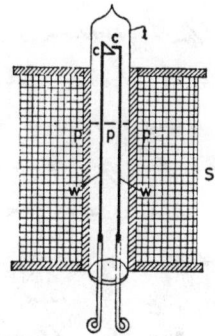

Colorado Springs June 19. 1899.

Sensitive automatic device for receiving circuits in telegraphy through the natural media, purposes of tuning e.t.c. The device in a simple form as illustrated in diagram below. In a small glass tube to are fixed two thin wires ww of soft iron or steel carrying contact points of platinum cc on the top of spool S wound with wire surrounded below t. The contact points are sharpened so that the wires can deviate considerably without the separation becoming too great.

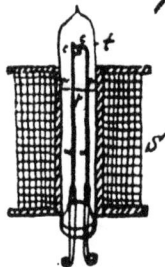

When the current passes through coil S the wires ww are separated and the distance between the contact points cc increases. The tube' is moderately exhausted. The dielectric between the points is strained up to sensitive process very nearly to the point of breaking down by means of a battery and when the distance reaches the current the dielectric gives by under the increased strain and the battery current passes through coil S separating the terminals and now breaking the leaky circuit. It is supposed in this instance that the contact points cc and coil S are connected in series with the battery; but the connection may be made in many other ways for the purpose of securing the device necessity is of automatically interrupting the current after the signal has been received. The contact points must be very close together and pointed. Stops $p p p p$ are provided to limit the movement of the wires ww and prevent vibration upon each action. An additional coil may be placed upon S for the purpose of adjusting the wire in the first coil will be at the required miromate distance from each other which is easily effected by graduating the strength of the current passing through the additional coil and an independent relay may be connected to the circuits in any convenient way for repeating the signals. The degree of vacuum may also be kept adjustable. In the first device coil S had 24 layers, 90 turns per layer = 2256 turns, No. 21 wire Res. 14.7 ohms.

Approximate estimate of some particulars of apparatus. With new jars the capacity will be about 0.174 mfd. that is with two sets of condensers in series as usual. Assume 20,000 volts on the supply transformer, the energy per impulse will be

$$\frac{4 \times 10^8 \times 0.174}{2 \times 10^6} = 34.8 \text{ watts estimated roughly.}$$

Suppose 1600 discharges through the primary per second, the condensers will deliver $34.8 \times 1600 = 55,680$ watts or a little over 74 H.P. With ten thousand volts they would still deliver $\frac{74}{4} = 18.5$ H.P. Now the vibration of the primary will be approximately:

$$T = \frac{2\pi}{10^3} \sqrt{\frac{7 \times 10^4}{10^9} \times 0.174} = \frac{2\pi}{10^5} \sqrt{0.7 \times 0.174} = \frac{2.2}{10^5} \text{ or } \frac{22}{10^6}$$

This gives $n = 45,500$ per sec. about. This vibration supposes only one primary turn.

The wave length calculated from this is about 4 miles or 21,120 feet and $\frac{\lambda}{4} = 5,280$ feet. As each turn has, on the average, 130 feet we shall want to make up the length of a quarter wave $\frac{5280}{130} = 40$ turns about. Or, if two primaries are used in series, the capacity remaining as before, the wave length will be doubled and 80 turns will be needed. Let it then be assumed that 80 turns are used, the self-induction of the secondary will be not far from 165×10^6 cm. The period of the secondary will then be:

$$T_1 = \frac{2\pi}{10^3} \sqrt{\frac{165 \times 10^6}{10^9} \times \frac{38}{9 \times 10^5}}$$

assuming no internal capacity or that it is overcome by suitable construction and only a ball of 30″ diam. or approximately 38 cm. radius on the free terminal of the secondary. We have then

$$T_1 = \frac{164}{10^7} \quad \text{and} \quad N = 61,000 \text{ approx.}$$

But this vibration will not be in harmony with the primary vibration. To effect this the self-induction of the secondary can be estimated. We have namely:

$$T = \frac{1}{45,500} = \frac{2\pi}{1000} \sqrt{\frac{38}{9 \times 10^5} L_s},$$

where L_s is the self-induction of the secondary required.

From this $L_s = \frac{10}{32}$ henry or $L_s = 312,500,000$ cm. Suppose the wire wound on the same spool or frame and the length to remain as before, the turns necessary can be estimated from $\frac{165 \times 10^6}{312 \times 10^6} = \frac{6400}{N^2}$, from which follows $N^2 = 12,102$ and $N = $ **110 turns.**

In addition to the wire already on hand this would cost about $ 250 but with 80 turns only $ 100 will be necessary. To keep the vibration of the secondary the same, the capacity on the free terminal will have to be increased. The capacity necessary will be C and we have:

$$\frac{1}{45,000} = \frac{2\,\pi}{1000} \sqrt{\frac{165 \times 10^6}{10^9}}\, C \quad \text{from which follows } C = 67.3 \text{ cm.}$$

A ball of this size is not to be had. If we employ a disc we have $\dfrac{2\,r}{\pi} = 67.3$, $\mathbf{r = 56}$ cm. A disc could scarcely be employed except with small pressures, there would be too much leakage.

All these estimates assume, of course, that the distributed capacity of the secondary is overcome in some way or other as by condensers in series, for instance. It is quite certain that the vibration of the secondary will be much slower.

Colorado Springs

June 21, 1899

Considerations of the various particulars of apparatus to be used, continued:

The present supply transformers can furnish 26 H.P.

Assume this energy consumption, that is, $26 \times 750 = 19,500$ watts and 1600 breaks or charges of the condensers per second. This gives per each break $\dfrac{19,500}{1600} = 12$ watts roughly. Let us further suppose that an excess of power is supplied so that the secondary receives clear 12 watts per each discharge of the primary. This means to say that the capacity on the end of the secondary will be charged 1600 times per second to a potential p. If C be the capacity on the free end we have $12 = \dfrac{p^2}{2}\, C$ from which $p^2 = \dfrac{24}{C}$. Assume C to be a sphere 38 cm. radius we have

$$\frac{24}{\dfrac{38}{9 \times 10^{11}}} = p^2 = \frac{9 \times 10^{11} \times 24}{38}, \quad \text{from this}$$

$$p = 3 \times 10^5 \sqrt{\frac{240}{38}} = 3 \times 10^5 \sqrt{6.32} = 3 \times 10^5 \times 2.51 = \mathbf{753,000} \text{ volts}$$

Approximate estimate of primary voltage necessary for above output.

To get e.m.f. lowest value it would be necessary to connect condensers in multiple, both sets. This would give a capacity of $0.174 \times 4 = 0.696$ mfd. Calling p_1 the primary e.m.f. necessary for this output, we have

$$\frac{0.696 \times p_1^2}{2 \times 10^6} = 12.$$

From this $p_1^2 = \dfrac{10^9}{29}$ and $p_1 = 6000$ volts approx. (26 H.P. expenditure, 1600 breaks per sec.)

With this e.m.f. assume 4 ohms res. of arc, the initial current would be 1500 amp. through the primary. From these assumptions the loss in the primary may be computed.

Colorado Springs

June 22, 1899

Wire for the new secondary ordered from Habirshaw No. 10 B.& S. rubber covered; all in all about 11,000 feet needed (more nearly 10,500 feet). This will do for 80 turns of an average length of 131 feet each.

No. 10 am. gauge 5.26 mm. square or $\dfrac{5.26}{645}$ sq.inch

100 feet will have: $\dfrac{5.26}{645} \times 1200 = 9.8$ cu.inches.

The weight of this, taking 5.13 ounces per cu.inch will be

$$\frac{5.13}{16} \times 9.8 = 3.14 \text{ lbs.}$$

Accordingly, 11,000 feet will weigh 345.4 lbs. This will give still less copper in the secondary than there is in the two primary turns. With secondary wire double we shall have 40 turns and with four wires (for quick vibration) 20 turns. The weight of copper should be equal and some of the No. 10 cord may be used on the first low turns.

Some arrangements were tried aiming chiefly at prolonging the vibration in the primary after each break. One of these was as illustrated in the diagram below:

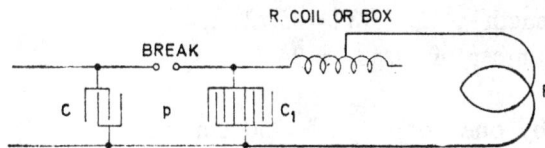

The condenser C_1 was placed in shunt to the primary P. Since there was no spark gap in this circuit and the magnifying factor was very large, the resistance being minute, the vibration continued much longer after each break as would be the case with the ordinary connection. A very curious feature was the *sharpness* of tuning. This seems to be due to the fact that there are two circuits or two separate vibrations which must accord exactly. The sparks were strong on terminals of the secondary always when $C = a\, C_1$, a being a whole number (no fraction), and particularly when $a = 2$ or 4.

This arrangement was carried out in New York on one of the later type oscillators and similar results were observed.

In this form there was a loss in circuit p since this part did not act upon the secondary in inductive relation to P. A modification consisted in including in circuit p one or more turns of the primary P or independent turns which acted inductively upon the secondary.

An arrangement intended for the same purpose was also tried. It consisted of providing two primaries, one independent of the break and merely shunted by a condenser,

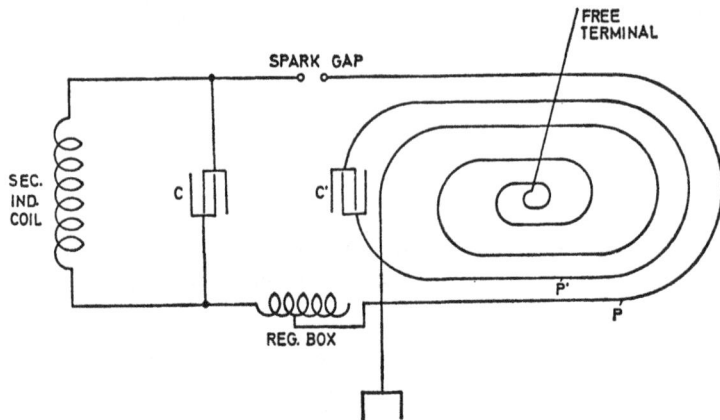

as illustrated. This plan was also experimented with in New York and it was found that it is good when the break number is *very small*. When the break is very rapid there is not much difference. In making the adjustments $C'P'$ was first tuned to the vibration of $C P$, then the secondary was adjusted.

This to be followed up.

Colorado Springs

June 23, 1899

Approximate self-induction of Regulating box brought from New York to be used in primary.

Dimensions: diam. of drum $12''=30.48$ cm.
        Length ,, ,, $18''=45.72$,,
        Number of turns $=24$

Area inclosed by one loop $\dfrac{\pi}{4}d^2=730$ cm. sq.

From this

$$L=\frac{4\pi N^2 S}{l}=\frac{12.57\times576\times730}{45.72}\ cm=$$

$$=0.275\times576\times730=115{,}600\ cm.$$

From this for approximate estimates we may take $\dfrac{115{,}600}{24}=4800$ cm per turn. This will be too much, as turns are far apart and thick. After Langevin's formula $L_8=\dfrac{\mathfrak{L}^2}{l}$. Here $\mathfrak{L}$ total length of wire is $30.5\times3.1416\times24=2300$ cm approx.

44

This would give value $L' = \dfrac{2\,300^2}{45.72} = \dfrac{5,290,000}{45.72}$

$$L' = 115,700 \text{ cm., remarkably close.}$$

Experiments with oscillator secondary 36 1/2 turns were continued. Many modified arrangements with auxiliary condensers — one of which is illustrated in the sketch below — were tried. All these chiefly aimed at prolonging the vibrations in the primary after each break and also at effecting sharper tuning of the circuits. In using auxiliary condensers in this way circuits are obtained containing no spark gap in which the damping factor was extremely small and the magnifying factor very great.

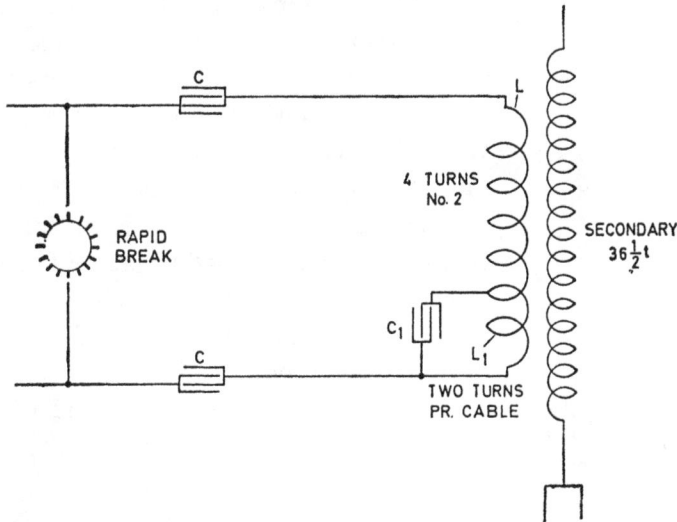

In this arrangement the relation $\dfrac{C}{2} L = C_1 L_1$

had to be attained for the best result. Or $L_1 = \dfrac{L}{4}$ $C_1 = 4 \times \dfrac{C}{2}$

Resonance was obtained with 15 jars on each side, 6 turns primary. With 4 turns primary there would have been necessary $15 \times \dfrac{36}{16} = 34$ jars or thereabouts. On thick cable about 68 jars. (for reference)

Note: Several rates of vibration were tried with such arrangements. Remarkable was the sharp tuning in some of them, one turn of the regulating coil being sufficient to entirely de troy the effect or to produce a great maximum rise of pressure. The jars broke down frequently, owing to sudden rise, as the handle of the regulating coil was turned.

Colorado Springs

June 24, 1899

The following plan of producing a conducting path of extremely low resistance suitable for resonating circuits and other uses offers the possibility of attaining results which can not be reached otherwise. It is based on my observation that by passing through a rarefied gas a discharge of sufficient intensity, preferably one of high frequency, the resistance of the gas may be so diminished that it falls far below that of the best conductors. So through just a bulb of highly rarefied gas an immense amount of energy may be passed and currents of a maximum strength such that they can not traverse a copper wire, owing to its resistance and impedance, may be made to traverse the rarefied gas. The plan now is to constitute a circuit composed of a rarefied gas column, heated by auxiliary means to a very high incandescence so as to offer an inconceivably small resistance to the passage of the current and use this column for the purpose to which it is suited. To illustrate the use of this idea in telegraphy, for instance the diagram below is shown in which S is a source of oscillating currents of preferably high frequency, C a condenser in shunt to same, L a coiled glass tube containing the rarefied gas which is kept at a high degree of excitation. The conductor L is connected, as in my system, to earth and a capacity preferably elevated. Through this path the currents of a distant transmitter are made to pass which are of the same frequency and cause a great rise of the e.m.f. on the terminals of conductor L, which may then be utilised to affect a receiving instrument in many ways.

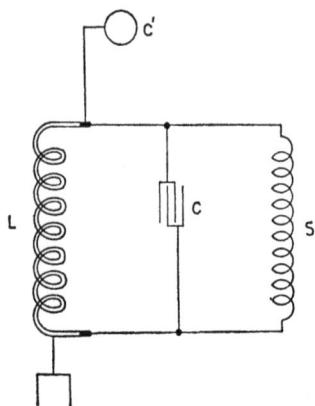

(To follow up).

Colorado Springs

June 25, 1899

The following plan seems to be well adapted for magnifying minute variations such as are produced by the action of a microphone, for example. Suppose that on a rotating or, generally speaking, moving surface of iron (polished or smooth) there is arranged a brush of soft iron, steel or at least having a surface of such magnetic materials whatever they be then there will be a certain amount of friction developed on the contact surfaces between the brush and moving surface and the brush will be dragged in the direction of movement of the surface. A spring may be used to pull the brush back against the friction and to maintain it in a position of delicate equilibrium. Let now the brush or surface be but slightly magnetized, then the friction between the magnetic surfaces will be enormously increased and the brush will be pulled forward with great force. A small variation in the magnetization of the surface will thus make great changes in the force exerted upon the brush, and the movements of the latter may be utilised for any purpose, as for instance in loud speaking telephones, or in perfecting a "wireless telephone" or such purposes. A simple form of apparatus is illustrated below: A is a rapidly rotating

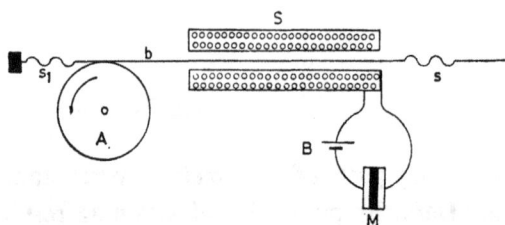

cylinder with a polished iron surface, if not all of iron; *b* is a small bar or brush bearing upon the cylinder and also of soft iron. This light plate or bar is held in a balanced position by differential spring $s \, s_1$ so as to bear lightly on the cylinder *A*. *S* is a solenoid energized through battery *B* in series with a microphone *M*. By speaking upon the latter the bar *b* will be vibrated back and forth and the movements of the bar may control any other apparatus, for instance a valve or other microphone.

Colorado Springs

June 26, 1899

In following up an old idea of separating gaseous mixtures by the aplication to them of an excessively high electromotive force, the following apparatus is to be adopted with a new oscillator.

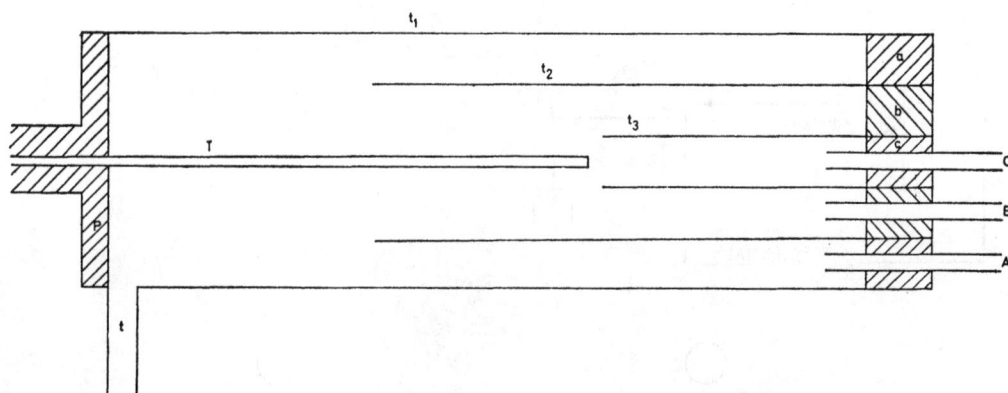

Note: in this apparatus it will be preferable to use a form of oscillator with mercury break supplied from a source of direct current, so that the force on *T* will be mostly uni--directional. Any other generator developing the necessary e.m.f. should, however, accomplish the same result.

Three tubes $t_1 \, t_2 \, t_3$ (assuming only three will be needed) are slipped one into the other, being held apart by insulating plugs *a b c*. In these plugs are fastened outlet tubes *A B C* to lead the several separated gases away to reservoirs into which they may be compressed. It is therefore to be understood that there is the desired degree of suction on the outlet pipes, or else the mixture is forced under desired pressure through a tube *t* serving to let in the mixture. The high tension terminal *T* is led in through an insulating plug *P* fastened into the largest tube t_1. The particles of the gas coming in contact with the active terminal are thrown away with great force and are projected at different distances according to their size and weight, hydrogen farther then most others. The latter element, if it be present, will therefore pass through tube *A*, that is mostly, the heavier and larger molecules through the other tubes. By repassing the gases drawn off through the apparatus again, any degree of purification or separation is obtained.

Arrangements of apparatus in telegraphy through the natural media aiming at exclusion of manager, in accordance with method experimented with in New York. This is not quite so good as the method used with condenser of commutating individual impulses, but great safety can be secured nevertheless. The idea was to provide more than one synchronized circuit and to make the receiver dependent in its operation on more then one such circuit. Experiments have shown that a great degree of safety is reached with two circuits. I think with three it is almost impossible to disturb the receiver when the vibrations have no common harmonics very near to the fundamental tones. Several arrangements experimented with are illustrated below. These are to be followed up.

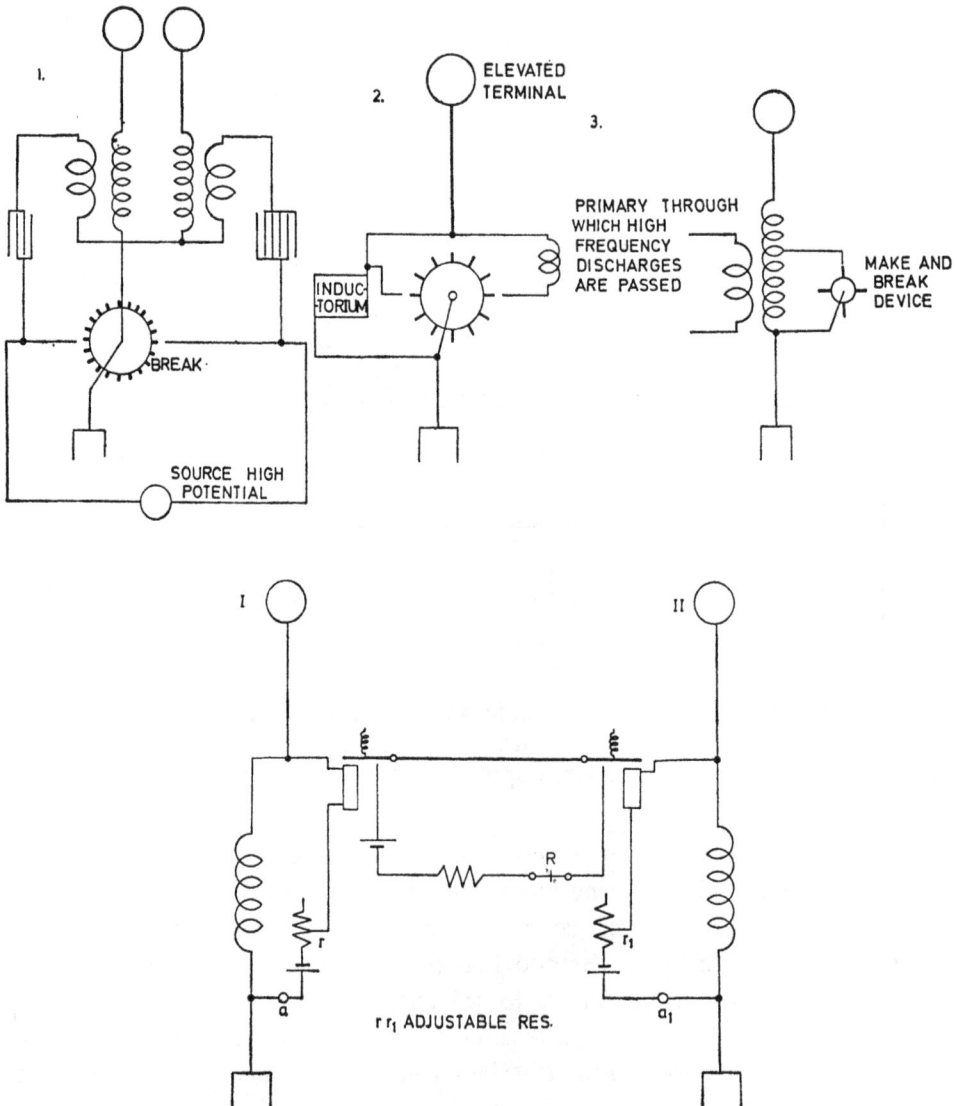

Figs. 1., 2. and 3. illustrate some arrangements of apparatus on the sending station by means of which two vibrations of different pitch are obtained. A greater number is omitted for the sake of simplicity. In case 1. are provided two sending circuits which should

be some distance apart and which are energized alternately by discharging condensers of suitable capacity through the corresponding primaries. In Figures 2. and 3. one sending circuit is arranged so that its period is altered by inserting some inductance as in 2., or by short-circuiting a part of the circuit periodically, by means of an automatic device. It is not necessary to use such a device; however, arrangements of this kind will be later illustrated. On the receiving station two synchronized circuits responding to the vibrations — each to one — of the sender. The receiver R responds only when both circuits I and II affect sensitive devices $a\,a_1$. The diagrams are self-explanatory.

Colorado Springs

June 28, 1899

Approximate estimate of the secondary with 20 turns on tapering frame, before referred to, from data of the secondary with 36 turns on the same frame. In the latter the wires 3 notches apart, in the former 7 notches.

Roughly, the capacity of the secondary with 20 turns will be, if C be that of the secondary with 36 turns:

$$C_1 = \frac{20}{36} \times \frac{3}{7}\, C = \frac{60}{252}\, C = \frac{20}{84}\, C = \frac{10}{42}\, C = \frac{5}{21}\, C$$

and the self-induction L_1 of the secondary with 20 turns compared with L — that of the secondary with 36 turns, will be

$$L_1 = \left(\frac{20}{36}\right)^2 \times \frac{36 \times 3}{20 \times 7}\, L = \left(\frac{5}{9}\right)^2 \times \frac{27}{35}\, L = \frac{675}{2835}\, L \qquad \text{Now } L = 383 \times 10^5$$

$$C = 1200 \text{ cm.}$$

Therefore $C_1 = \dfrac{5}{21} \times 1200 = \mathbf{290\ cm.}$ and $L_1 = \dfrac{383 \times 675}{2835} \times 10^5 = 9 \times 10^6 \text{ cm.}$ From this

$T = \dfrac{2\pi}{10^3} \sqrt{\dfrac{290}{9 \times 10^5} \times \dfrac{9 \times 10^6}{10^9}} = \dfrac{107}{10^7}$ approx. as period of sec. system (roughly) and

$n = \dfrac{10^7}{107} = 93,458$ per sec. Now the length of wire for 20 turns, about 139 feet per

turn, will be 139×20 feet. This gives $\lambda = 11,120$ feet or $\dfrac{\lambda}{4} = 2780$ feet and this would

correspond to $n = 90,000$ approx.

Adding a ball of 38 cm. capacity would give a total capacity

$$290 + 38 = 328 \qquad \sqrt{328} = 18.11 \qquad \sqrt{290} = 17 \text{ approx.}$$

hence by adding a ball the secondary vibration will be reduced by a ratio of: $\dfrac{17}{18.11}$ or it

will be $\dfrac{17}{18.11} \times 93,460 = 88,000$ approx. This would be too quick a vibration to best suit

the apparatus as then we would have only 4 jars on each side of the primary.

With the additional coil of 1500 cm. capacity added in series with secondary on free terminal, the capacity would be 1500+290=1790, that is about 6 times as much as before. The vibration will then be slower $\sqrt{6}$=2.5 approx. times slower, about 37,400 per sec. **This better suited.**

Colorado Springs

June 29, 1899

The first good trial of a new wound secondary with 36 turns was made today. The wire was No. 10 cord, the turns being wound in every third groove. The distance of wires is approx. 1 7/8".

Vibration under the conditions of the first experiments: Approximate self-induction of secondary about 5×10^7 cm. Additional coil connected to free end of secondary, the coil having 240 turns, spool 6 feet long, 2 feet diam. Estimated self-induction of coil roughly 10^7 cm.

A=2900 sq.cm.

N=240 turns

l =183 cm.

$$l=\frac{2900\times240^2\times4\pi}{183}=\frac{576\times29\times10^4\times12.57}{183}=$$

$$=1140\times10^4=114\times10^5 \text{ for rough approximation } =10^7$$

The wave length should be (ignoring capacity):

$4\times[5280 \text{ (sec)}+1440 \text{ (spool)}]=4\times6720=26,880$ feet or about 5 miles.

To give this wave length the primary vibration should actually be 187,000 : 5= =37,400 per sec. (same number before found).

The capacity C_p in primary to be found:

$$\frac{1}{37,400}=\frac{2\pi}{1000}\sqrt{LC_p}=\sqrt{\frac{7\times10^4\times C_p}{10^9}}=\frac{2\pi}{10^3}\sqrt{\frac{7\times C_p}{10}}$$

$$\frac{1}{374}=\frac{2\pi}{10^3}\times0.84\sqrt{C_p}, \quad \sqrt{C_p}=\frac{10^3}{2\pi\times0.84\times374}=\frac{1000}{1975} \text{ or } 0.5 \text{ approx.}$$

$C_p=0.7$ mfd

This would require $\dfrac{0.7}{0.003}$ jars=233 jars with two primary turns in multiple and $\dfrac{233}{4}$ or about 58 jars total with two primaries in series. As so many jars were not available evidently only a higher vibration was obtainable. This explains why first results unsatisfactory.

Arrangements of apparatus experimented with in carrying out condenser method. (This for Curtis application)

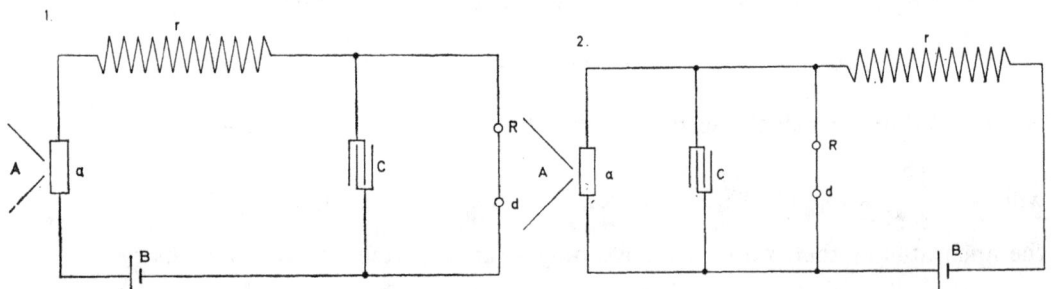

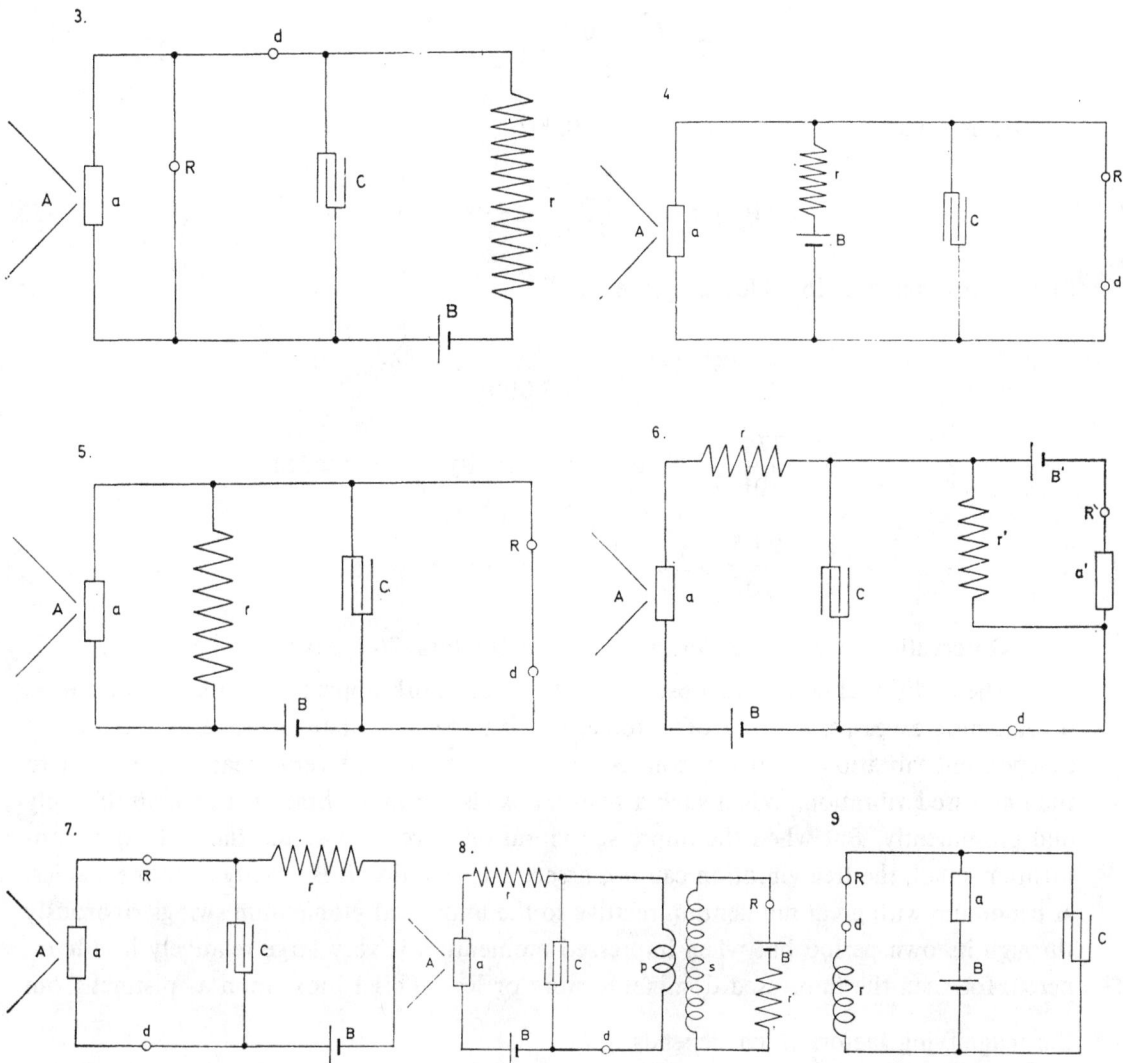

Colorado Springs

Simple formulas to be used in rough estimates of the quantities frequently wanted.

In formula $T = \dfrac{2\pi}{10^3}\sqrt{LC}$, L is in henry but usually is wanted in cm. We may therefore use

$$T_1 = \frac{2\pi}{10^3 \times 10^4 \times \sqrt{10}}\sqrt{LC} \quad \text{or approx. when } L \text{ is in cm.}$$

$$T_1 = \frac{2}{10^7}\sqrt{LC} \quad C \text{ in mfd.} \qquad (1)$$

From $L = \dfrac{T_1^2 \times 10^{14}}{4C}$ we have

$$C = \frac{T_1^2 \times 10^{14}}{4\,L} \quad \cdots \cdots \cdots \cdots \quad (2)$$

Introducing jars for C we have $C = n \times 0.003$

$$\text{therefore} \quad n = \frac{T_1^2 \times 10^{17}}{12\,L} \quad \cdots \cdots \cdots \cdots \quad (3)$$

Introducing again λ in miles in place of T_1

$$T_1 \text{ being} = \frac{\lambda}{187,000} \quad \text{we have:}$$

$$\lambda = \frac{374}{10^4}\sqrt{LC} \quad \text{or, since usually} \quad \frac{\lambda}{4} \quad \text{is needed}$$

$$\frac{\lambda}{4} = \frac{93.5}{10^4}\sqrt{LC} \quad \cdots \cdots \cdots \cdots \quad (4)$$

Observations made in experiments with oscillators, 36 1/2 turns and additional coil:

The additional coil is, as observed in the New York apparatus, an excellent means of obtaining excessive electromotive force. But it is peculiar that to properly develop the independent vibration of such a coil its momentum should be very great with respect to the impressed vibration. When such a relation exists the free vibration asserts itself easily and prominently. But when the impressed vibration is very large and the coil's own momentum small, the free vibration can not assert itself readily. It is exactly as in mechanics. A pendulum with great momentum relative to the impressed momentum swings rigorously through its own period but when impressed momentum is very large relatively it is hampered, for then the impressed dominates more or less. This I look upon as distinct from the magnifying factor which depends on $\frac{pL}{R}$.

It was evident that in such excitation of the additional coil there should be, for the best result, three vibrations falling together: that of the coil, that of the secondary and that of the combined system. In view of the above it is of advantage to place inductance between the secondary and additional coil to free the latter, when impressed vibration is too powerful to allow the intended vibration of the coil to take place readily.

From experiments it further appeared as though it would be of advantage to have some self-induction in the primary spark gap. This is to be ascertained. The use of condensers in series with the supply secondary is sometimes of advantage but little so when the vibration of the secondary is in resonance with the primary. Then there is less short circuiting of the secondary of the supply transformer and sparks are loud and sharp.

Colorado Springs Notes

July 1—31, 1899

To this to be added two applications filed with Curtis and other patent matters, mostly foreign.

Colorado Springs July. 1. 1899.

Various ways of connecting apparatus when
applying condenser method of magnifying effects. The
charging or discharge of condenser is controlled by the effects
transmitted through the medium and the condenser discharges are passed through the primary of the oscillatory
transformer. The diagrams below show various arrangements
with the instruments in the secondary of the transformer.

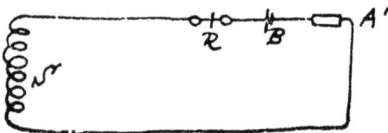

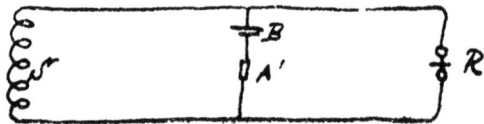

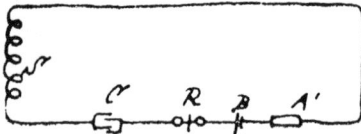

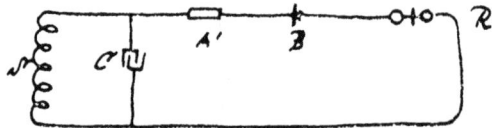

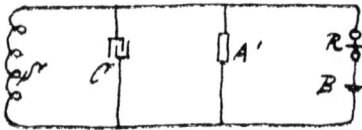

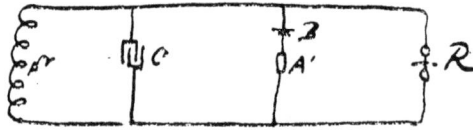

In these arrangements the primary is not shown. The wire
is assumed to be connected in the circuits in any way but in these
the charging or discharging of the condenser is controlled by a sensitive
device affected by the feeble effects which are to be magnified. In
the above diagrams S is the secondary of oscillatory transformer, B battery to strain
sensitive device in secondary; A' sensitive device, R fine relay (magnetic), C condenser
in secondary. The primary circuit which is not shown includes: a sensitive device, battery,
condenser and discharge break device. This method with two sensitive devices very good.

Various ways of connecting apparatus when applying condenser Method of Magnifying effects. The charging or discharging of the condenser is controlled by the effects transmitted through the media and the condenser discharges are passed through the primary of the oscillatory transformer. The diagrams below show various arrangements with the instruments in the secondary of the transformer.

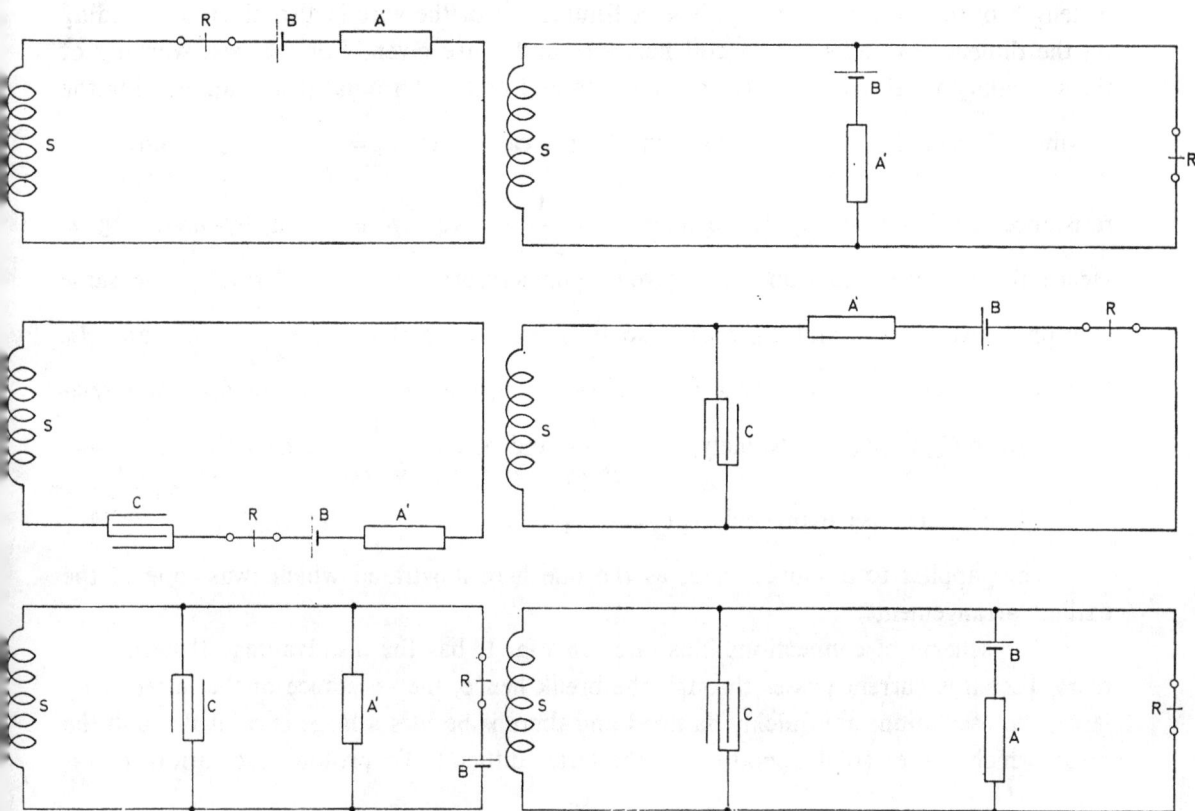

In these arrangements the primary is not shown. The same is assumed to be connected in the circuit in any way but so that the charging or discharging of the condenser is controlled by a sensitive device affected by the feeble effects which are to be magnified. In the above diagrams S is the secondary of oscillatory transformer, B battery to strain sensitive device in secondary; A' sensitive device, R fine relay (magnetic); C condenser in secondary. The primary circuit which is not shown includes: a sensitive device, battery, condenser and make-and-break device. This method with two sensitive devices is very good.

Considerations regarding best conditions of working apparatus in experimental station particularly with reference to stationary waves in the ground which will be investigated.

First assumption on which to base calculations of other elements is made by deciding on the wave length of the disturbances. This in well designed apparatus determines $\dfrac{\lambda}{4}$ or length of secondary wound up. The self-induction of the wire is also given by deciding on the dimensions and form of coil hence L_s and λ are given. For the best working of the secondary we should have the capacity on end or free terminal just counteracting the self-induction of the secondary. This will be the case when $C_s = \dfrac{L_s}{R_s^2 + p^2 L_s^2}$. Since the resistance should be negligible we have $C_s = \dfrac{1}{p^2 L_s}$. Now p is given by assuming λ. Hence from above equation C_s is given. Furthermore, to get best results the same relation between p, L_p and C_p must also be preserved so that $C_p = \dfrac{1}{p^2 L_p}$, C_p and L_p being the capacity and inductance of primary. From these considerations follows a relation between C_s, C_p, L_s, L_p. Namely, $C_s = \dfrac{1}{p^2 L_s}$ and as $p^2 = \dfrac{1}{C_p L_p}$ we have $C_s = \dfrac{1}{L_s \dfrac{1}{L_p C_p}} = \dfrac{L_p C_p}{L_s}$ or otherwise expressed, $L_s C_s = L_p C_p$ .... I.

This applies to a simple case, as the one here illustrated which was one of the earliest arrangements.

The scheme of connections illustrated in Fig. 1. has the disadvantage that the primary discharge current passes through the break hence, the resistance of the latter being large, the oscillations are quickly damped and there is besides a large current through the break which makes good operation of the latter difficult. To prolong oscillation in pri-

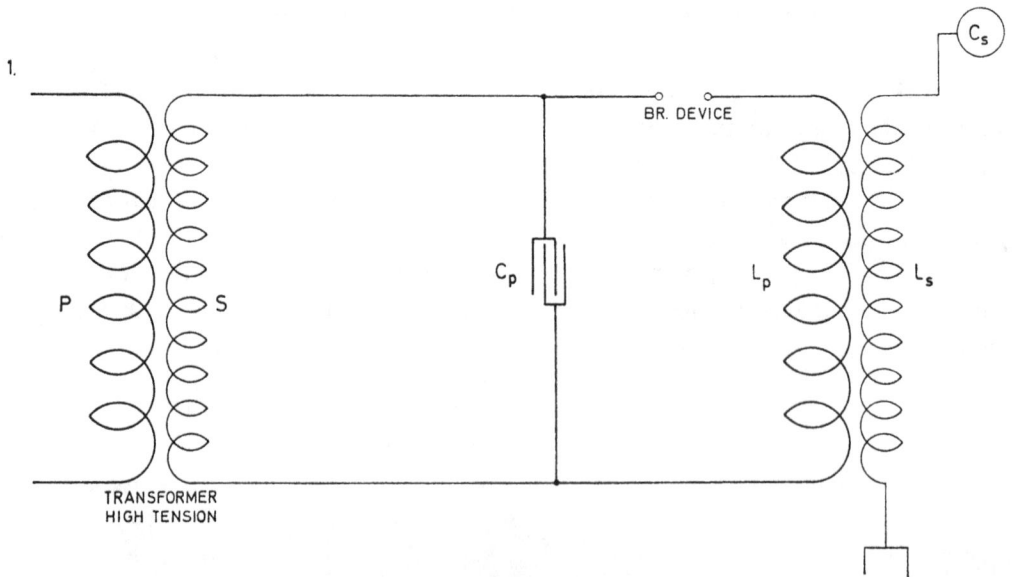

mary and increase economy one of the schemes before considered may be resorted to. One of these is illustrated in Fig. 2. which follows: in this arrangement the currents through the break device are much smaller and the oscillations started by the operation of break

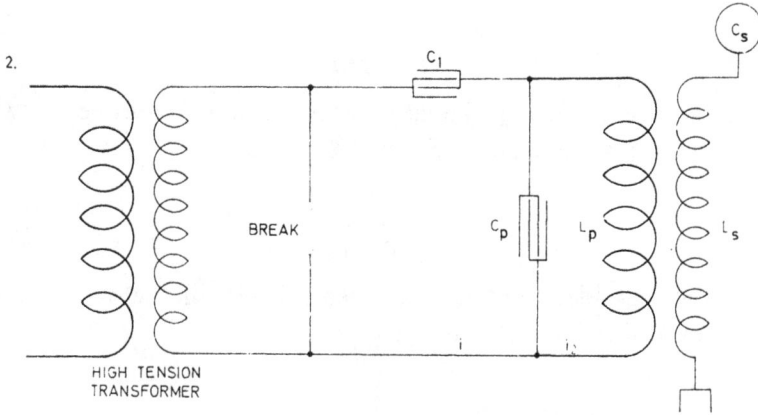

in circuit $L_p C_p$ continue much longer. We can now determine the magnitude of currents i, i_2. Capacity $C_p = \dfrac{L_p}{R_p^2 + p^2 L_p^2}$ from foregoing or approximately $C_p = \dfrac{1}{p^2 L_p}$ as before, neglecting resistance. We have then $\dfrac{i_2}{i} = \sqrt{\dfrac{R_p^2 + p^2 L_p^2}{R_p^2}}$ or, since the first term is negligible in numerator, we have $\dfrac{i_2}{i} = \dfrac{p L_p}{R_p}$ or $i_2 = i \dfrac{p L_p}{R_p}$. From this relation it is evident that, other difficulties or disadvantages not considering the arrangement as illustrated in Fig. 2. should be superior to that in Fig. 1. It secures two chief advantages: 1) less current through the break and more through the primary and 2) longer and better oscillation in circuit including primary because it is easy to constitute such a system without break with an extremely small resistance or frictional waste.

Fig. 3. illustrates an arrangement similar to that shown in Fig. 2. but with a condenser in each branch. The same considerations made in regard to Fig. 2. hold good for this and in both cases, if there is to be resonance and best conditions attained, the circuit including the break should have the same period and be in phase with primary circuit $L_p C_p$ and secondary $L_s C_s$. Referring to Fig. 2. as the simpler, this is the case when the

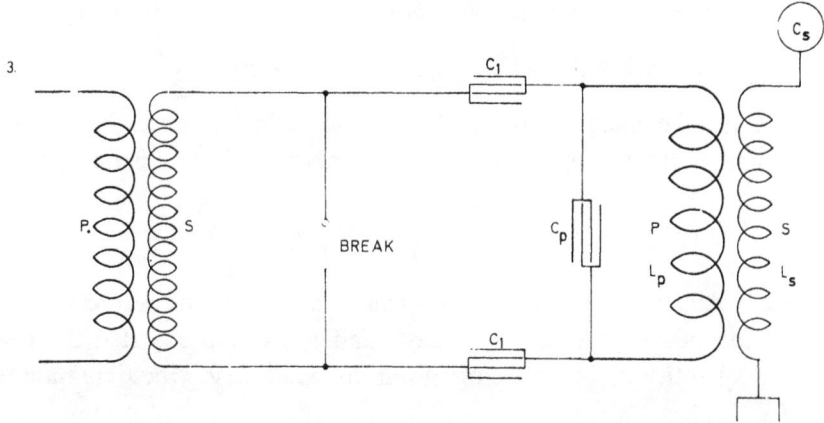

relation between C_1 and self-induction in this circuit is such that they annul each other at that frequency.

Fig. 4 — a further modification shows the system with inductance L_1 in circuit. To satisfy above conditions we must have

$$C_1 = \frac{L_1}{R_1^2 + p^2 L_1^2},$$

R_1 being the resistance including the arc. Since in most cases, even with the arc included, R_1 will be negligible against pL_1, we have again

$$C_1 = \frac{1}{p^2 L_1}.$$

From all the above considerations we get a general relation between the constants of all the three circuits which is expressed by:

$$\left. \begin{array}{c} \dfrac{1}{C_s L_s} = \dfrac{1}{C_p L_p} = \dfrac{1}{C_1 L_1} \\[2mm] \text{or } C_s L_s = C_p L_p = C_1 L_1 \end{array} \right\} \ \ldots\ldots\ldots\ldots\ldots \text{ II.}$$

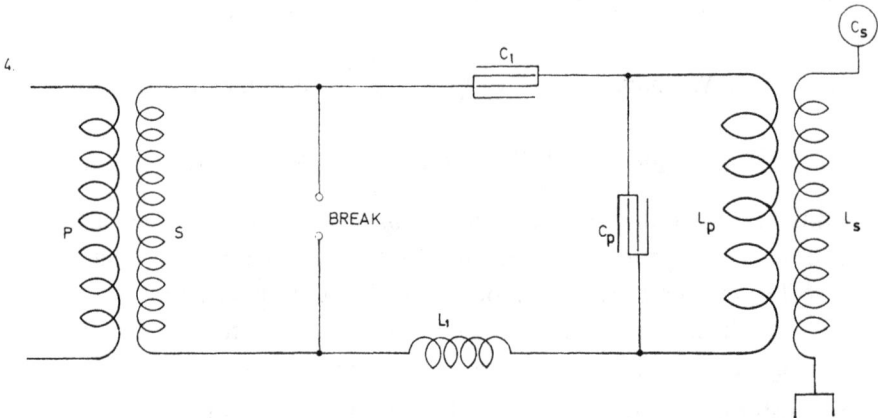

We have seen from the preceding that of the quantities considered p was given by arbitrary selection of the wave length, L_s necessarily by the wire and design of secondary and, lastly, C_s as following from the two preceding quantities. One more quantity is, however, given by practical considerations and that is C_1. Namely, this capacity C_1 must be sufficient to take up the entire energy of the transformer, if all things are rightly proportioned. Now let P' be the e.m.f. on the secondary of the supply transformer, then there will be stored each time in condenser C_1 a quantity of energy $\dfrac{P'^2 C_1}{2}$ and if this value be taken for the average energy stored each time and if, furthermore, the frequency of the make and break be p_1, that is $p_1 = 4\pi n$, n being number of charges per second, or

$$n = \frac{p_1}{4\pi}, \ \text{ then } \ \frac{P'^2 C_1}{2} \cdot \frac{p_1}{4\pi} = M \ \text{ or } \ C_1 = \frac{8\pi M}{P'^2 p_1}$$

M is here the total power of the transformer expended and expressed in watts, P' the pressure of secondary (average as defined) and p_1 as before stated the frequency of the break. The quantity C_1 thus being given in each case, these remain still to be determined: L_p, L_1 and C_p.

Now it is evident that when the relation between p_1, L_p and C_p exists, which is here implied, the current passes through the system as if there would be no inductance, hence insofar as the circuit including the break, C_1 and L_1 is concerned the system $L_p C_p$ will comport itself as if it consisted of a short wire of inappreciable resistance, the primary being generally made of stout short conductor — therefore, in estimating the quantities of the circuit $C_1 L_1 b$ the compound system $L_p C_p$ may be neglected since it will have little influence upon the period under the conditions assumed, and we may then put $C_1 = \dfrac{1}{p^2 L_1}$ when resistances are, as always before, negligible.

Since C_1 is known we can determine L_1 because

$$L_1 = \frac{1}{p^2 C_1} = \frac{1}{p^2 \dfrac{8\pi M}{P'^2 p_1}} = \frac{P'^2 p_1}{8\pi M p^2}$$

Presently then all the quantities are known for determining the constants of circuit $L_p C_p$ from the two equations:

$$\left. \begin{array}{l} L_p C_p = L_s C_s \\ L_p C_p = L_1 C_1 \end{array} \right\} \ \ldots\ldots\ldots \ \text{III.}$$

Colorado Springs

July 3, 1899

In experiments with the secondary as last described, fairly good resonance was obtained with 15 jars on each side of the primary. A length of wire No.10—170 feet — was covered with intense streamers. The capacity — total — was $7.5 \times 0.003 = 0.0225$ mfd. L_p was approximately estimated $36 \times 7 \times 10^4$ cm. (six primary turns in series). From this $T = \dfrac{4.836}{10^5}$ as calculated. This gives $n = \dfrac{1}{T} = 20{,}700$ per sec. approx. With this vibration λ was nearly 9 miles or $\dfrac{\lambda}{4} = 2.25$ miles. Actually, there was only one mile of secondary wire but owing to the large capacity (distributed) in the secondary the vibration was much slower than should be inferred from the length of wire. We may estimate the ideal capacity, which associated with the inductance of the secondary would give a vibration of the above frequency. Since there was resonance we have:

$$T = \frac{4.836}{10^5} = \frac{2\pi}{10^3} \sqrt{\frac{5 \times 10^7}{10^9} C_s}$$

Taking the inductance of the secondary as being 5×10^7 cm. and from this

$$C_s = \frac{23.34}{20{,}000} = \text{roughly } \frac{1}{1000} \text{ mfd. more exactly } 0.0012 \text{ mfd. or } \mathbf{1080} \text{ cm.}$$

But we may approximately estimate capacity in another way taking the wires in pairs as a condenser. This would give $C_s = \dfrac{A}{4\pi d} \times 40$, there being 40 pairs since there are 40 turns wire. Now A=surface of half wire: 131 feet, about 4000 cm. long; half circumference about 0.4 cm. gives A=1600 cm. sq. The distance of wires d was=5 cm. From this: $C_s = 40 \times \dfrac{1600}{4\pi \times 5} = 1040$ **cm. approx.** (probably accidentally close.)

Last night the 50,000 volts transformer brought from New York broke down. This happened upon connecting to the lower end of the secondary a condenser composed of two adjustable brass plates of 20″ diam., one being connected to the ground, the other to the secondary. The plates were about 5″ apart. The experiment was repeated with the transformer after repairing it and all was found in good order.

Experiments were now continued with the secondary of 40 turns wire No. 10 just one mile long. In connection with this secondary a coil was used wound on a drum 2 feet in diam. and 6 feet long, with wire No. 10 (cord), there being 260 turns. Approx. estimate of capacity after a previous similar estimate: 6 feet=$6 \times 12 = 72″ = 72 \times 2.54 = 183$ cm. Half circumference of wire 0.4 cm. Area=$183 \times 0.4 = 74$ sq.cm. (about)=a, distance of wires about 1 cm.=d. This would give roughly the capacity $C_1 = \dfrac{a}{4\pi d} = \dfrac{74}{4\pi}$ for one pair of wires; there being 260 pairs, total capacity would be according to this $\dfrac{260 \times 74}{4\pi} = 1532$ cm. $=C_t$ (for coil). Now, the capacity in the secondary was previously found 1080 cm. Hence the total capacity of the system would be $1080+1532 = 2612$ cm. $=C'$. This, of course, gives only an idea and the determination in this manner is far from being exact. Consider what the period would be with the capacity and secondary and coil together as a whole. Since the coil will have only about 12,000,000 cm. the inductance of the secondary will be the chief governing factor. Taking this at 5×10^7 cm. we would have a total inductance $5 \times 10^7 \times 12 \times 10^6$ or $(50+12) \times 10^6 = 62 \times 10^6$ cm. This would give

$$T = \frac{2\pi}{10^3} \sqrt{\frac{62 \times 10^6}{10^9} \times \frac{2612}{9 \times 10^5}} = \frac{2\pi}{3 \times 10^7} \sqrt{164{,}944} =$$

$$= \frac{2\pi \times 406}{3 \times 10^7} = \frac{2549.68}{3 \times 10^7} = \frac{849.9}{10^7} \text{ or } \frac{850}{10^7} = \frac{85}{10^6}$$

and n=11,800 roughly. *To this the primary to be adjusted* for first approximate trials. Now the primary has six turns. Since one turn is approx. 7×10^4 cm. we may put:

$$\frac{85}{10^6} = T = \frac{2\pi}{10^3} \sqrt{\frac{36 \times 7 \times 10^4}{10^9} C_p}$$

From this a rough idea of the capacity in primary may be gained. We get C_p=0.0717 mfd. roughly. Taking capacity of one jar=0.003 mfd. we would want total $\dfrac{0.0717}{0.003} = 24$ jars approx. or 48 jars on each side of primary. This vibration would be impracticable under the present conditions as the transformer could not charge this number of jars. Although for *stationary waves* in ground it would be desirable to use such a low frequency the vibration will have to be quickened. An octave would require only 12 jars on each side. This was tried and results *were good* although the octave vibration had only 1/4 of the energy as the fundamental would have had. To get the true vibration we shall want at least 8 *turns* in the primary with present transformer to keep the capacity in the primary within the limits given by the output of transformer. This would then give $48 \times \frac{36}{64} = 27$ jars on each side. While this might do, still for best conditions not more than 16 jars should be used on each side of primary. This just taxes the transformer to full capacity within safe limits.

Conclusion: *Used about 10 turns in the primary.*

Colorado Springs

July 4, 1899

Observations made last night. They were such as not to be easily forgotten, for more than one reason. First of all a magnificent sight was afforded by the extraordinary display of lightning, no less than 10—12 thousand discharges being witnessed inside of two hours. The flushing was almost continuous and even later in the night when the storm had abated 15—20 discharges per minute were witnessed. Some of the discharges were of a wonderful brilliancy and showed often 10 or twice as many branches. They also appeared frequently *thicker* on the *bottom* than on top. Can this be so? Perhaps it was only due to the fact that the portion close to the ground was nearer to the observer. The storm began to be perceptible at a distance as it grew dark and continuously increased. An instrument (rotating "coherer") was connected to ground and a plate above ground, as in my plan of telegraphy, and a condenser was used to magnify the effects transmitted through the ground. This method of magnifying secures much better results and will be described in detail in many modifications. I used it in investigating properties of Lenard and Roentgen rays with excellent results. The relay was not adjusted very sensitively but it began to play, nevertheless, when the storm was still at a distance of about 80—100 miles, that is judging the distance from the velocity of sound. As the storm got nearer the adjustment had to be rendered less and less sensitive until the limit of the strength of the spring was reached, but even then it played at every discharge. An ordinary bell was connected to earth and elevated terminal and often it also responded. A small spark gap was bridged by a bright spark when the lightning occurred in the neighbourhood. By holding the hands across the gap a shock was felt indicating the strength of the current passing between the ground and the insulated plate. As the storm receded the most interesting and valuable observation was made. It happened this way: the instrument was again adjusted so as to be more sensitive and to respond readily to every discharge which was seen or heard. It did so for a while, when it stopped. It was thought that the lightning was now too far and it may have been about 50 miles away. All of a sudden the instrument began again to play, continuously increasing in strength, although the storm was moving away rapidly. After some time, the indications again ceased but half an hour later the instrument began to record again. When it once more ceased the adjustment was rendered more delicate, in fact very considerably so, still the instrument failed to respond, but half an hour or so it again began to play and now the spring was tightened on the relay very much and stil it indicated the discharges. By this time the storm had moved away far out of sight. Byl readjusting the instrument and setting it again so as to be very sensitive, after some time it again began to play periodically. The storm was now at a distance greater than 200 miles at least. Later in the evening repeatedly the instrument played and ceased to play in intervals nearly of half an hour although most of the horizon was clear by that time.

This was a wonderful and most interesting experience from the scientific point of view. It showed clearly the existence of *stationary waves*, for how could the observations be otherwise explained? How can these waves be stationary unless reflected and where can they be reflected from unless from the point where they started? It would be difficult to believe that they were reflected from the opposite point of the Earth's surface, though it may be possible. But I rather think they are reflected from the point of the cloud where the conducting path began; in this case the point where the lightning struck the ground would be a nodal point. It is now certain that they *can be produced* with an oscillator

(*This is of immense importance*)

Measurement of inductance of oscillator secondary 36 1/2 turns on tapering frame repeatedly referred to. The approximate dimensions and form of same are indicated in

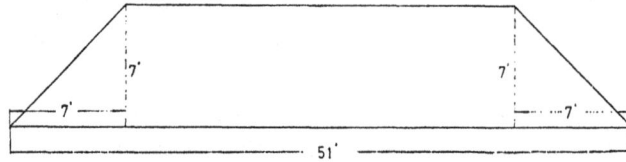

the sketch. On the base it was about 51 feet and the sides were inclined at an angle of 45°. The sides were formed of light lattice work notched for the reception of the wires. The first turn of the secondary began some distance from the ground so that the average turn was smaller than it ought to have been, judging from dimensions, that is nearly 145 feet. Nevertheless more wire was actually coiled up owing to the fact that there was some loss in the corners, and the wire not being perfectly straight added still further to the length so that 6 coils of wire were rolled up, their lengths being: $1000+1000+1005+1002+762+546=5315$ feet total, No. 10 B. & S. wire. Deducting ends left gave very nearly 5280 feet or a mile $= 1610$ meters approx. The wire was wound on by the help of a stand rolled on the floor and supporting the reel. The resistance of the wire was 5.55 ohms. The readings were as follows:

ω motor	E	I		
2115	212	6.2	from this average	$I=6.17$
2100	211	6.16	values were	$E=211.33$
2120	211	6.16		$\omega=883.42$

$$\frac{E}{I} = 34.25 \qquad \left(\frac{E}{I}\right)^2 = 1173 \qquad R^2 = 30.8 \qquad L = \frac{\sqrt{\left(\frac{E}{I}\right)^2 - R^2}}{\omega}$$

gives **$L=3826\times10^4$ cm. or 0.03826 henry.**

With 40 turns placed at same distance we may take approximate inductance to be about 42×10^6 cm. Note: It was before assumed 5×10^7 but the turns were a trifle closer.

Readings were taken today with the synchronous 8 pole motor to ascertain ω as closely as possible for future measurements with following results:

Time	Speed generator (10 pole)	Speed motor	Speed reduced gen.
5.35	1700	2155	1724
5.40	1700	2165	1732
5.50	*1730*	2160	*1728*
6.10	*1710*	2150	*1720*
6.20	1705	2170	1736
6.25	*1717*	2150	*1720*

Speed of generator on station was taken by Mr. L. and on motor in exp. station by myself. The readings on generator were more liable to be underestimated. General results show that high values obtained on generator agree well with values obtained on motor. The latter *can* go a little faster without load. This I have observed with such motors before. The reason for this may be found in the fact that the generator fluctuates about a certain average value, a greater momentum is impacted to the motor when the speed of the generator is *above* than when *below* that value. This will result in the motor making a few more revolutions than would strictly follow from the average value. Or, the counter--electromotive force is out of phase giving higher e.m.f. on motor then on generator, thus changing amount of positive or negative slip. Taking average from three evidently best readings we get on generator 1719, on motor 1722 which agrees fairly. Accordingly, *with this generator* ω will be generally $2\pi \times 29 \times 5 = 911$ or, for ordinary estimates $\omega = \mathbf{900.}$

From older notes:

Consider generation of hydrogen for balloons in the ordinary way:

$$H_2SO_4 + Zn = ZnSO_4 + H_2 \qquad \left.\begin{array}{l} H = 1 \\ S = 32 \\ O = 16 \\ Zn = 65 \end{array}\right\} \qquad H_2SO_4 = 98$$

From this:

$98\,(H_2SO_4) + 65\,(Zn) = 2H$ in lbs. Now weight of hydrogen 0.00561 lbs. per cubic foot, for filling 10 ft balloon we would want $523 \times \dfrac{2}{3} = 350$ cu. feet hydrogen, namely capacity of 10 foot balloon would be 523 feet but it ought to be filled only to about 2/3. This quantity of hydrogen will weigh about $350 \times 0.00561 = 1.96$ lbs.

Result: We want $\left.\begin{array}{l} 100 \text{ lbs } H_2SO_4 \\ 65 \text{ ,, } Zn \end{array}\right\}$ to fill balloon of only 10 feet diam.

Now consider and compare process which some years ago occurred to me and which consists in decomposing a hydrocarbon as by an electric current heating a wire to incandescence. To get a rough idea take, for instance, a hydrocarbon of the general composition C_2H_4 (not to speak of combinations richer still in hydrogen). In such a combination we have for the quantity of hydrogen contained in it an extremely small weight. For instance, from $\left\{\begin{array}{l} C = 12 \\ H = 1 \end{array}\right\}$ 28 units of weight total we get 4 unity of weight hydrogen, much more than possible in former method. In case therefore a very small weight is essentially required this method is *most excellent*. Now as to electrolytic generation: 1 amp.-hour gives 37.3 milligram of hydrogen, 1000 amp.-hours $\dfrac{37,300 \times 2.2}{10^6}$ lbs., 0.00561 lbs per cu. foot gives 1.2 cu. feet per 1000 amp.-hours! Ridiculously small!

On a previous occasion the capacity of a coil was estimated by considering it as a series of parallel conductors and in this manner a tolerably close estimate was obtained. Applying this to the secondary of 40 turns we would have:

$$C = \frac{0.01206}{\log \dfrac{d}{r}} \text{ from a practical formula } \left\{\begin{array}{l} d = \text{distance between} \\ r = \text{diameter of} \end{array}\right\} \text{ wires}$$

This gives C in microfarad per 1 km., according to the authority. Here the length of wire total is 5280 feet roughly or $5280 \times 12 \times 2.54 = 160{,}934$ cm. Taken as a pair of conductors this would give length $\dfrac{160{,}934}{2} = 80{,}467$ cm.

$$\text{Now} \quad \frac{d}{r} = \frac{4.23}{0.127} = \frac{4230}{124} = 33.3 \qquad\qquad d = \frac{10''}{6} = 4.23 \text{ cm.}$$

$$\log \frac{d}{r} = 1.522444 \qquad\qquad r = \frac{0.1''}{2} = 0.127 \text{ cm.}$$

from these data we would have per 1 km. in microfarad ·

$$C_1 = \frac{0.01206}{1.522444} = 0.00792 \text{ mfd. per km.}$$

Now the length of the pair of parallel wires as supposed would be 80,467 cm. or 0.80467 km. or for the secondary the capacity would be $C = 0.80467 \times C_1 \times 2$, double since both sides must be counted, or $C = 0.00792 \times 2 \times 0.80467 = 0.01584 \times 0.80467 = 0.012746$ mfd. or $9 \times 10^5 \times 0.012746 = 1274.6 \times 9 = 11{,}471.4$ cm. Evidently this is not applicable, the capacity could be at the utmost 3000—4000 cm. judging from the vibration of the secondary.

Colorado Springs

July 7, 1899

General conclusions arrived at after all these and previous experiences with electrical oscillators of this kind. It is important as a rule and sometimes imperative to overcome distributed capacity. In large machines it also becomes necessary to overcome the too great self-induction since it prevents obtaining a very high frequency which is generally of great advantage. The high e.m.f. being for the chief purposes aimed at — that is power transmission and transmission of intelligible messages to any point of the globe — essentially necessary, it is important to ascertain the best manner to obtain it. As has been already stated this result may be reached in two ways radically different — either by a high ratio of transformation or by resonant rise. For power transmission it seems that ultimately the former method must prevail, but where a small amount of energy is needed the latter method is unquestionably the better and simpler of the two. By placing the secondary in a very close inductive relation to the primary, the self-induction is diminished so that the self-induction in a highly economical machine of this kind would not seem to be an impediment in the way of obtaining a very high frequency, at least not one which could not be more or less overcome by scientific design of the machine. But the distributed capacity is a troublesome element in such a machine and all the more so as the e.m.f. increases. When the pressure reaches a few million volts almost all the energy is taken in charging the condenser or capacity distributed along the wire. The difficulty becomes greater still when it is realized that in an economical machine the turns must be close together, which increases the drawbacks resulting from the distributed capacity. Now one way of reducing the internal capacity is to place between the turns, and *in series* with them, condensers of proper capacity, but this is not always practicable. This will be *later considered* more in detail. By such means the full rise of pressure on the terminal or terminals of the secondary may be obtained, which is impossible with distributed capacity of any magnitude. Very often only a small rise at the terminals can be obtained as all the charge remains "inside". Now as to obtaining the required pressure by a resonant rise there are again two ways: either to place a secondary in *loose* connection to a primary thereby enabling the free vibration of the secondary to assert itself, or using a secondary in *intimate* connection with the primary and then raising the pressure by an additional coil — extra coil — or inductance *not in inductive relation* to the primary. The latter method I have found preferable when a very high e.m.f. is desired. More particularly for purposes of telegraphy to any point of the globe which is one of the objects, I conclude that: 1) ratio of transformation should be as large as practicable with reference to the preceding; 2) magnifying factor of coil as large as possible; 3) minimum internal capacity; 4) high self-induction in coil for *sharp tuning*. Experiences up to present indicate flat spiral form of coil in sections as best suitable.

Rough estimate of period of vibration to be adopted with Westinghouse transformer 40,000 — 60,000 volts.

Required: that only one turn of primary be used because of 1) high ratio of transformation to be attained and 2) facility of regulation with the Regulating coil brought from New York. Now the total output of W.E. Transformer will be, say, 50 H.P. (though the machine may be strained to many times that output). From this follows the number of jars which it will be possible to use. We have $50 \times 750 = \frac{1}{2} \times 60{,}000^2 \times 300 \times C$, assuming now 150 cycles per second, a little more than is likely to be the case. From this follows

$$C = \frac{75 \times 10^3}{36 \times 10^8 \times 3 \times 10^2} = \frac{75}{36 \times 3 \times 10^7}$$ farad or in centimeters we would have the capacity

of condenser which the transformer will be able to charge wi hout considering resonant conditions or other causes which may enable the transformer to charge many more jars

$$- \; C = \frac{9 \times 75 \times 10^{11}}{108 \times 10^7} = \frac{9 \times 75 \times 10^4}{108} = C = 62{,}500$$ cm. total. Now taking the capacity

of one jar as 0.003 mfd. or 2700 cm. this would give only $\frac{62{,}500}{2700} = \frac{625}{27} = 23$ jars total,

or in two series 46 jars on each side of primary. The capacity of new jars will be probably 0.0025 and a correspondingly greater number may be taken. With 40,000 volts we would be able to take $\frac{36}{16} \times 23 =$ nearly 52 jars *total* or 104 on each side. Assuming 60,000 volts and say 48 of new jars on each side this would give capacity in the primary most suitable to the transformer $24 \times 0.0025 = $ **0.06 mfd.** and the inductance of the primary being say 7×10^4 cm. The period T would be $= \frac{2\pi}{10^3} \sqrt{\frac{7 \times 10^4}{10^9} \times 0.06} = \frac{12.874}{10^6}$ and $n = 77{,}660$ per second.

Further conclusions relative to the best working conditions and constructional features of such oscillators derived from observations made in these and previous experiments. Beginning with the primary, the capacity should, as stated before, be best adapted to the generator which supplies the energy. This consideration is, however, of great importance only when the oscillator is a large machine and the object is to utilize the energy supplied from the source in the most economical manner. This is the case particularly when the oscillator is designed to take up the *entire output* of the generator, as may be in the present instance. But generally, when the oscillator is on a supply circuit distributing light and power the choice of capacity is unrestricted by such considerations. In most cases the advantages secured by using a very high frequency are so pronounced that the primary circuit will have to be designed with this feature in view. The resistance of the primary circuit should be in any event as small as it is practicable to make it. I also think that generally, the inductance should be as small as practicable for that frequency which is supposed to be arbitrarily selected beforehand. When, however, the break number is comparatively small, that is, much smaller than the number of free vibrations, it is of great advantage to have the inductance great in order to give a greater momentum to the circuit and to thus enable it to vibrate longer after each break. But if the break number is of a frequency comparable with that of the free vibrations, the inductance should be as small as possible for however small it be, the circuit will generally vibrate long enough. One more reason why the inductance should not be large in such a case is that, in the primary, it is unnecessary to raise the pressure by making $\frac{pL}{R}$ very large. Necessarily this factor *will* be large in a well designed circuit, but should be so chiefly owing to an extremely small resistance and not owing to a high self-induction. By making the inductance smaller a greater capacity may be used and this will give a greater output, a feature which is sometimes of importance. Of course, as the capacity becomes large the difficulties in the make and break device increase, but with a properly designed mercury break these difficulties are in large measure overcome. I conclude from the above facts that the best way to construct a primary in such a machine is to use thin sheet of copper or at any rate a stranded cable. I have settled upon using copper sheet in the smaller machines since long ago, this giving the best result. By using sheet a very small inductance is obtained and *more length* of conductor can be wound on for same frequency, at the same time the opportunity for radiation is excellent and the construction is simple and cheap. For the same section sheets heat *much less* than cable and the difference in this respect is so marked that I have been tempted to believe that there is a special reason for it, not yet satisfactorily explained. The *actual length* of the primary conductor, relative to that length which is obtained by dividing the velocity of light by $2n$, n being the number of vibrations of the primary per second — is of little importance since the primary is generally but a very small fraction of that length, but I believe to have observed that it is preferable, in a slight degree, to make the conductor of such a length that, if l be this length and n the frequency, $2\,Knl$ should $= v$, the velocity of light, and K should be a *whole* number and not a *fraction*. At least this seems to hold good in circuits made very long, expressly for the purpose of ascertaining whether there is truth in this idea which was arrived at by considering the ideal conditions of such a vibrating circuit. In this abstract case l should be rigorously equal to that *half wave* length which is obtained by computation from the velocity of light. In practice it is invariably observed that l is smaller and K is not as it should be $= 1$ but is often a large number, this simply following from the fact that the velocity of propagation in a circuit with considerable inductance and capacity is generally much smaller than that of light and often considerably smaller. It is to be stated that for a number of reasons it is of advantage, whatever be the actual length of the primary conductor, to arrange it so that it is *symmetrical* with respect to the condenser and the make-and-break device, one of the chief objects being to secure the maximum difference of potential on the terminals of the condenser. This consideration leads to the adoption of at least *two condensers* in *series*, the primary generally joining the outer coatings while the inner ones are bridged by the break device.

Coming now to the secondary quite different considerations apply. First, we must decide whether the secondary high electromotive force is to be obtained exclusively or entirely by transformation as in the commercial transformers with iron core, or not. In the first case obviously similar rules of economic design as followed in ordinary transformers will have to be respected. The secondary will have to be placed in the closest possible inductive relation to the primary and this will give an economical machine and one of relatively high frequency, since the inductances of the circuits by mutual reaction will be considerably reduced. But it is at once seen, that in a machine as here chiefly considered for the purposes followed from the outset, the connection between the primary and secondary can never be as close as in ordinary transformers, and the connection must be all the less intimate as the pressure on the secondary is increased since the wires must necessarily be placed at a greater distance from each other. From this it follows that in such a machine the free vibration of the secondary can never be quite ignored even if the electromotive force is not extraordinarily high. Now directly as the free vibration of the secondary becomes an important element to consider in the design, the careful adjustment becomes obviously imperative. It goes without saying that $\dfrac{pL}{R}$ should be as large as possible in all cases where resonant rise is one of the objects. But here is where we find in practice, and particularly in a large machine, difficulties not easily overcome. Both the inductance and capacity grow rapidly as turns are added, so much so that very soon it is found the secondary period becomes longer than that of the primary. The chief drawback is, as has been already pointed out, the distributed capacity but also the inductance though in a lesser degree. While the inductance in a certain sence has a great redeeming feature and is *necessary*, yet it stands in the way of obtaining a very high frequency in a large machine. To get a high electromotive force we must have many turns or turns of great length and this means great inductance and this again entails the drawbacks of *slow vibration*. Thus, in a large machine we encounter those difficulties which meet us in the design of too large a bridge, for instance, difficulties which are based on the very properties of matter and seemingly insuperable. Make a wire rope of twice the section and it will not be able to carry a longer piece of its own, since the weight is increased in the same proportion as the section and the strain per unit of the latter remains the same. Fortunately for us in electrical machinery, of this kind at least, this limit is immensely remote owing to the wonderful properties of this agent. Still the difficulties encountered on account of the capacity and inductance, and equally on account of the insulation are such as will require great deal of persistent effort to be effectively done away with in these oscillators, if the results aimed at are to be achieved in a thoroughly practical manner. Much attention will be devoted to this part of the problem in perfecting the machines which are necessary for the successful carrying out of the projects of transmitting power as well as effecting communication with any point irrespective of distance. But the machines for these two purposes will necessarily differ in design, since in one case — the first — a great amount of energy is imperative, while in the other only a high electromotive force and immense rate of *momentary energy delivery is required.* The two most promising lines of development are evidently these two: either to obtain the necessary electromotive force in a secondary alone, or in an extra inductance, not affected inductively by the primary or even by the secondary but merely excited by the latter, the rise of pressure being due to a great magnifying factor. The latter method has been found to be by far the best when a high e.m.f. but not a great amount of energy is required and there is scarcely any limit to the maximum pressure so producible. But it does not seem to me as though this method should be applied in power transmission contemplated, but this will be decided in the future.

Recognizing in the distributed capacity the chief drawback, I have since long thought on ways of overcoming it and the best seems to be so far, to make the secondary in sections of the determined length and to join them all in series through condensers of the proper capacity. By this means the greatest possible length of conductor may be utilized by a given frequency and placed in good inductive connection with the primary. Theoretically it is possible to entirely do away with the effect of distributed capacity in this manner and to use an excessively large inductance and consequently a large magnifying factor as well as magnifying transformation ratio. The wave length in such a theoretical case will be then exactly that which follows from the velocity of propagation of light.

Various arrangements experimented with for the purpose of studying effect of diminished inductance of secondary.

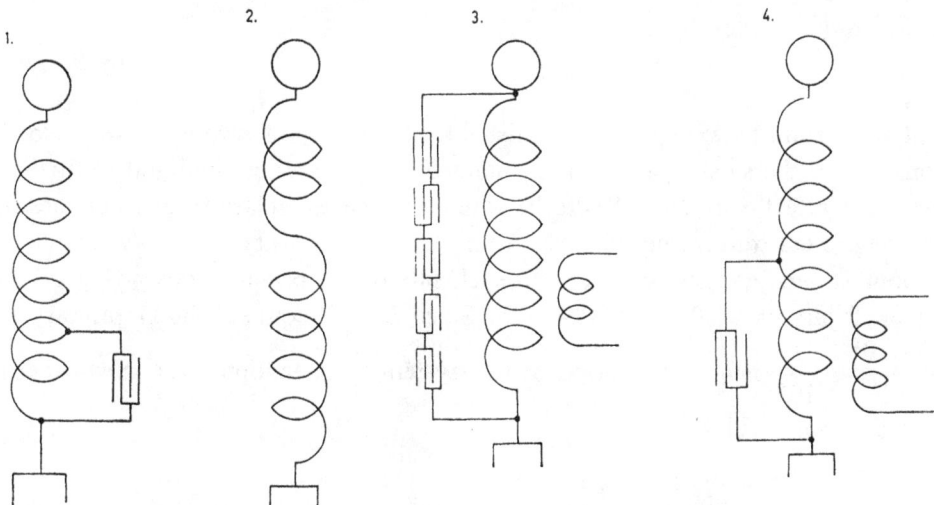

In Diagram 1. one part of the secondary was used as primary, this part being shunted by the primary condenser; in 2. a few turns of the secondary, those remote from the primary, were wound opposite; in 3. a series of condensers capable of standing the entire pressure of the secondary were used to shunt the same; in 4. a larger part of secondary inductance was counteracted by a condenser; in 5. the entire secondary inductance was counteracted by a condenser; in 6. the inductance of the secondary was reduced by placing

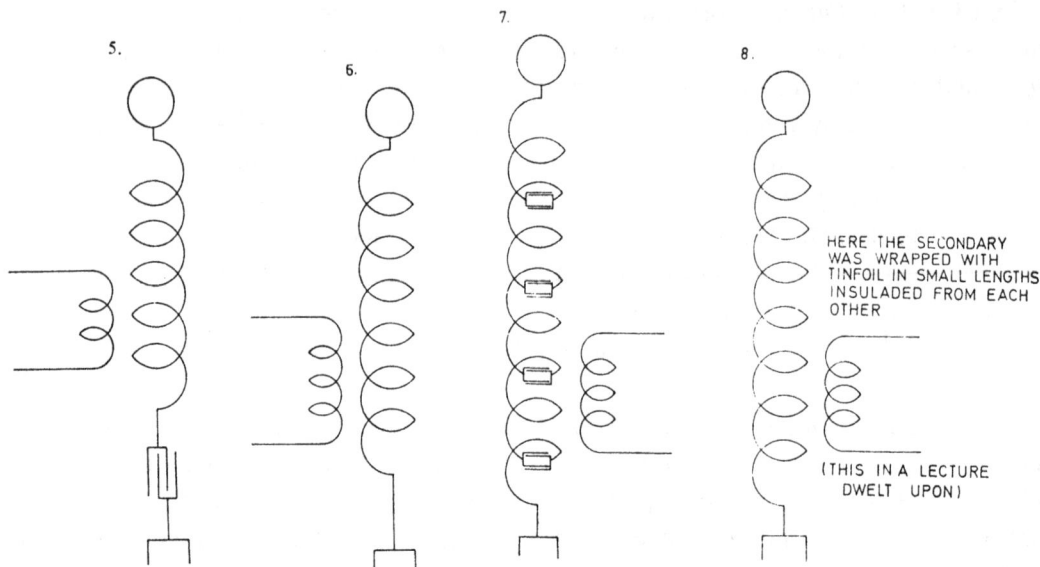

HERE THE SECONDARY WAS WRAPPED WITH TINFOIL IN SMALL LENGTHS INSULADED FROM EACH OTHER

(THIS IN A LECTURE DWELT UPON)

the turns far apart; in 7. inductance was counteracted by a series of condensers inserted between the turns and also distributed capacity was reduced and in 8. (not shown) static screeening was resorted to.

Colorado Springs

July 9, 1899

To ascertain to what extent the distributed capacity of secondary wire No. 10 was responsible for the small spark length obtained on the free terminal and to further study this capacity effect, wire No. 31 is to be wound on the secondary frame. For these experiments the present transformer brought from New York is to be used — which can charge just about 16 jars, as now operated, on each side of the primary, this giving primary capacity = 8 jars or $8 \times 0.003 = 0.024$ mfd. Since L with connections (1 primary turn) is $L = \dfrac{7 \times 10^4}{10^9} = \dfrac{7}{10^5}$ henry, the period of the system to be adopted for new experimental coil of this wire will be:

$$T = \frac{2\pi}{10^3}\sqrt{\frac{7}{10^5} \times \frac{24}{1000}} = \frac{2\pi}{10^7}\sqrt{168} \qquad \sqrt{168} = 12.9$$

$$68 : 2$$

$$2400 : 24$$

$$159$$

or T approx. $= \dfrac{6.28 \times 13}{10^7} = \dfrac{82}{10^7}$

This would give $n = 10,000,000 : 82 = 122,000$ per second.

180

160

From this again we get $\lambda = 186 : 122 = 1.524$ miles, or length of secondary roughly $=$ $= \dfrac{\lambda}{4} = 0.38$ miles or wire length $= 2000$ feet approx. When two primary turns in series are to be used we shall have $\dfrac{\lambda}{4}$ or length of wire $= 4000$ feet.

This to follow up.

The effects of distributed capacity in some experiments with the secondaries constituted of wire No. 10 (or cord) were so striking that it seemed worth while to carry on some investigations with very thin wire and consequently very small capacity in the secondary. It was decided to use wire No. 31 in these experiments. The diameter of this wire was only $\dfrac{1}{11.4}$ of that of wire or cord No. 10 hence the capacity of the new secondary, assuming all other things to remain the same, would be only $\dfrac{1}{11.4}$ of the capacity of the old secondary. The capacity of the new coil would be, however, reduced or increased in proportion to the length of the new wire relatively to the old and would be furthermore regulated by the distance of the turns. It was resolved to adopt 122,000 per second (see note before) in the primary as compared with 21,000 per second with the old secondary which was obtained with 15 jars on each side of the primary. For this vibration (122,000) the length of the new secondary should be about 2000 feet, this being the length of a quarter wave. Calling now the distance between the turns of the new coil d, its capacity as compared with that of old secondary of a length of 5280 feet would be:

$$C_1 = \dfrac{1}{11.4} \times \dfrac{20}{53} \times \dfrac{d}{d_1} C,$$ C being capacity of old secondary and d_1 the distance of

turns in same. We may call $\dfrac{d}{d_1} = D$ a number which will modify the capacity according to the distance of the turns, and we then have the capacity of the new coil $C_1 = \dfrac{1}{11.4} \times$ $\times \dfrac{20}{53} \times DC.$ Now L being the inductance of old coil and L_1 that of the new, we have, disregarding effect of the smaller diameter of wire for the present, $L_1 = \left(\dfrac{14}{36}\right)^2 DL$ for the inductance will evidently be changed in accordance with the same number D. This relation follows from the fact that 2000 feet of new secondary with 143 feet average length of turn give $\dfrac{2000}{143} = 14$ turns or nearly so, and there were 36 turns in old secondary, the length being preserved the same in both coils. From this it follows that the new system will vibrate with a period $T=?$ which will be $\sqrt{\left(\dfrac{14}{36}\right)^2 \dfrac{20}{53 \times 11.4}} \; D^2$ times the period of

the old system, or $\dfrac{14}{36} D \sqrt{\dfrac{1}{30}}$ or $0.07D$ times. From this we may calculate D.

The old primary system was 21,000 per second, the new 122,000 per second, hence we have $0.07\,D \times 21{,}000 = 122{,}000$ or $D=83$. This means to say that without any additional capacity on the free terminal the turns would have to be put at a distance only $\dfrac{1}{83}$ that of the former secondary. This would not be realizable because the sparks would pass between the turns. Capacity must be, therefore, added on the free terminal. Calling this capacity c and the old C, the total capacity on the new coil would be $C' = \dfrac{20\,DC}{11.4 \times 53} + c$ and the inductance would be $L'_1 = L_1 = \left(\dfrac{14}{36}\right)^2 DL$. Hence, for estimating D and c we have the equation:

$$\frac{1}{122 \times 10^3} = \frac{2\pi}{10^3} \sqrt{\left(\frac{14}{36}\right)^2 D \, \frac{38 \times 10^6}{10^9} \left(\frac{20 \times 1200\,D}{11.4 \times 53} + c\right) \frac{1}{9 \times 10^5}} \qquad \begin{array}{l} L = 38 \times 10^6 \\ C = 1200 \end{array} \left.\begin{array}{l} \text{old} \\ \text{coil} \end{array}\right.$$

$$\frac{1}{122 \times 2\pi} = \frac{14}{36 \times 3 \times 10^4} \sqrt{D(40\,D + c)} \quad \text{or} \quad \frac{108 \times 10^4}{3416\,\pi} = \sqrt{D(40\,D + c)}$$

Taking here now $D=1$ we get approx: $\dfrac{11{,}664 \times 10^7}{1166 \times 10^4} = 10{,}000 = (40+c)$ or $c=$ roughly 10,000 cm. for same distance of wires in new coil as in old.

It was of interest to determine the period of the combined primary and secondary system of the experimental oscillator and the following method was adopted. A coil wound with thin wire, turns separated by a string to reduce distributed capacity, was placed at a distance of a few feet from the vibrating system and so that it was about equally affected by the primary and secondary. One end of the coil was connected to earth and the other end was left free. An idea was already obtained beforehand as to the frequency which was likely to be found but more wire was wound on the coil to enable the adjustment to be effected by taking off turns. Turns were then taken from the lower end of the coil until a maximum of spark length from free terminal was obtained. The coil then gave a spark 5″ long. This took place when there were 1140 feet of wire on the coil, this length being found by measurement of resistance $\begin{cases} \text{res. of coil 38.4 ohms} \\ \text{res. of 12 feet 0.405 ohms} \end{cases}$

Neglecting the capacity of the coil this length should be $= \dfrac{\lambda}{4}$ or the quarter of a wave length. Of course, it was less but it was surmised that it would not be very much less. The total length of wave was then $\lambda = 4560$ feet. From this would follow

$$n = \frac{186{,}000 \times 5280}{4560} = 215{,}370 \text{ per second.}$$

Now the capacity in the primary circuit was ten new jars on each side, this making a five jars total or $5 \times 0.0025 = 0.0125$ mfd. or $\dfrac{125}{10^4}$ mfd. From this the period of the system ought to have been

$$T = \frac{2\pi}{10^3} \sqrt{\frac{125}{10^4} L \left(1 - \frac{M^2}{NL}\right)}$$

N = inductance of secondary
L = inductance of primary
M = mutual inductance

But on a previous occasion $1 - \dfrac{M^2}{NL}$ was found to be nearly 0.6 so that

$T = \dfrac{2\pi}{10^3} \sqrt{\dfrac{125}{10^4} \times 0.6 \times L}$ Taking now $L = 7 \times 10^4$ cm. or $\dfrac{7}{10^5}$ henry we have

$$T = \frac{2\pi}{10^3} \sqrt{\frac{125}{10^4} \times \frac{6}{10} \times \frac{7}{10^5}} = \frac{2\pi}{10^8} \sqrt{42 \times 125} =$$

$$= \frac{2\pi}{10^8} \sqrt{5250} = \frac{72.46 \times 2\pi}{10^8} = \frac{455}{10^8} \quad \text{approx.}$$

and from this would follow $n_1 = \dfrac{10^8}{455} = 219{,}800$ per second. This is fairly close within the limits of ordinary errors of measurement and observation.

On the bases of above estimate of n, we may also determine the inductance of primary cables or cable as modified by the influence of the secondary since we have the equation

$$\frac{1}{215{,}370} = \frac{2\pi}{10^3} \sqrt{\frac{125}{10^4} L'} \quad \text{which gives} \quad L' = \left(\frac{10^4}{21{,}537 \times 2\pi}\right)^2 \times \frac{1}{125} = \frac{10^8}{185{,}537 \times 10^5} \times \frac{1}{125}$$

$$L' = \frac{10^8}{2319 \times 10^4 \times 10^5} = \frac{1}{23{,}190} \quad \text{henry or} \quad \frac{10^9}{23{,}190} = \frac{10^8}{2319} = \textbf{43,120 cm.}$$

This is not far from the truth since

$$7 \times 10^4 \times 0.6 = 70{,}000 \times 0.6 = \textbf{42,000 cm.}$$

Estimate of capacity of a large coil (secondary) by Lord Kelvin's formula for concentric cable, to see how far it is applicable to a coil. Kelvin gives $C = \dfrac{KS}{4\pi r \log \dfrac{r'}{r}}$. Here S is the surface of inner copper conductor, r' radius of inside hole of conductor outside of cable, r radius of copper conductor inside (Diag. 1.). Assuming now a cable wound up having n turns at a distance d, each turn may be considered as having a conductor on either side at distance d. If we draw a circle of radius d around the conductor and imagine the inner surface conducting, we have the capacity of such a system according to above formula. Now in the case of a coil of this ideal surface we utilize only a small part which is approximately, when d is very large compared to r, $\dfrac{2r\pi}{2\pi d} = \dfrac{r}{d}$.

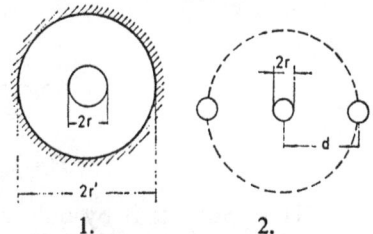

1.   2.

73

Now if there are n turns there will be $(n-1)$ such systems as illustrated in Diagram 2. Taking air as insulation and neglecting the effect of the small thickness of other dielectric, we would then have the total capacity: $C_1 = \dfrac{S_1}{4\pi r \log \dfrac{d}{r}} \times (n-1) \times \dfrac{r}{d}$. Now $S_1 = 2\pi r l$, l being the length of one turn. The values calculated out in this particular case of a secondary of 37 turns (36 1/2+connecting wires=approx. 37 turns) are as follows:

$n = 37$

$S_1 = 3600$ sq.cm.

$r = 0.125$ cm.

$d = 4.2$ cm.

$K = 1$

$l = 143$ feet $= 4360$ cm.

$\dfrac{d}{r} = 33.5$, $\quad \log \dfrac{d}{r} = 1.525045$

Substituting these values we have:

$$C_1 = \frac{3600 \times 36}{4\pi \times 1.525 \times 4.2} = \frac{9 \times 3600}{\pi \times 1.525 \times 4.2}$$

$$= \frac{32,400}{3.1416 \times 1.525 \times 4.2} =$$

$$= \frac{32,400}{4.791 \times 4.2} = \frac{32,400}{20.12}$$

$$C_1 = \text{approximately} = \frac{32,400}{20} = \textbf{1620 cm.}$$

This is not far from the value found. It shows that an approximate estimate may be made in this manner.

Colorado Springs

July 10, 1899

The following consideration conveys an idea of the drawbacks of distributed capacity in the secondary. Suppose the total capacity of 12 such large turns would be 1200 cm. as may be frequently the case, or 100 cm. per one turn and now let us ask to what potential this condenser may be charged — assuming further the energy to be carried away for the performance of work — by expending 1 H.P. In this case we would have $\dfrac{1000\,P^2}{9 \times 10^{11} \times 2}\,2n = 750$ watts. Here P is the potential, n the number of vibrations per second.

In our case n may be 20,000 and then we may put

$$750 = 40,000 \times \frac{1000\,P^2}{9 \times 10^{11} \times 2} = \frac{2\,P^2}{9 \times 10^4} \text{ or}$$

$$P^2 = \frac{9 \times 10^4 \times 750}{2} = 9 \times 10^4 \times 375 = 3375 \times 10^4 \text{ or}$$

$$P = 100\,\sqrt{3375} = \text{nearly } \textbf{5800 volts.}$$

This shows that by only charging the internal capacity to the insignificant pressure of 5800 volts we would have to expend 1 H.P. Of course, normally the power is small although the capacity is charged up to a much higher potential, but the consideration shows why with a large distributed capacity a very high pressure can not be obtained on the free terminal. All the electrical movement set up in the coil is taken up to fill the condenser and little appears on the free end. This drawback increases, of course, with the frequency and still more with the e.m.f.

In accordance with the preceding, an experimental coil of No. 31 wire (No. 30 not being on hand) was wound on the secondary frame. In the first experiment 14 1/2 turns were coiled up. The results were disappointing and for some time mystifying. The induced e.m.f. ought to have been 14 1/2 times the primary less 40% of total as before stated, but it did not seem so. Finally it was recognized that, as the capacity of the new secondary was very small, the free vibration of the coil was very high hence no good result could be obtained. The capacity in the primary was now reduced until to all evidence resonance was obtained, but the results were much inferior to what might have been expected, probably because $\dfrac{pL}{R}$ was small owing to large resistance. One of the reasons was, however, that the capacity in the primary was too small to allow a considerable amount of energy to be transmitted upon the secondary. To better the conditions, one of the balls of 30'' diam. was connected to the free terminal; this allowed a greater number of jars in the primary to be used but the capacity of 38.1 cm. was by far too small to secure the best condition of working. The resonating condition in secondary was secured with approximately 7 jars on each side of the primary and when the ball was connected with about 14 jars.

The capacity of the secondary was estimated to be 40 cm. and the inductance approx

$$15 \times 10^6 \text{ cm.} = \frac{15}{10^3} \text{ henry.}$$

Note: In these estimates I consider not the actual distributed capacity, but an ideal capacity associated with the coil.

This gave period of secondary roughly:

$$T = \frac{2\pi}{10^3}\sqrt{\frac{10}{10^3} \times \frac{40}{9 \times 10^5}} = \frac{2\pi}{10^7}\sqrt{\frac{600}{9}} = \frac{20\pi}{3 \times 10^7}\sqrt{6} = \frac{2\pi}{3 \times 10^6} \times 2.45 =$$

$$= \frac{4.9 \times \pi}{3 \times 10^6} = \frac{15.4}{3 \times 10^6} \quad \text{or about} \quad \frac{1}{195 \times 10^3} = T.$$

From this $n = 195,000$ per second. This vibration was far above that of the primary circuit working under favorable conditions, that is with the full number of jars. As the thin secondary did not yield any satisfactory results a coil was now associated with it. It was one used in some experiments before, having 260 turns of cord No. 10 (okonite) wound on a drum 2 feet in diam. and 6 feet long. The total length of wire was 1560 feet and the capacity of the coil (as above) 1530 cm. This coil was connected to the free terminal of the secondary and the free end of the coil was placed vertically on the top of the same and in the prolongation of its axis. Fairly good resonant rise was obtained on the free end, the streamers being 2 1/2—3 feet long. The primary circuit had 9 jars on each side, this giving $4.5 \times 0.003 = 0.0135$ mfd. primary capacity and as the inductance of the primary was $\dfrac{7 \times 10^4}{10^9}$ H the period of the primary was about $\dfrac{2\pi}{10^6}$ and $n = 160,000$ per second. According to this $\dfrac{\lambda}{4}$ should have been about $\dfrac{1.16}{4}$ miles or 1531 feet. Assuming the coil and the secondary of 14 1/2 turns vibrated together as one, λ would have been $4\,(2000 + 1560) = 4 \times 3560$ feet and $\dfrac{\lambda}{4} = 3560$ feet. But the experiment disproved this and showed clearly that the coil vibrated *alone*, the secondary merely exciting it, since the $\dfrac{\lambda}{4}$ on primary bases was 1531 feet, while the spool had 1560 feet, very nearly the same. For best results secondary should have been tuned to the same period as that of the coil. *Conclusion* of *experiments with thin wires* was: results must be inferior as free vibration is important.

In using wire No. 30 in the experiments proposed instead of wire No. 10 in the present secondary, we shall have to consider the following: the largest number of jars would be 154 on each side of the primary and with *one primary turn* which it is advantageous to use on account of facility of adjustment. We would have, as before found,

$$n = 43,500 \text{ approx. Thus } \left.\begin{matrix} C_p \\ L_p \end{matrix}\right\} \begin{matrix} \text{primary capacity and} \\ \text{inductance are } \textit{fixed.} \end{matrix}$$

The wave length will be:

$$\frac{186,000}{43,500} = 4.2 \text{ miles of } \frac{\lambda}{4} = 1 \text{ mile roughly.}$$

Now No. 30 wire has diam. 0.01
„   10 „   „ 0.1   } Suppose the same number of turns as

before, wound on the secondary frame and placed at twice the distance, the distributed capacity will be not far from $\dfrac{1}{2} \times \dfrac{1}{10}$ or $\dfrac{1}{20}$ of that of the old secondary. The inductance of the new coil — neglecting effect of small diameter of No. 30 wire — will be one half of the old, hence the new system will vibrate $\sqrt{20 \times 2} = \sqrt{40} = 6.3$ times quicker than the old secondary system. Since the old system vibrated 21,000 per sec. the new will vibrate 6.3 times that or 132,000 times per second, this number being $= n$. We should now, for the best suitable conditions, make the new system vibrate only about 1/3 that number since this would be approximately the vibration of the primary system, as above stated. Now taking the capacity of the old secondary 1200 cm., that of the new would be only $\dfrac{1200}{20} = 60$ cm.; we would have, therefore, to bring the pitch down to make the capacity 9 *times* as large or $9 \times 60 = 540$ cm. or we would have to put about 480 cm. on the free terminal of the new coil.

Further consideration in using a new secondary of No. 30 wire. The length of 1 mile of this wire will have a resistance on the basis of 9.7 feet per ohm from tables $\dfrac{5280}{9.7} = 544$

ohms approx. To get a rough idea of how the coil might work suppose that on the primary there would be 40,000 volts and that there were 36 turns of secondary, then there would be theoretically an e.m.f. on the terminals of secondary $40,000 \times 36 = 1,440,000$ volts and, deducting with reference to mutual inductance 40%, we would have $1,444,000 \times 0.6 = 864,000$ volts impressed or induced e.m.f. Suppose now we had a capacity of 480 cm. on the free terminal as before estimated, the charge stored in this condenser would be:

$$\left.\begin{array}{c}\text{Coulombs}\\\text{or}\\\text{amp. --- sec.}\end{array}\right\} : \quad \dfrac{864,000 \times 480}{9 \times 10^{11}} \quad \text{per one charge and there being}$$

$43,500 \times 2 = 87,000$ charges per second and taking a theoretical case we

would have $\dfrac{864,000 \times 480 \times 87,000}{9 \times 10^{11}} = \dfrac{86 \times 87 \times 48}{9 \times 10^4} = 4$ amp. nearly. The current in the

secondary would be quite large for the small section of the wire. Suppose 50 H.P., expended we may say roughly: $E_s i_s = 50 \times 750 = 37,500$ watts. With the above pressure we would have: $864,000 \times i_s = 37,500$ or $i_s = 0.043$ amp. only. In this

case the energy lost per second would be $544 \times \dfrac{43^2}{10^6} = \dfrac{1849 \times 544}{10^6} = 1$ watt approx.

The loss in this case, in spite of the great amount of energy transformed, would be ridiculously small owing to the great e.m.f. despite comparatively high resistance of the secondary. But this only seems so, for 544 ohms would be extremely small resistance for such e.m.f. The theoretical case above considered is hardly realizable, it would require an immense amount of energy.

Additional coil used before in some experiments, also in trials with a secondary
of No. 31 wire.

Approximate inductance of coil

Data: diam. 2 feet$=24''=61$ cm.$=d$

$$\text{Area} = \frac{\pi}{4} d^2 = \frac{\pi}{4} \times 3721 = 2922 \text{ sq.cm.}$$

length $71'' = 71 \times 2.54 = $ approx. 180 cm.

From this $L = \dfrac{4 \pi N^2 A}{l} = 12.57 \times \dfrac{260^2 \times 2922}{180}$ $\quad N = 260$ turns, this gives

$$L = 13,925,000 \text{ cm. approx.} = 0.0139 \text{ H}$$

Period of primary was before found $\dfrac{1}{16 \times 10^4}$, hence

$$\frac{1}{16 \times 10^4} = \frac{2\pi}{10^3} \sqrt{\frac{13,900,000}{10^9} C}$$

From this a rough idea of the capacity of the coil may be had, resonance being observed when the primary had $T = \dfrac{1}{16 \times 10^4}$ as above. From above equation we find

$$C = \frac{10}{142,336} \text{ mfd. or in cm.}$$

$$C = \frac{9 \times 10^5 \times 10}{142,336} = \frac{9 \times 10^6}{142,336} = 63.2 \text{ cm.}$$

This is a much smaller value than would be expected from previous approximate estimates. The correctness of the value found depends, among other things, chiefly on the correct determination of the period, for resonance might have been also obtained with a lower or upper harmonic, but this is not very likely to be the case as the vibration was very intense. There were approximately 5000 volts on primary turn. Resistance of coil was 1.56 ohm. From this $\omega = 2\pi \times 16 \times 10^4 = 1,004,800$ or 10^6 approx; $\omega L = 0.014 \times 10^6 = 14,000$, $\dfrac{\omega L}{R} = \dfrac{14,000}{1.56} = 8910$ as factor for magnifying e.m.f. impressed; **very large.**

Colorado Springs

July 11, 1899

Some considerations on the use of "extra coils". As has been already pointed out an excellent way of obtaining excessive electromotive forces and great spark lengths is to pass the current from a terminal of an oscillating source into such a coil, properly constructed and proportioned, and having preferably a conducting body — best a sphere — connected to the free terminal. In free air the *highest economy* is obtained with a well polished sphere, but for the greatest spark length — if this be the chief object — no capacity on the free terminal should be used, but all the wire should be carefully insulated so that streamers can not form except on the very end of the wire which as a rule should be pointed. This, however, is not always true. When the apparatus delivers a notable amount of energy the curvature of the end of wire or terminal attached to it should be such that the streamer breaks out only when the pressure at the terminal is *near the maximum*. Otherwise, very often, when a finely pointed terminal is employed the streamer begins to break out already at a time when the e.m.f. has a small value and this reduces, of course, the spark length and power of the discharge. By careful experimentation and selection of terminal the most powerful spark display is easily secured which the particular apparatus used is capable of giving. When the conditions are such that for the most powerful discharge a terminal of some, relatively small, curvature is needed, the curvature of the terminal can be beforehand calculated so that the discharge will break out at any point of the wave desired, when the e.m.f. at the terminal has reached any predetermined value. The smaller the curvature of the terminal the smaller an electromotive force is required to enable the discharge to break out into the air. In fact, the curvature of the terminal may serve as an indication of the value of the e.m.f. the apparatus is developing and it is often convenient to determine the e.m.f. approximately by observing how large a sphere will just prevent the streamers from breaking out, and for such purposes I have found it useful to provide the laboratory with such metallic spheres of different sizes up to 30″ diam. It is of course necessary to guard in such experiments against errors which might be caused by any modification of the constants of the vibrating circuit through the addition to the system of a body of some capacity. The latter should be such as to insure the maximum rise of the pressure. With apparatus of inadequate power the pressure may be very much diminished by the addition of capacity merely because there is not enough energy available to charge the same to the full pressure. It is a notable observation that these "extra coils" with one of the terminals free, enable the obtainment of practically *any* e.m.f. the limits being so far remote, that I would not hesitate in undertaking to produce sparks of thousands of feet in length in this manner. Owing to this feature I expect that this method of raising the e.m.f. with an open coil will be recognized later as a material and beautiful advance in the art. No such pressures — even in the remotest degree, can be obtained with resonating circuits otherwise constituted with two terminals forming a closed path. It is also a fact that the highest pressure, at a free terminal, is obtained in that form of such apparatus in which one of the terminals is connected to the ground. But such "extra coils" with one terminal free may also be used with ordinary transformers and by using one such coil on each of the terminals of the transformer, practically any spark length may be reached. Of course, it is desirable that the frequency of the currents should be high, as with the common frequencies of supply circuits the lengths of the wires in the coils become too great. In the diagrams below the two typical arrangements with such an "extra coil" or coils are illustrated in which Diagram 1. illustrates their use with an ordinary transformer, which may have an iron core or not, and Diagram 2. shows typically the connection as

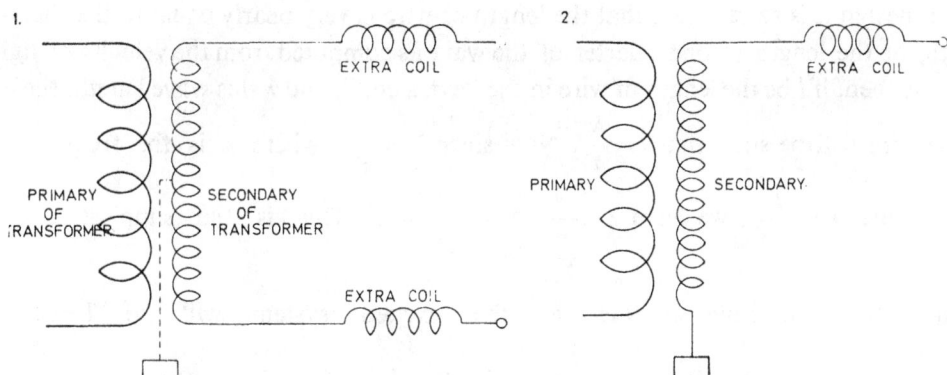

1.

EXTRA COIL

PRIMARY OF TRANSFORMER   SECONDARY OF TRANSFORMER

EXTRA COIL

2.

EXTRA COIL

PRIMARY   SECONDARY

I use it in my "single terminal" induction coil. As has been stated on a previous occasion in connection with this subject, to enable a considerable rise of pressure to take place in a circuit, the same must be tolerably free from inductive influences of other circuits. It follows from this that, although with a secondary in *loose* connection with a primary a very high pressure is obtainable, yet the pressure will never be as high as when an "extra coil" *not* in inductive connection with the primary is employed to raise the pressure, because the secondary always reacts upon the primary thus dampening the vibration, while the "extra coil" does not react in *such* a *manner*, the rise of pressure being simply due to the factor $\dfrac{pL}{R}$.

The object of the considerations which follow is to establish simple relations between the quantities which are known or adopted beforehand, so as to enable the experimenter to construct such coils without previous trials.

Calling E_0 the impressed e.m.f. and E the pressure measured with reference to the free terminal — that is the maximum pressure, p the product $2\pi n$ as usual, L the inductance of the "extra coil" and R its resistance, we have the known relation $E=\dfrac{pL}{R}E_0$. Obviously the maximum rise will take place when the period of the excited system or "extra coil" is the same as that of the oscillating system impressing the movement, for although the results obtained with a lower or upper harmonic, and particularly the former, may be sometimes so remarkable, as to be mistaken for effects of the true vibration, they are nevertheless always inferior, and I as a rule try the first upper and undertones to be sure of the result, when there exists any doubt in this respect. In ordinary practice the first element which is given will be the frequency, hence the wave length must be assumed as the first fixed quantity. But as has been already stated on another occasion, in an apparatus *designed* to give the *best result* the actual length of the wire should be that which is obtained on the basis of a velocity of propagation v equal to that of light. I have already remarked before that this is generally not true, the actual length of wire being always smaller than the theoretical length, and I propose to put together data derived from many experiments with coils wound with different wires and varying in size, from which it will be possible to always obtain, beforehand, with any particular wire, insulation and size of coil etc. the length required — by multiplying the theoretical length with a coefficient dependent on these and other particulars of this kind, different in special cases. Such coefficients will be certainly useful to the practitioners. The chief element determining the length of the wire is the distributed capacity and I shall presently suppose, that by proper design it is so reduced, that the length of wire is very nearly equal to the theoretical length, or the length of one quarter of the wave as computed from the velocity of light. In this case then, if l be the length of wire in the "extra coil", and λ this wave length, the length of the wire will be such that $l=\dfrac{\lambda}{4}$. Now since $\lambda=\dfrac{v}{n}$, where n is the frequency and furthermore $n=\dfrac{p}{2\pi}$, we have $\lambda=\dfrac{v}{\dfrac{p}{2\pi}}$ or $=\dfrac{2\pi v}{p}=\lambda$. This is the simplest case. The period of the exciting as well as the excited system will be $T=2\pi\sqrt{LC}$

where C is the capacity in farad $\Big\}$ of each of the
and L the inductance in henry $\Big.$ „extra coils"

$$\text{or } n=\frac{1}{T}=\frac{1}{2\pi\sqrt{LC}}=\frac{p}{2\pi}\ ; \quad \text{or } p=\frac{1}{\sqrt{LC}}\ \text{ or } LC=\frac{1}{p^2}$$

This condition, as is well known, must be fulfilled, whatever be the length of the wire, to enable the maximum rise of pressure to take place, and also, p must be the same number for both systems, obviously. Now evidently in designing the apparatus, in any case, the experimenter will know approximately what e.m.f. he would wish to secure, and consequently he will have an idea how much difference of pressure he will have between the turns, and this will give him again an idea how he must place the windings most advantageously. He will furthermore recognize at once that the simplest form of such a coil, and also the cheapest, will be one with *one* single layer and, settling upon this form, he will get an approximate estimate as to the diameter of the coil to be adopted. This will, of course, depend much on the kind of wire he uses and particularly on the insulation, since the better the insulation the closer will be two points of the coil between which there will exist a certain maximum pressure. Granted now the diameter of the coil to be constructed is settled upon, it will be at once seen that, assuming turns are wound on until resonance is obtained, the inductance of the coil and the capacity of the same will vary in the like manner. If more turns are wound on the drum their number will be proportionate to the length of the coil, therefore to the length of the wire, and the inductance will be proportionate — on the one hand, to the square of the turns or respectively to the square of the length of the wire, and on the other hand, it will be inversely proportionate to the length of the coil. Inasmuch as this length is likewise proportionate to the length of the wire, the inductance, on the whole, will be proportionate to the ratio $\frac{l^2}{l}$ or to l, that is to say, the inductance of the coil, as more turns are wound on, will grow as the length of the wire. And so will also, and obviously, the distributed capacity. And furthermore, and for the same reasons, under the conditions considered, both the capacity and the inductance of the coil will vary inversely as the distance of the turns which I shall designate τ. This is clear since, as far as the inductance of the coil is concerned, the number of turns will be inversely as τ, and the length of the coil *directly* as τ, hence the inductance will be inversely as τ; and, as regards the distributed capacity, it will be, of course, inversely proportionate to τ. Hence we can express both the unknown quantities L and C_1 (distributed capacity) in terms of l and τ. But it should be remembered that in the equation $LC = \frac{1}{p^2}$, C is capacity associated with the coil on the free terminal and not the distributed capacity. In case, therefore, we should make the capacity on the free terminal very large in comparison with the distributed capacity of the coil, or if capacity be associated with the coil in other ways, as by *shunting* the coil with a condenser as in sketch 3. — or in any other way, but so that the distributed capacity *may* be *neglected*, then the design of the coil is much simplified, for then one of the constants, preferably C, can be adopted beforehand and the other constant calculated. It will be in such case better to adopt first a capacity, because it is easy to get an idea of what kind of condenser to use when the

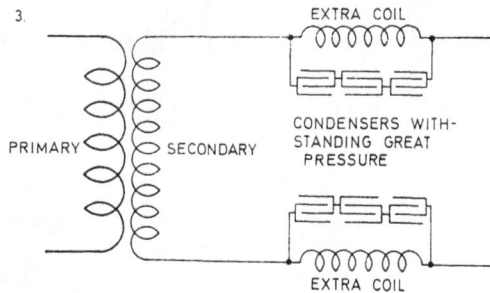

pressure of the terminals of the coil is approximately known. A practical way is, sometimes, to adopt a construction before suggested, securing a negligible internal capacity, consisting of the placing of condensers in series with the turns of the coil, and then merely calculating one of the constants, assuming first a value for the other constant, whichever is the more convenient of the two. I have also found it practicable in some cases to avail myself of some methods of tuning allowing exact observations as to the rise of pressure in the excited circuit, and to tune a small number of turns first to a much higher harmonic and, after completing this adjustment, to calculate the dimensions of the coil for the fundamental vibrations from the experimental data secured. But in most cases such resources are not readily available and an approximate idea must be gained in other ways. There are a number of different considerations which, when followed out in connection with the preceding, will lead to the establishment of simple relations between the quantities primarily adopted and will enable an experimenter to construct such a coil to suit source, without previous experiment and some of these I propose to consider on other occasions. Presently I shall indicate a way which, in some cases to which the calculated data were applied, has given satisfactory results.

The ideal capacity C which should satisfy the equation $LC = \dfrac{1}{p^2}$ is always a function of the distributed capacity C_1 and furthermore a linear function, so that $C = K_1 C_1$, where K_1 is a constant, the value of which is to be deducted from the results of many varied experiments carried on to this end. But this capacity C_1 is, as has been found in many experiments with coils of widely different dimensions, directly proportionate to the length of wire and to the diameter of the same and furthermore to the diameter of the drum. The latter will be understood when it is considered that the greater the diameter of the drum the greater is the potential difference between the turns and, consequently, the greater is the amount of the energy stored in the coil with a given length and diameter of the wire. Finally the quantity C_1 is inversely proportionate to the distance of the turns τ. As to the dielectric constant it is only then important to consider when the turns are quite close together so that the entire space between the turns is filled with the dielectric. When the turns are far apart this constant may be taken $=1$. From this it follows that the capacity C interpreted as above may be expressed by the following equation: $C = K \dfrac{D\,d\,l}{\tau \times 9 \times 10^{11}}$ when the dielectric constant can be neglected. Here D is the diameter of the drum, d diameter of the wire, l length of wire, τ the distance between the turns. C is the "ideal" capacity which will satisfy equation $LC = \dfrac{1}{p^2}$ expressed in farads. The constant K is determined by practical experiments. When the turns are very close the dielectric constant must be introduced and C will be multiplied with the latter quantity. Now the inductance of the extra coil to be constructed is $L = \dfrac{4\pi A N^2}{l_1 \times 10^9}$ henry. Here A = area of coil in cm. square,

N = number of turns, and l_1 length of coil in cm. Now $A = \dfrac{\pi}{4} D^2$, $l_1 = N(\tau + d)$. Hence

the inductance will be $L = \dfrac{4\pi \cdot \dfrac{\pi}{4} D^2 N^2}{N(\tau + d)\,10^9} = \dfrac{\pi^2 D^2 N}{(\tau + d)\,10^9}$ henry. Taking these values

for L and C we have, with reference to above, $\dfrac{1}{p^2} = \dfrac{\pi^2 D^2 N}{(\tau + d)\,10^9} \times \dfrac{d\,l\,D}{\tau \times 9 \times 10^{11}} K$ or

$$p^2 = \frac{(\tau + d)\,\tau \times 9 \times 10^{20}}{\pi^2 D^3\,d\,l\,N\,K}$$

Since in the preceding the diameter of the drum is assumed, from practical considerations it will be convenient to find the number of turns N. The quantities D and τ are, of course, interconnected since by assuming D and deciding on the pressure to be obtained beforehand, τ is practically given. The diameter of the wire will in most cases also be selected beforehand so that then merely N is to be determined to satisfy the condition of resonance for any frequency specified. Now $l=\pi D N$ hence substituting this we have from above:

$$p^2 = \frac{(\tau+d)\,\tau \times 9 \times 10^{20}}{\pi^2\,D^3\,d\,N\,\pi\,D\,N\,K} \quad \text{or} \quad p^2 = \frac{(\tau+d)\,\tau \times 9 \times 10^{20}}{\pi^3\,D^4\,d\,N^2\,K} \quad \text{and from this we get}$$

$$N^2 = \frac{(\tau+d)\,\tau \times 9 \times 10^{20}}{\pi^3\,D^4\,d\,p^2\,K} \quad \text{or} \quad N = \frac{3 \times 10^{10}\,\sqrt{(\tau+d)\,\tau}}{D^2\,p\,\sqrt{\pi^3\,K}\,\sqrt{d}} \ldots\ldots\ldots$$

This formula may serve to give an approximate idea of how many turns are to be wound on in cases when the length of the wire, owing to the capacity in the excited circuit, is smaller then $\dfrac{\lambda}{4}$ $\Big($or respectively smaller than $\dfrac{\lambda}{2}$ if the circuit is not one of the kind illustrated in diagrams above — that is, one in which the potential on one terminal is many times higher than on the other, but an ordinary circuit, in which there is a symmetrical rise and fall of pressure at both the terminals$\Big)$, but the equation assumes that K, the dielectric constant, is $=1$ or nearly so.

From a number of experiments the value for K with wire No. 10 as used in preceding experiments was found to be nearly $=\dfrac{52}{10^6}$. Introducing this value in equation for N and reducing the constant quantities we find $N=\dfrac{747 \times 10^9}{D^2\,p}\sqrt{\dfrac{\tau\,(\tau+d)}{d}}$.

To see how close this formula will give the value of N in a special case take, for instance, the secondary with 40 turns experimented with before. In this particular case we have the following data:
diameter of coil, average, 40 feet$=480''=1220$ cm. approx.

$$D=1220 \text{ cm.}; \quad D^2=1488 \times 10^3; \quad \tau=5 \text{ cm.}; \quad d=0.254 \text{ cm.}$$

Resonance in secondary from previous tests took place with the primary period being $T=\dfrac{4.836}{10^5}$, this was also the secondary period and from this $n=20{,}700$ approximately and this gives $p=130{,}000$ very nearly. On the basis of these data we would have:

$$N=\frac{747 \times 10^9}{D^2\,p}\sqrt{\frac{\tau\,(\tau+d)}{d}} = \frac{747 \times 10^9}{1488 \times 10^3 \times 13 \times 10^4}\sqrt{\frac{5 \times 5.254}{0.254}}$$

$$= \frac{747 \times 10^2}{1488 \times 13}\sqrt{103.4} = \frac{747 \times 10^2}{19{,}344} \times 10.16$$

$$= \frac{747 \times 1016}{19{,}344} = \frac{758{,}952}{19{,}344} = \mathbf{39.24} \text{ turns}$$

This comes indeed very close, the turns being actually 40 for the condition of resonance as experimentally shown.

This will be followed up.

Self-induction coil for condenser method in conjunction with oscillating transformer. Adapted to Thomas clockwork and condenser 1/2 mfd. mica on hand (one of the two small condensers). *Capacity* given 1/2 mfd., break also given: wheel of clockwork breaking and making contact has 180 teeth, turns about 20 a minute. This gives breaks $\dfrac{180 \times 20}{60} = 60$ per second. Here at each make and break we have a wave in the condenser, and tuning

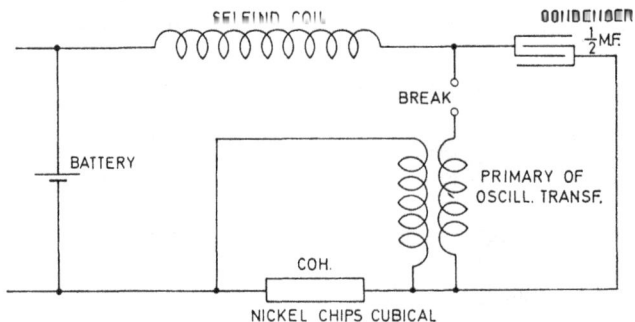

may be effected either by making $n=60$ or $n=30$. Best result seemingly, from former experiments with oscillators, seems to make $n=$ the number of breaks. We have then

$$T = \frac{2\pi}{10^3} \sqrt{L \times \frac{1}{2}}$$

$$T = \frac{1}{n} = \frac{1}{60} = \frac{2\pi}{10^3} \sqrt{L \times \frac{1}{2}}$$

From this $L = \dfrac{2 \times 10^3}{144} = \begin{matrix} 2000 : 144 = 14\,\text{H} \\ = \quad 560 \end{matrix}$

This would not be realizable with a condenser of $\dfrac{1}{2}$ mfd. Therefore a larger condenser or quicker break necessary — will vibration always take place through the high resistance of sensitive device? $\dfrac{4L}{C} > R^2$ Take $L=1$ henry, we have $\dfrac{4}{\dfrac{1}{2 \times 10^6}} > R^2$, $8 \times 10^6 > R^2$ or roughly $3000 > R$. This shows that although resistance may be very large, vibration will still take place.

Follow up.

Considerations regarding working of oscillator without spark in secondary. This is a considerable advantage because of economy and also facility of exact synchronous adjustment. When spark used the latter difficult as capacity is changed by varying distance of terminals, also because spark establishes short circuit temporarily. In general, the process is very complicated and the tuning only partially successful. But using spark allows obtaining of great suddenness and using short wave lengths. The shortness of waves gives high e.m.f. and, therefore, great effect at distance. Without spark it is difficult to obtain high e.m.f. with *short waves*. Long waves on the other hand are less absorbed and allow exact tuning. Following plan seems to offer particular advantages that seemed to work well in New York oscillator.

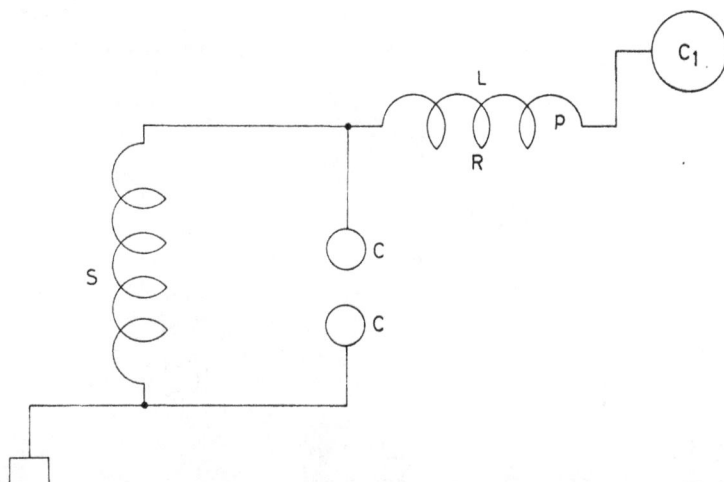

S is the secondary of oscillator. To this is connected a coil L with capacity C_1. The secondary is shunted by a condenser $C\,C$. This condenser can be of large spheres when practicable. No spark should go over the spheres $C\,C$ and streamers should be prevented. Now the adjustment may be such that system $L\,C_1$ is any upper harmonic. In this system $\dfrac{Lp}{R}$ should be as large as possible. The free vibrations of $L\,C_1$ can be transmitted upon earth through condenser $C\,C$.

Further considerations in regard to producing most effective movement without spark gap in secondary. 2) A way which was experimented with in New York about a year and a half ago and worked exceedingly well and also later with both, was to produce a very quick primary vibration and induce currents in secondary of a few turns which has one of its ends to earth and the other connected to a large capacity. Connections were as illustrated.

Fig. 1) supply direct current 220 volt. Mercury break 1600 per second. The secondary S_1 with small condenser C and spark gap d, primary P 2—3 turns.

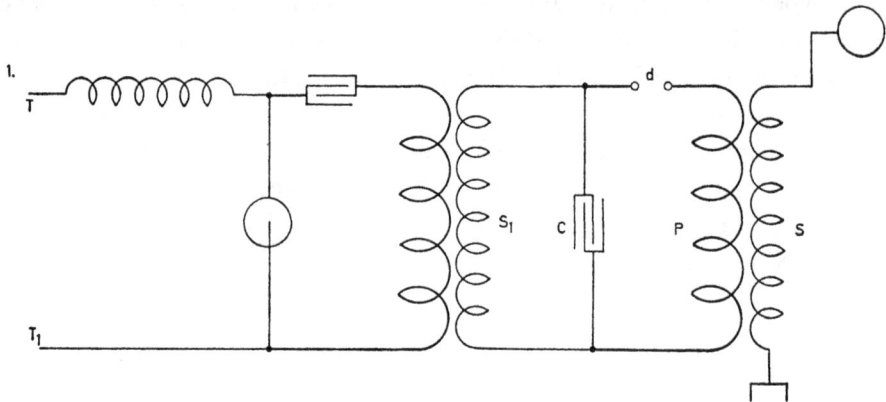

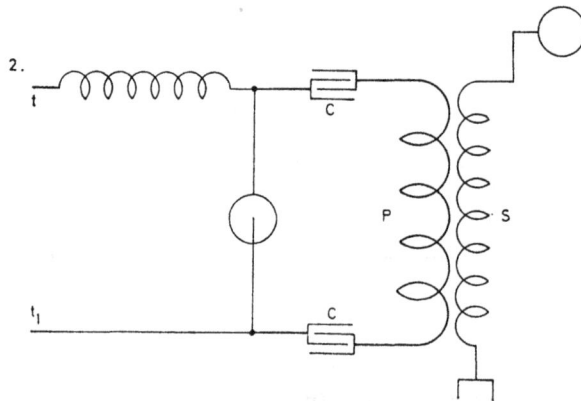

Fig. 2) supply circuit about 600 V; small condensers $C\,C$, $\dfrac{1}{2}$ mfd. each, 1 turn primary. Both of the arrangements worked well, that illustrated in 2) more economical but waves longer.

3). Another way (and seemingly best) is to provide a secondary which consists of a number of elements comprising condenser and coil each of a high frequency of vibration and all joined in series. The primary vibration should be quick corresponding to that of each of the elements of the secondary. In this manner any e.m.f. may be secured, the secondary may be of any length yet it will vibrate quick.

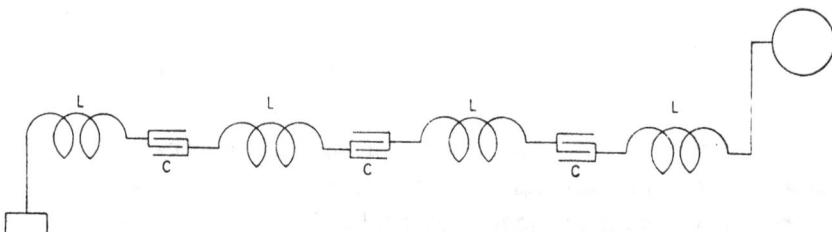

(To be followed up).

Some arrangements in telegraphy involving the Dynamo principle (first brought to Page some years ago). Present apparatus built two years ago in New York, worked very well. Consider which the best of the following modifications: In case 1. sensitive

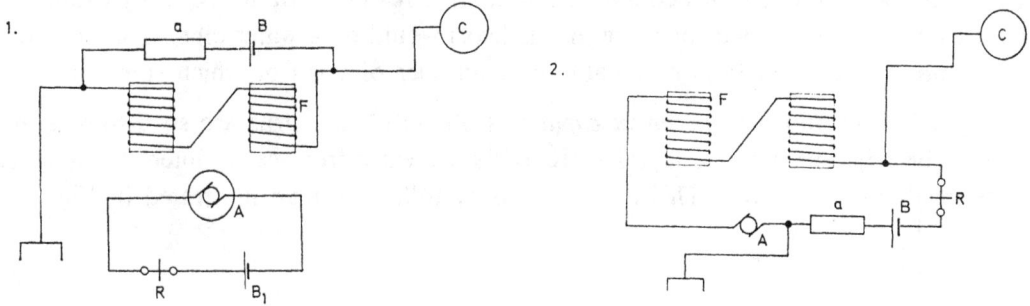

device *a* with battery around field *F*. In armature circuit independently a receiver *R* and battery B_1. The receiver may be a relay, and in addition, to insure greater sensitiveness another sensitive device as *a*, may be joined in convenient manner in the armature circuit. In case 2. the armature and field circuits are joined in series and battery and receiver are in shunt to both, also sensitive device *a*. In both cases the sensitive device may be also in series with the field or field and armature though arrangements 1 and 2 seem preferable. In arrangement 3. a shunt dynamo is shown, the sensitive circuit being also in shunt to the terminals of the dynamo. In addition, to regulate excitation of dynamo a shunt of high self-ind. is placed around the sensitive device *a*. Such a shunt may also be used with good effect in Fig. 2.

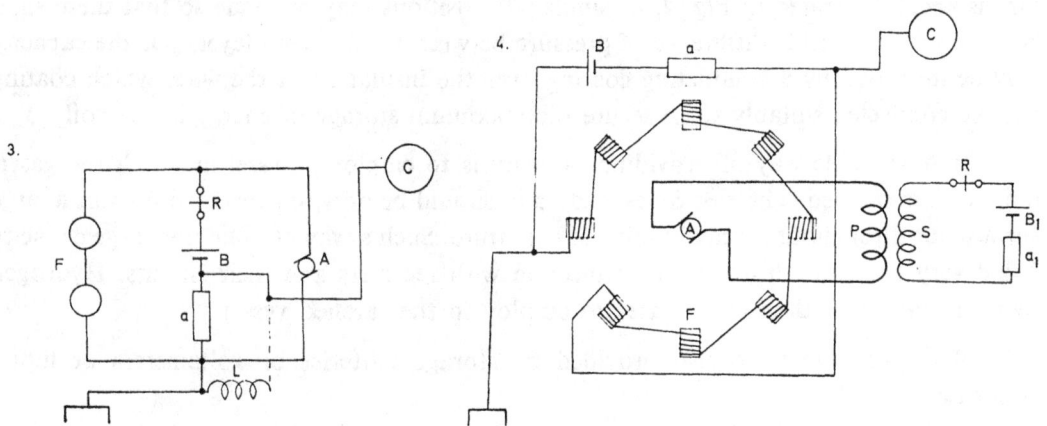

Fig. 4 illustrates one of the dispositions with an alternate and — preferably — high frequency dynamo. The letters are self-explanatory. The sensitive device a_1 may be omitted.

Colorado Springs

July 16, 1899

In order to produce the greatest possible movement of electricity through a region of the earth in accordance with the plan involving use of a single terminal oscillator, as here experimented with, it is desirable to obtain in some way a large capacity on the free terminal. This is connected with difficulty as spheres get to be too large with moderate tensions and when the tensions go into the millions, streamers can not be easily overcome. The streamers involve loss of pressure just as leaks would on a water pipe which is closed at one end. Large capacity is obtainable in a number of ways of which some are:

1) a coil wound for *maximum capacity* (internal). The turns are so disposed that between the adjacent turns of layers there exists a great difference of potential, as much as the insulation can stand. This is best done by following plan illustrated in Fig. 1 in

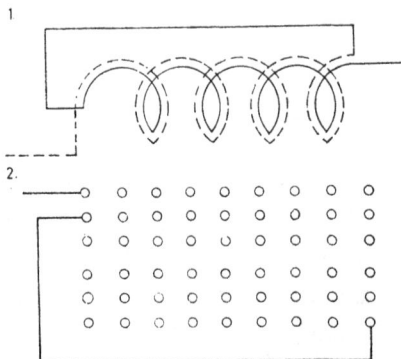

1

2.

which there exists between each two turns one half of the total difference of pressure which is active on the terminals of the coil. But other arrangements may be followed as, for instance, illustrated in Fig. 2, or similar dispositions may be made so that there shall be the greatest possible difference of pressure between the adjacent layers. Or the capacity may be increased by a conducting coating over the insulation of the wire, which coating may be connected suitably so to secure the maximum storage of energy in the coil;

2) A valuable way of providing capacity is to employ a vessel in which the gas is more or less rarefied. The electrodes leading in should be of wire gauze and present a large surface but throughout of small radius of curvature. Such a way of obtaining large capacity I find very good in telegraphy in connection with receivers and their circuits. Hydrogen seems to be better than other gases to employ in the rarefied vessel.

3) Capacity may be also provided by storage batteries or voltameters or liquid condensers.

4) Another way is by local condensers arranged on end of wire near the free terminal. This is illustrated in annexed diagrams. In Fig. 1. and 2. two of the many arrangements are shown. In Fig. 1. a condenser is placed at the free end of the secondary *S* of the oscillator, the other end being connected to earth. One coating is directly connected to the end of the secondary, that is to *b*, the other coating to a point *a* which has a suitable difference of potential with respect to *b*. By the operation of the oscillator energy is stored in the

condenser C, which energy must all be supplied through the secondary, thus producing a large movement in and out of the earth. A modified arrangement is shown in Diagram 2.

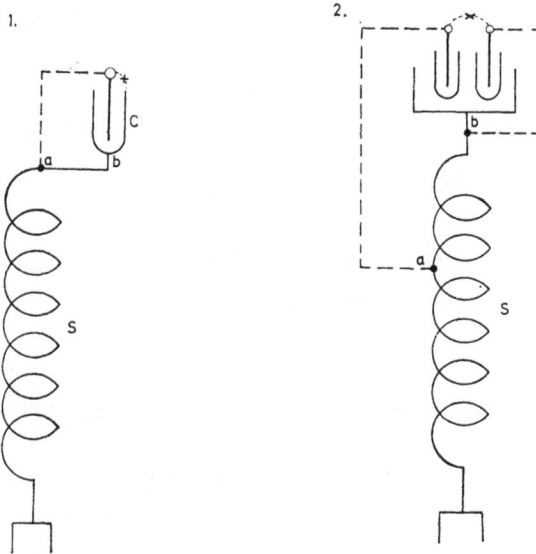

An arc may be maintained on the places marked x. Proper rules of tuning are observed to secure best result.

Colorado Springs

July 17, 1899

Some arrangements of apparatus experimented with. Modifications of former plans.

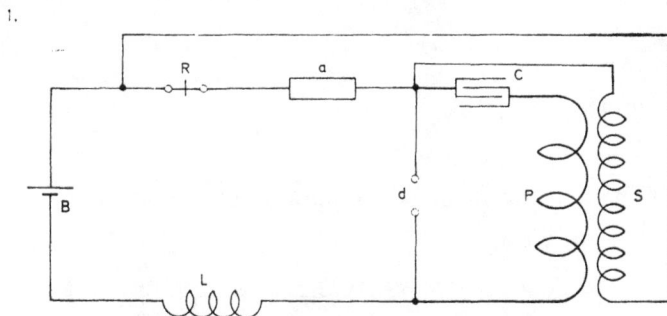

Here relay is placed in series with sensitive device but in secondary. In this way relay is not affected by break. The charge of condenser may be regulated by varying L or by resistances in series with L or with S.

2.

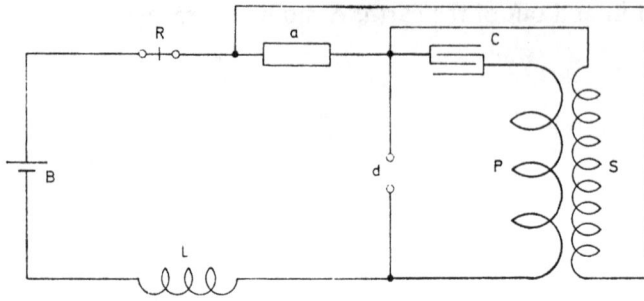

The relay is affected by the break in this disposition but the action was good in some instances; probably secondary *S* was more effective in breaking through the sensitive device *a*.

3.

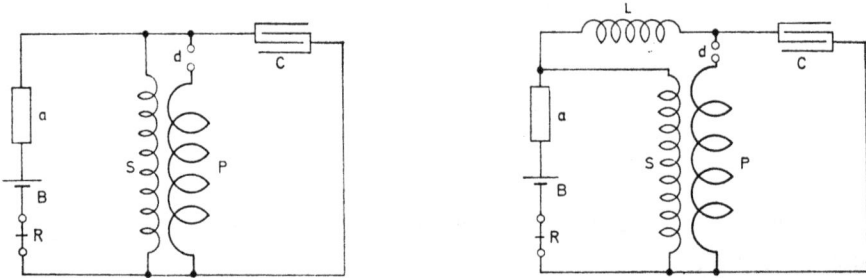

This disposition is simple and secures good results but one disadvantage has been found in the short circuit of the secondary through the condenser, which is necessarily too large for the high tension secondary since it fits the primary *P*. The above defect is reduced largely by the introduction of regulable self-induction *L*, or similarly a resistance is used instead of *L*. In all these dispositions of apparatus the effect upon the sensitive device is rendered accumulative.

Colorado Springs

July 18, 1899

Other arrangements of apparatus experimented with.

1.

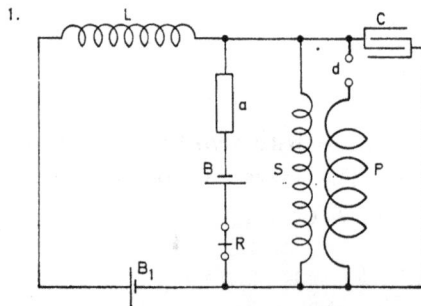

90

In this scheme the excitation of the condenser and therefore also of the sensitive device a was regulated by an adjustable self-induction and additional battery B_1. The battery B can be in the same or opposite direction working through the device a. The former was apparently preferable.

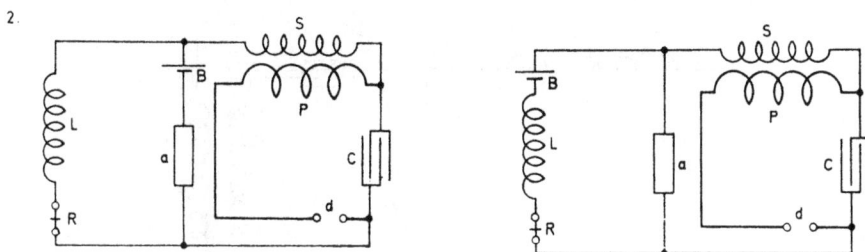

Of these two connections the first was advantageous as the battery was not working except when the sensitive device was excited.

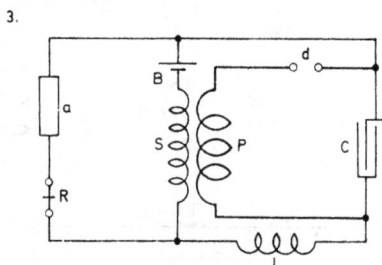

This was a plan (3) similar to one previously experimented with, only the battery B was placed so as to be able to charge the condenser.

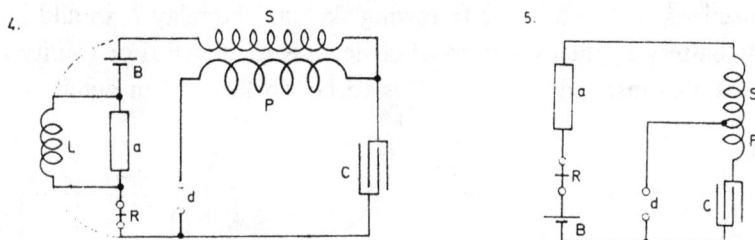

To determine excitation of sensitive device to the point of breaking down a self--induction coil L (very high) was placed around it (4). This coil was also tried with connections changed as indicated by dotted lines.

Here again (5) a part of secondary was used as primary. The arrangement worked well probably because as in some instances previously the secondary was open and the rise of pressure considerable upon a small excitation of device a. This is suitable for a device of great resistance.

Some simple dispositions in the practical uses of apparatus as now available.

1.

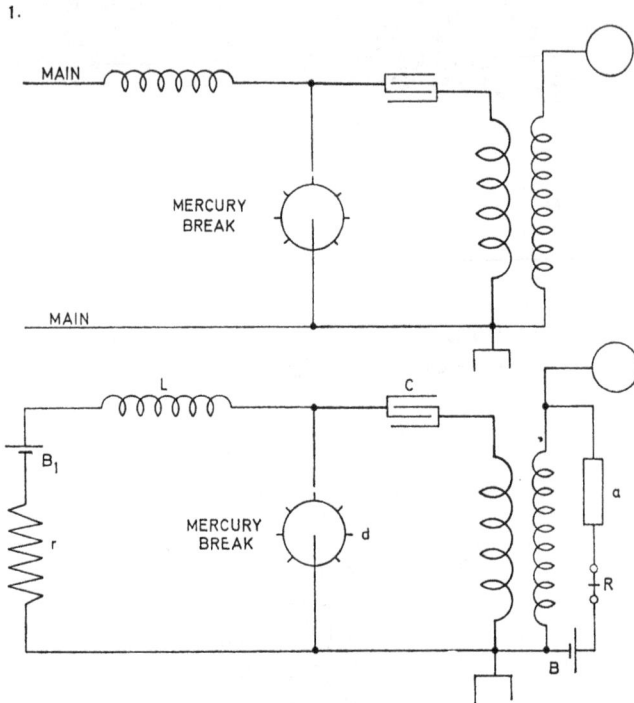

The connections in the oscillator as now manufactured are as shown in first sketch. In this way the apparatus is used as a sender. The connections are now by a throw of a switch changed in such a way that all can be used in receiving the message. One of the simplest connections is shown in the following sketch. The relay R should have small self--induction. By battery B_1 the excitation of device a is regulated. For facility of adjustment a resistance r is also inserted. The switch is to be worked out in detail.

2.

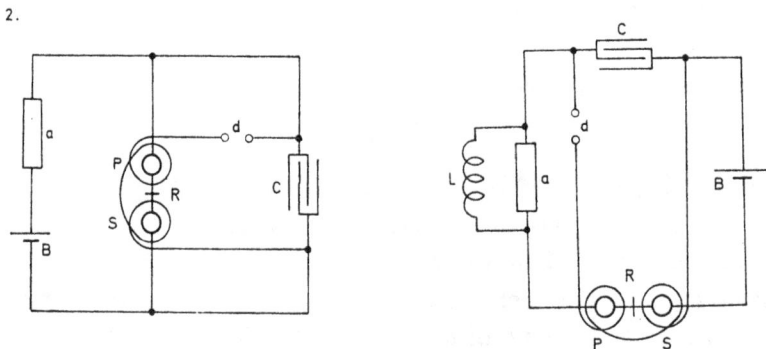

In using the method of exciting the device a by means of oscillating transformer the construction of a special apparatus may be obviated by winding the primary directly upon the relay so that the relay itself is the transformer. This is schematically indicated in the sketches in which the letters indicate the same. In the first the battery should be an open circuit, in the latter a closed circuit type.

July 20, 1899

Galvanometer from Colorado College set up on lead plate and four rubber supports. Lead plate 50 lbs. Resistance roughly 2530 ohms. The filament is very short, vibration quick, altogether not best quality but possible suitable for approximate determinations of ratios and resistance measurement etc.

1 dry cell 1.43 volt

$$i = \frac{e}{10,000 + \dfrac{2530 \times 100}{2630}} = \frac{1.43 \times 2630}{2630 \times 10,000 + 2530 \times 100}$$

$$i = \frac{1.43 \times 263}{2,655,300} \quad \text{this gave 13.5 deflection.}$$

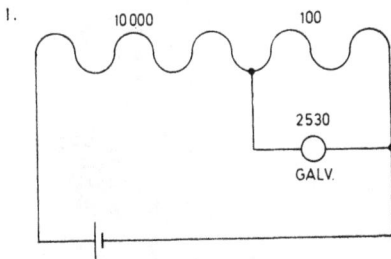

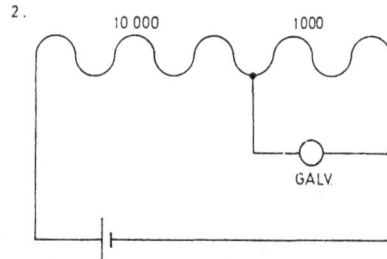

Now the current in galvanometer was

$$I = \frac{100}{2530} i \quad \text{or} \quad I = \frac{10}{253} \times \frac{1.43 \times 263}{2,655,300} = \frac{376.09}{67,179,090} ;$$

this gave 13.5 degrees of scale, hence K for one degree current

$$K = \frac{376.09}{67,179,090 \times 13.5} = \frac{1}{\dfrac{67,179,090 \times 13.5}{376.09}} = \frac{1}{2,411,438} \quad \text{approx. } \textit{Will do for use}$$

intended.

To test proportionality on scale the galvanometer was connected across 1000 ohms. This gave a deflection of *98°* on scale. The currents computed were in both cases as $\frac{10,716.7}{100,961.9}$. This showed the necessity of increasing range of reading by placing scale further. For small deflections proportionality quite rigorous.

93

Colorado Springs

July 21, 1899

Various arrangements with two sensitive devices for the purpose of increasing efficiency of receivers in telegraphy. Also for directing currents.

Two sensitive devices, disposed as indicated and so constituted that they break down or respond easier to impulses of one direction than to those of the other, allow the commutation of alternating currents. For this purpose a device may be employed, described before, consisting of a glass tube and two metallic plugs, the glass tube being about half filled with nickel chips or fillings of other metal. In Fig. 1. it is supposed that the devices $a\,a_1$ have this quality which may be given, for instance, by a battery in each circuit as shown in Fig. 2. In both sketches 1) and 2) a relay $R\,R$ is shown having one of its legs in one of the circuits and the other leg or coil in the second circuit. In this manner the impulses coming, for instance, from a distance as in telegraphy can be made to exercise an accumulative effect. The alternating impulses are led in through terminals $t\,t$.

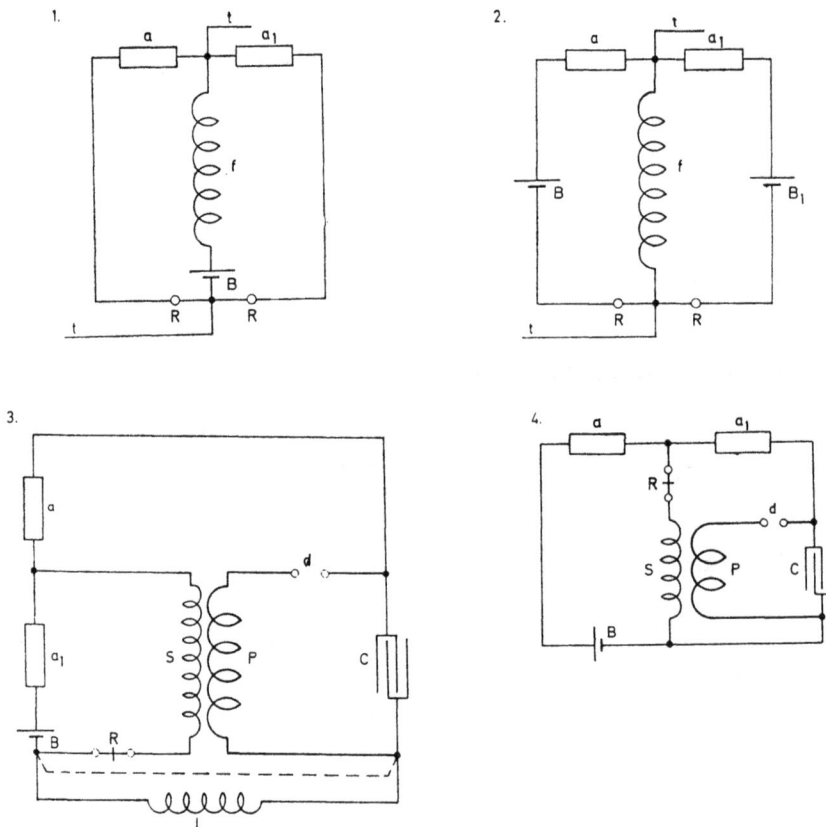

Fig. 3. illustrates one of the connections of apparatus experimented with with good success. The method of excitation and magnification by means of an oscillating transformer is used and the relay and secondary were connected either as shown in the plain lines or in the modification indicated by the Botted lines. The letters have the same meaning as in previous instances. In Diagram 4. similar connections are shown, merely the self-induction coil L has been done away with and secondary S adjusted accordingly.

Figs. 5.—8. show again modified dispositions. Fig. 5.: the condenser is in the bridge and the legs of the receiver are placed one in each of the two circuits. In Fig. 6. the condenser is placed around a battery which is graduated by a resistance (not shown) so that the secondary S strains the devices $a\,a_1$ to the point desired. Fig. 7. shows a similar plan with the secondary in shunt to one of the sensitive devices and in Fig. 8. two sensitive devices one in one and the other in the second circuit of which each contains a condenser and its own primary.

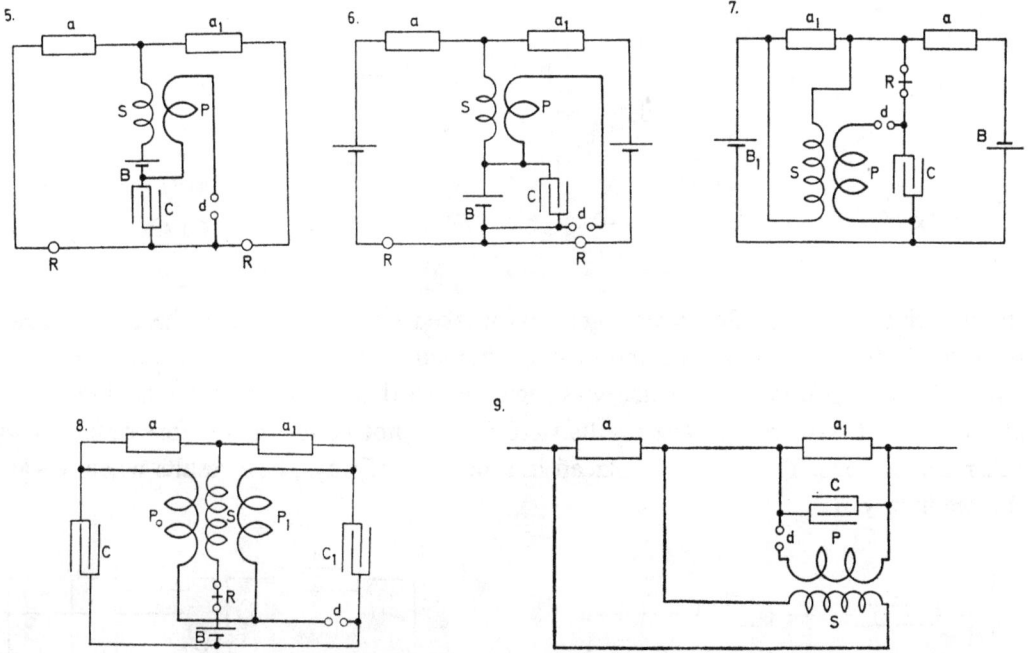

Fig. 9. shows a plan followed in numerous variations and which is capable of excellent results.

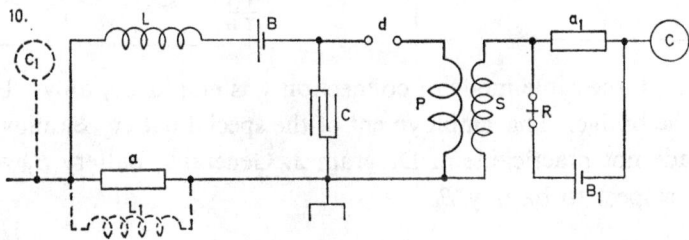

Fig. 10. again shows a form of connections which works extremely well. It is suitable for use in connection with single terminal oscillators. The capacity or elevated terminal may be at C and also at C_1. Both cases result good.

Further modes of connecting apparatus with two sensitive devices for telegraphy and such purposes. A connection as in sketch showed quite satisfactory results. The sensitiveness was probably due to the fact that the secondary was open as in a previous instance. Several plans of working with two or more devices in multiple were experimented with. The idea was to introduce greater regularity and reduce resistance of the path through

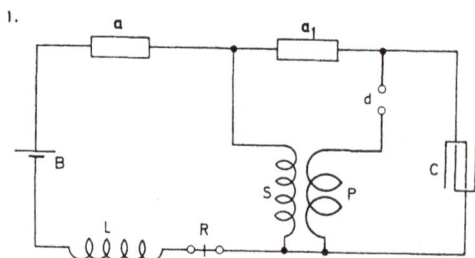

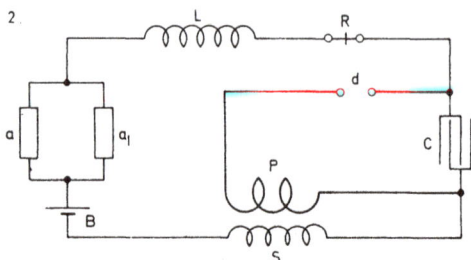

the sensitive apparatus. Some arrangements worked well, for instance the one illustrated in sketch 2. In 3. both devices a and a_1 were shunted by a high self-induction L, the inductive and ohmic resistance of which was regulated so that devices $a\,a_1$ would break down at the slightest disturbance. The results were fair but not better than before obtained with other dispositions. The relay was placed in a number of ways, best results when it was in the secondary S.

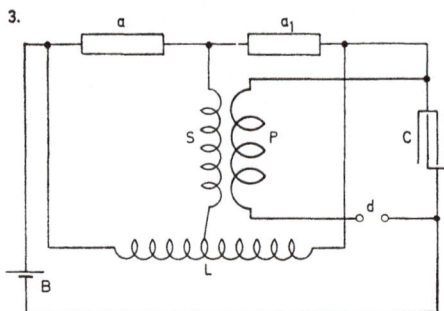

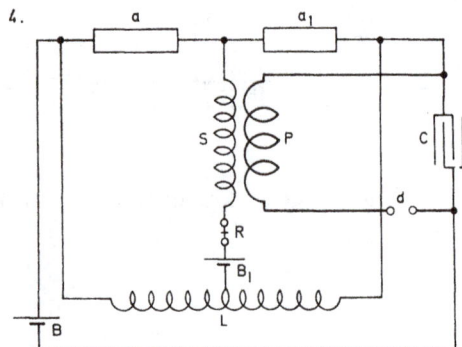

In Diagram 4. the same mode of connection was employed, only a battery and relay were placed in the bridge. The employment of the special battery B_1 allowed some adjustments to be made not practicable in Diagram 3. Generally, battery B_1 was differentially connected with respect to battery B.

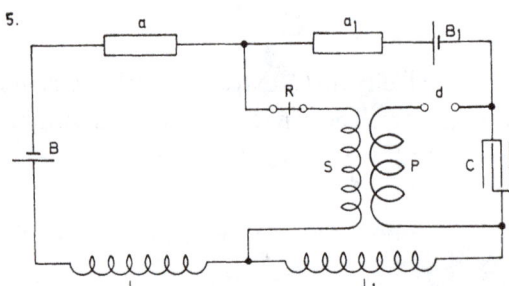

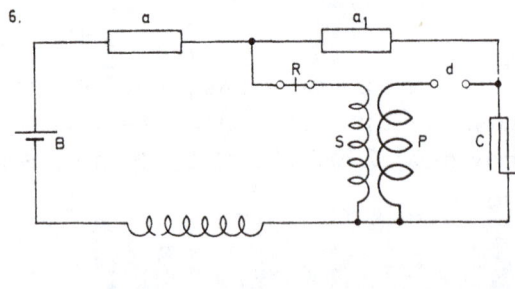

Diagram 5. again shows a form of connections slightly modified, two bateries being used in series to strain the sensitive devices $a \, a_1$. Results about the same.

In Diagram 6. the connections were as previously shown in Diagram 1. only the relay was placed in the secondary S. This worked *excellently*.

In Fig. 7. the same connections were retained only a high self-induction was connected around the devices and regulated so that devices were rendered very sensitive.

Fig. 8. shows an arrangement to be used in connection with present oscillator employed as sender. The switching connections are to be simplified. The oscillator with mercury break and its induction coil work *much better* than interrupters driven by clockwork and small special coils. *Important*.

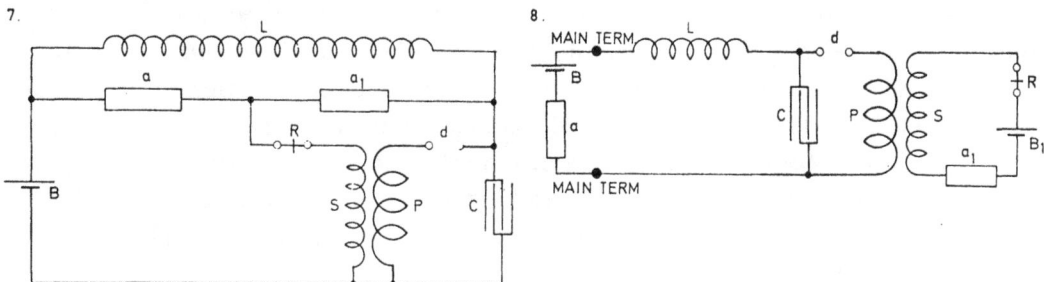

The employment of the mercury break particularly makes the apparatus efficient, probably because of perfect regularity of working which can not be secured by spring contacts or brushes. The efficiency is also in a measure due to the small resistance of primary circuit because of the large copper section and good mercury contact. It is highly important that in preceding dispositions of apparatus the break is of high frequency, the condenser large and best insulated and the conversion in coil *efficient*.

Colorado Springs

July 23, 1899

In investigating the propagation through the media, and more particularly through the ground, of the electrical disturbances produced by the experimental oscillator, as well as those caused by lightning discharge, to which work a few hours were so far devoted every day, a form of sensitive device used in some experiments in New York was adopted,

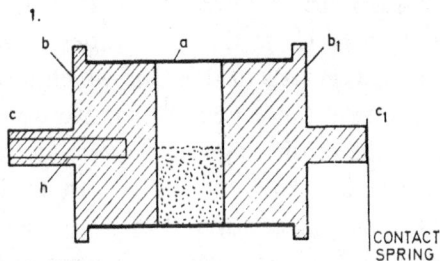

as the best suitable fot these purposes. This device, and the manner of preparing it, it is now necessary to describe. In one form it comprised a glass tube 3/8″ inside diameter having two brass plugs fitted in its ends. The plugs had their inner surfaces highly polished and the distance between them was from 1/8″—1/2″. The tube is illustrated in Diagram 1. in which a is the glass tube and b b_1 the plugs of metal with narrow projections C C_1 for support and contact, respectively. The space between the plugs was filled about 1/3 full with coarse chips of nickel. These chips were made by a milling tool or punch so as to be as much as possible equal in size and shape, this being of considerable importance for the good performance of the instrument. The plug b had a small reamed (tapering) hole h in the center extending to some distance into the plug so as to enable its being placed on a small arbor fitting into the hole and rotated by clockwork at a uniform rate of speed. In some cases when the working of the device was excellent the speed was 16 revolutions per minute. But often the instrument was rotated very much faster in which case it was merely necessary to increase the e.m.f. of the battery which was used to strain the device to the point of breaking down. A beautiful feature of this kind of device is that by regulating the speed its sensitiveness may be regulated at will and in this respect it is preferable to similar devices which are stationary, the contact after being established being broken by tapping. The device acts exactly like a cell of selenium, its resistance diminishing when the disturbances reach it, being automatically increased in consequence of rotation and separation of the chips when the disturbances cease to affect the latter. The rotation of the device replaces here the property of recovery which the selenium possesses, otherwise the similarity is complete. To insure a quite satisfactory working and permanent state I prepare this form of device in the following manner:

The glass tube, plugs and the chips to be used are first thoroughly cleaned with pure absolute alcohol and dried. Next, one of the plugs, as b, is slipped into one end of the glass tube and the required amount of chips is put in the other, plug b_1 being finally inserted closing the tube *nearly* hermetically, but not quite so. Now the device is placed upon a cylinder of metal with a hole in the center, to allow the small part of one of the plugs b or b_1 to slip in, with some space between, and permit the plug to rest upon and in good contact with the upper surface of the metal cylinder which is then slowly heated, as by being placed upon an electrical stove or a plate supported above an alcohol lamp. When the lower plug is brought to the required temperature, sealing wax is run around the rim projecting for this purpose, beyond the glass tube. The metal cylinder is now allowed to cool down slowly until the sealing wax is in some degree solidified when the instrument is turned over and placed with the other plug on the cylinder and the operation of sealing the joint between glass and plug repeated. During this preparation the chips are of course at an elevated temperature and all moisture is expelled so that, when the instrument is ready, a thoroughly dry atmosphere exists within the same, this being essential for good performance. The atmosphere is, however, at a pressure slightly below that of the surrounding air. When the device is carefully prepared it works remarkably well, and in comparative tests showed itself superior to this kind of device of the form ordinarily advocated. During a few days I carried on tests of this kind which brought out the good qualities of this kind of instrument. In one instance two of them were compared with a third device of the ordinary form in which the sensitive grains were immersed in an atmosphere considerably rarefied and contact was broken with a tapper. In all three instruments the grains of nickel were of the same size and shape. One of the terminals of each of the devices was connected to a ground wire, while the other terminals were each joined to a piece of wire extending to a small height, these pieces of wire being the same in all particulars. All the three devices were strained as far as was practicable by batteries so as to be at the point of breaking down and sensitive to a high degree. Although the pieces of wire extending into the air were only of a length of a few feet, all the instruments recorded the discharges of lightning up to about 30 miles as the storm moved away. At this point it was found necessary to set the instrument with the "tapper" so that it was still more sensitive when it responded but in an irregular manner, while the other two devices continued to record regularly up to a distance of about fifty miles when the disturbances ceased, probably owing to the cessation of the storm. I inferred from these experiments, carried on for some time with the view of selecting and adopting the best form of such a device, that when the particles of metal are rotated they are, as it were, suspended in the air and in this condition more susceptible to the influence of the disturbances than when they are kept stationary. It seems, however, that when rotated, the particles are not so liable to stick together and cause irregularity of action such as observable in the ordinary form of such a device. As to the amount of chips, if more are put in the instrument must be rotated at a higher speed or else the battery straining the dielectric must be weaker. Through this kind of instrument much stronger currents can be passed without damaging it and making it further unfit for work. Another form of such instrument particularly suitable for experi$f$ mentation is illustrated in Diagram 2. It consisted of a brass plug b with a fibre tube -

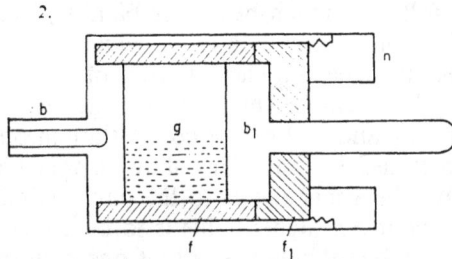

2.

into which was fitted another brass or metal plug b_1 which was held in place by a fibre washer f_1 and metal nut n. In other devices of similar construction the space between the plugs was adjustable. This form of instrument was particularly suitable for testing the properties of sensitive grains g. Before testing the grains and the instrument as well were thoroughly dried. To get an idea of the resistance of such devices when in either state, excited or not, the resistance of many was measured under varying conditions. A fair idea is conveyed by saying that, unexcited, they measured more than 1,000,000 ohms while the resistance would sink down to 300 or even 50 ohms or still less when excited. When highly sensitive they would respond to sound waves at a considerable distance.

Experiments with oscillator 35 turns in secondary on tapering frame No. 10
B. & S. wire.

This is the first test of the Westinghouse transformer installed a few days ago. It was tried yesterday evening but only for a short time to merely get an idea how it will behave. The e.m.f. used was 7500 volts or less. Today a pressure of 15,000 volts on the secondary was used. Best resonating action was obtained with *one* primary turn and a few turns in the regulating coil. The spark gaps were as long as obtainable in the box, that is, about 7 turns of the screw on each side, possibly an inch or so. An approximate estimate places the primary inductances at 75,000 cm. or $L_p = \dfrac{75}{10^6}$ H. The primary capacity was 88 jars in each of the sets in series. The capacity of one jar being approximately 0.0035 mfd.

The total capacity C_p is $=0.154$ mfd. From this calculated, and neglecting as in most cases before the reaction of the secondary, we get $T = \dfrac{214}{10^7}$ or $n = 46{,}730$ per second.

Observations: A spark gap being established between the free terminal of the secondary and an earthed wire, strong streamers were seen on the *latter*. This shows very rigorous action and demonstrates that the potential of the neighbouring parts of the ground must be considerably affected. *Very strong* sparks on lightning arresters as the secondary discharge is playing over the gap. This is certainly *extraordinary* as the ground is now excellent on the secondary. The arc, horizontally passing about 32″ long is very powerful, thick and giving a vivid light, the noise is deafening. The arc passes sometimes on a downward course. Is it attraction or due to surgings of the air in consequence of violent explosions? When large balls 30″ diam. are placed in the gap the spark length is nevertheless small. This shows the secondary can not supply the great amount of energy necessary for charging the large balls to full pressure. This may be due simply to the imperfect inductive connection with the primary or to the small amount of power now available from the supply transformers, as there are only two of them, and the Westinghouse transformer works only at 1/4 of the normal pressure. This would mean roughly 1/16 of its total performance. On some points of the balls small streamers are observed; must be due to roughness or points on the places. The balls will have to be gone over and all the surface polished up. It would be impossible for streamers to break out from balls of such size unless the pressure is a few millions of volts, which cannot be the case at present. A curious feature is to see the sparks deviate and follow wooden beams or planks placed nearby. I rather think this is merely due to an effect of the currents of air which are prevented from circulating freely on the side of the plank or beam. The intensity of the vibration in the primary is evidenced by sparks passing between the turns of the regulating self-induction coil in the primary. Between the beginning and end of the coil, although only a few of the turns are inserted, the sparks are sometimes 3″ long. This shows a very high e.m.f. on the primary and I almost think there must be a mistake as to the pitch estimated which, judging from these sparks, would seen to be much higher. This is to be investigated closer. Experimentation shows that it is very decidedly better to adopt one turn of primary instead of two and if a lower frequency is desired rather to increase the primary capacity. With one turn the explosions are more violent and the regulation is much more convenient. In these experiments the jars do not seem to be much strained, which indicates well. At times sparks will break through inside of the secondary between the turns and to the ground. The sparks are very strong from small wires attached to the free end of the secondary more so than from thick wires. When a coil was connected to the free secondary end the vibration could not be well established, evidently the coil was "out of tune" and by its capacity and inductance interfered with the free vibration of the secondary. The sparks went from the gap-box to the ground though the box was well insulated; there is danger of inflaming the building by this or by the secondary sparks following the wooden structure. The experiments were continued with 7500 volts as yesterday but the working was unsatisfactory. This showed finally that yesterday one of the jars in one set was bad and there was only one set acting, the other set being short circuited; that is why an e.m.f. of 7500 volts was sufficient yesterday. The Westinghouse transformer gains in e.m.f. as jars are put on, the maximum rise seems still remote, this argues well for the economy of the transformer. The incandescent lamps are all destroyed in consequence of the intense secondary vibration, the filaments being broken by electrostatic attraction towards the glass. Lamps were spoiled at a distance of *40 feet* from the secondary free terminal! This action is likely to give trouble in future experiments. A curious observation is that all horses shy. It is due to sound or possibly to current action through the ground to which horses are highly sensitive either owing to greater susceptibility of the nerves or perhaps only because of the iron shoe establishing good ground connection. I am not quite certain that the secondary vibration is fundamental although for a lower or higher tone it is too powerful. The external gaps used in some trials seem to improve the action somewhat in rendering the discharges of the primary more *sudden*. If time should permit the vibration will be investigated by a rotating mirror to be prepared.

Experiments with the *secondary 35 turns* were continued today. In connection with the secondary an "extra coil" was used. The same was before described having 260 turns on a drum 2 feet in diam. of cord No. 10. The inductance of the coil as before found was approximately 13,900,000 cm. or $\dfrac{139}{10^4}$ henry. The coil was connected with the lower end to the free terminal of the secondary while the upper end was left free, a few feet of wire extending into the air. Resonance, as evidenced by streamers (maximum) on free end of coil, was obtained exactly as in a previously recorded experiment with 9 jars in each set of the primary condensers, or slightly less, at any rate 8—9 jars caused the largest display of the streamers on the free end. The streamers were only three feet long as the energy from the supply circuit was limited, the intention being to first study the peculiarities and behaviour of the transformers before taking them to their full output. A simple computation showed that the resonant rise on the free end was chiefly, if not wholly, due to the rise in the coil itself, the free vibration of the secondary being comparatively of small moment. To study the harmonics the capacity in the primary circuit was doubled but the effect, as expected, was very small. Now the capacity in primary was again doubled, it being expected that the streamers would be of considerable power under these conditions. This was the first undertone and it should have been fairly strong. But singularly nothing to speak of was noted although the adjustments were carefully gone through again and again. It was thought that owing to a very small arc in the primary the oscillation did not readily establish itself but this was not highly probable though such has taken place in experiments which I made before. The arc was necessarily small as the capacity was very large in the primary. The tuning was very sharp with twice the capacity in primary so that a little variation in the self-induction regulating coil made the streamers change very considerably. I expect that it was sharper still with four times the primary capacity so that, after all, the resonant condition may have been missed. This might have been the case easily as all the variation from *no* streamers to their *maximum* would have taken place by going through only one *quarter* of *one turn*, and *possibly less*, of the regulating coil. This sharpness of tuning noted here and in previous instances in some arrangements again impresses me with the value of such dispositions in telegraphy, when it is of great importance to isolate messages. It seems possible to secure in such or similar ways an almost absolute privacy. The experiments with only two circuits show this sufficiently. Continuing the experiments one of the balls of 30″ diameter was connected to the free end of the coil and now the resonating condition was secured with 23 jars on each side of the primary. Summing up the results the vibration with coil alone, without ball, with 9 jars on each side of primary was, approximately, taking the primary capacity equal to $\dfrac{9}{2} \times 0.003 =$

$$= \frac{0.027}{2} = 0.0135 \text{ mfd.}$$

$$T_1 = \frac{2}{10^3} \sqrt{0.0135 \times \frac{7}{10^5}} = \frac{2\pi}{10^3} \sqrt{\frac{945}{10^9}} =$$

$$= \frac{2\pi}{10^7} \sqrt{94.5} = \frac{2\pi}{10^7} 9.72 = \frac{61.0416}{10^7} \text{ or approx.} = \frac{61}{10^7}$$

and $n = 164,000$ per second.

Now when the ball was attached the primary capacity was

$$\frac{23}{2} \times 0.003 = \frac{0.069}{2} = 0.0345 \text{ mfd.}$$

and the period

$$T_2 \text{ was} = \frac{2\pi}{10^3} \sqrt{0.0345 \times \frac{7}{10^5}} = \frac{2\pi}{10^3} \sqrt{\frac{2415}{10^9}} =$$

$$T_2 = \frac{2\pi}{10^7} \sqrt{241.5} = \frac{2\pi}{10^7} \times \frac{2\pi \times 15.54}{10^7} = \frac{97.6}{10^7} \text{ and } n = \textbf{102,460 per sec.}$$

The ball slows the vibration of the coil very much down. From a series of observation with capacities of varying value useful estimates may be made and the quantities of moment calculated. This mode of proceeding seems to offer features of considerable value in experimentation and it will be followed up. A curious observation in these experiments was that maximum rise was obtained always with the regulating coil practically all out.

How is this to be explained?

Experiments with the *secondary 35 turns were resumed.* The probable causes of the curious phenomenon that maximum resonant rise (on the coil attached to the terminal of the secondary, as before described) took place when the self-induction regulating coil was practically all cut out — were considered. Evidently when the coil was cut out there was more energy available for the excitation of the primary turn and therefore the secondary was more strongly energized, this giving a higher electromotive force on its terminals. Owing to this the impressed e.m.f. on the coil attached to the free terminal of secondary was greater and therefore the coil was more strongly excited. Assuming then that the secondary free vibration did not take place, this explanation would be acceptable but for one thing: the maximum rise on the coil with 260 turns did not occur, when *all* the turns of the primary regulating coil were cut out, but at a point when there remained still a few turns in series with the primary. The phenomenon must be therefore interpreted differently. To all appearances the secondary free vibration *did* occur, and there was a certain inductance in the primary which gave the highest e.m.f. on the excited coil on the free terminal of secondary. But now the latter was in fairly close inductive relation with the primary hence its own vibration was more or less modified by that of the primary. In altering the primary vibration, that of the secondary must have been, therefore, correspondingly altered. Now, the secondary excited the coil with 260 turns and, to insure the maximum rise on the free terminal of the coil, the secondary vibration ought to have been of exactly the same pitch as the free vibration of the coil. From this it is plainly seen that if the primary vibration was such as to favour a rise in the secondary of the pressure at the free terminal then the impressed e.m.f. on the coil with 260 turns was greater; but this evidently, judging from the actually observed results, took place when the secondary vibration was "out of tune", more or less with the free vibration of the coil. Thus it happened that by raising the secondary e.m.f. up to a certain point there was an increased resonant rise on the excited coil. But when, by further cutting out turns of the regulating primary coil, the secondary vibration was modified more and more and brought "out of tune" with the free vibration of the coil excited by the secondary, the resonant rise on the terminals of the excited coil was diminished. Now, with a certain small number of turns of the regulating coil still included in the primary, the relation between these opposing elements determining the resonant rise was such as to insure the maximum. I have no doubt that this is the correct explanation of the phenomenon observed. At first I thought that the *length* of the *primary* might have something to do with it, as I have observed before something to this effect, but now I must reject this view as improbable. From the preceding it is now quite evident that in cases when the free vibration of the secondary can assert itself, the primary capacity and self-induction has to be such that maximum e.m.f. is obtained on the secondary — then the excited coil must be such as to vibrate in accord with the secondary or (inasmuch as the secondary vibration is affected by the primary) the free vibration of the excited coil must be the same as that of the combined primary and secondary system. When the vibration in the secondary is exactly the same as the free vibration of the excited coil the maximum rise will be obtained on the coil, in any event, but for the best result the secondary must also be tuned to the primary so that greatest impressed e.m.f. is secured on the coil.

In cases where the secondary is in such intimate inductive connection with the primary then the latter condition need not be considered and it is only necessary to adjust the coil so that it will have the same period as the oscillation in the secondary. In fact, I believe this will be, in the end, the best condition in practice for, if the transformer be efficient, the connection between the primary and secondary must be a very close one. In such a case the high impressed e.m.f. on the excited coil will be obtained only by transformation and not by resonant rise.

A gratifying observation was made today which was the following: the water pipe to which the secondary lower end was connected, and which conveyed the currents to the ground, was disconnected from the secondary and the latter connected to a separate ground plate at a distance from all other ground connections. Everything was carefully examined to be quite sure that there was no other ground connection in the secondary. Nevertheless, when the secondary discharge was made to play, *strong sparks* went *continuously over the lightning arresters.* There was no other possible way to explain the occurence of these sparks than to assume that the vibration was propagated through the ground and following the ground wire at another place leaped into the line! This is certainly extraordinary for it shows more and more clearly that the earth behaves simply as an ordinary conductor and that it will be possible, with powerful apparatus, to produce the stationary waves which I have already observed in the displays of atmospheric electricity. The mere observation of the sparks speaks well for the power of the apparatus used and clearly shows that it is competent to carry to a great distance even as it is when used as a transmitter in telegraphy. Assume even that the pressure would diminish as the square of the distance from the source, still the performance would be remarkable. Such an assumption seems to be justified when we consider that the density of the current passing over the earth's surface will diminish as the square of the distance from the center of the disturbance and consequently, the effective pressure at least, ought to diminish correspondingly. Now, in the experiments above described the distance between point *a*, where the lower end of the secondary was grounded to point *b*, where the sparks jumped from the ground to the line or vice versa, was 60 feet. Hence on the above assumption we can

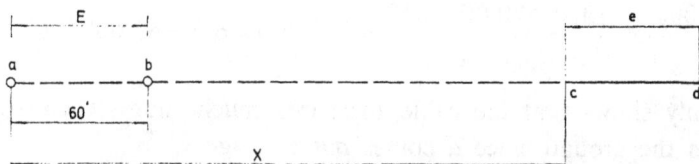

make the ratio $\dfrac{E}{e} = \dfrac{x^2}{60^2}$, where x will be in feet. At the lowest estimate E was 10,000 volts

and assuming an ordinary instrument be used as a receiver, requiring $\dfrac{1}{10,000}$ of a volt $= e$

between c and d, we would have $x^2 = 60^2 \times \dfrac{10,000}{\dfrac{1}{10,000}} = 36 \times 10^{10}$ or 600,000 feet or $x = 114$

miles nearly. But by making the distance between c and d much greater the transmission radius may be *greatly* extended.

Capacity of secondary 35 turns used in preceding experiments.

The average length of one turn may be put at approximately 135.1 feet $= 4120$ cm. The wire is No. 10 B. & S. diam $= 0.102'' - 0.26$ cm. Surface of wire in sq. cm. $= \pi \times 0.26 \times$ $\times 4120 \times 35 = \pi \times 37,500$ cm. approx. The capacity was compared with that of $1/2$ mfd. standard condenser and was found to be **C = 3600 cm.** This measurement was I expect correct within $1/2 \%$.

Note: The measurement was made by connecting the cable with one end, or with *both* ends to the source, the other terminal of which was connected to the earth. *Result* was the *same*.

Now it is of interest to compare the value found by measurement with that which

the cable would have when stretched out. Its capacity would then be $C_1 = \dfrac{l}{2 \log_e \dfrac{l}{r}}$ con-

sidered as a cylinder remote from the ground. Taking the values as above

$$C_1 = \frac{4120 \times 25}{2 \log_e \dfrac{4120}{0.13}} = \frac{4120 \times 35}{2 \log_e 31,700} = \frac{2060 \times 35}{\log_e 31,700} = \frac{2060 \times 35}{4.50106 \times 2.3} =$$

$$= \frac{2060 \times 35}{10.35244} = \textbf{6965 cm.}$$

If we consider the cable in its capacity with reference to the earth or conducting

plane its capacity would then be $C_2 = \dfrac{l}{2 \log_e \dfrac{2\,d}{r}}$. Since we have found the measured

value we may find from it the distance d; we have $3600 = \dfrac{4120 \times 35}{2 \log_e \dfrac{2\,d}{r}}$ from which

$\log_e \dfrac{2\,d}{r} = \dfrac{4120 \times 35}{2 \times 3600} = 20.3$ and this gives

$$\log \frac{2\,d}{r} = \frac{20.3}{2.3} = 8.82607, \text{ hence } \frac{2\,d}{r} = 67,000,000 \text{ cm.}$$

$$d = \frac{67,000,000}{2} \times \frac{13}{100} = \frac{670,000 \times 13}{2} = 670,000 \times 6.5 = 67,000 \times 65 = \textbf{44,550 meters!}$$

The result only shows that the cable measures much more when straight and at some distance from the ground since d comes *out so large*.

104

It may be of further interest to compare the capacity as found with capacities which would be obtained if the surface of the cable were converted into the surface of a disc or sphere, for instance. Taking first the latter and calling its radius r' we have $4\pi r'^2 = \pi 37,500$ and $r'^2 = \dfrac{37,500}{4}$, $r' = \dfrac{\sqrt{37,500}}{2} = \dfrac{193.65}{2} = \mathbf{96.825}$ **cm.** and this should be the theoretical capacity C_1 of the sphere. Taking now a disk of radius R, its capacity C_2 would be $C_2 = \dfrac{2R}{\pi}$. Now on the above assumption of an equal surface we would have $2\pi R^2 = \pi 37,500$ $R^2 = 18,750$ and $R = \sqrt{18,750} = 136.93$ cm. Hence $C_2 = \dfrac{273.86}{\pi} = \mathbf{87.2}$ **cm.** This is still smaller than the capacity of a sphere of the same surface. Suppose now the surface of the cable were converted into small spherical surfaces of a diameter equal to that of the cable and inquire what capacity would be obtained in this manner with the given surface. The surface of one of the small spheres being $4\pi r^2$ and calling their number n we have $4\pi n r^2 = \pi \times 37,500$ and

$$n = \frac{37,500}{4\,r^2} = \frac{37,500}{4 \times 0.13^2} = \frac{37,500}{4 \times 0.0169} = \frac{37,500}{0,0676} = 554,734 = n$$

The total capacity C_3 of all these small spheres, neglecting mutual screening action, will be n times the capacity of one of the spheres and since the latter is $=0.13$ we would have total capacity $C_3 = 554,734 \times 0.13 = \mathbf{72,115}$ **cm.!** A very large value indeed, which would have been still greater if the diameter of the spheres would have been smaller. These primitive considerations show that to get the largest possible capacity with a given surface we must use the latter in the form of minute surfaces of the smallest possible curvature. This makes it obvious why exhausted bulbs show under certain conditions such comparatively large capacity. And this explains the virtue of bulbs when used in telegraphy for the purpose of supplanting a large elevated plate or wire leading to a great height. Such a bulb or tube, particularly when filled with *hydrogen*, is (as I have found) very effective and I look to a valuable use in the future of such rarefied vessels in connection with telegraphy through the media or the like. The greatest capacity with a given surface will, of course, be obtained with spheres of the smallest possible diameter, as the spheres of hydrogen. The next best form to give to the surface would be a cylinder of great length and minute diameter. The above considerations make it plain why thin wires have such a comparatively large capacity. Taking two such wires of the same length L and diameters δ and δ_1, we would have their capacities as

$$\frac{L}{2\log_e \dfrac{2L}{\delta}} : \frac{L}{2\log_e \dfrac{2L}{\delta_1}} \quad \text{or as} \quad \log_e \frac{2L}{\delta_1} : \log_e \frac{2L}{\delta} .$$

Now suppose, for the sake of illustration, that $\dfrac{2L}{\delta}$ were $=10$, and $\dfrac{2L}{\delta_1} = 1000$, then the above capacities would be as $\dfrac{1}{2.3 \times 3} : \dfrac{1}{2.3 \times 1}$ or as $\dfrac{1}{3} : 1$. In other words, making the diameter of one wire 100 times as great as that of another, but keeping the length of the wires the same, the capacity of the thick wire or cylinder would be only three times that of the thin one.

Experiments with the *secondary of 35 turns* continued. The secondary was tuned alone and more carefully, the result being that maximum rise of pressure was obtained with *one turn of the primary*, the two cables being connected in multiple, and 78 jars on each side of the primary. When best action was obtained there were a few turns in the regulating coil in series with the primary cables, the total self-induction in the primary being estimated to be $L_p = 85,000$ cm. or $\dfrac{85}{10^6}$ henry.

The capacity in primary was:
$$\begin{cases} 16 \text{ old jars} = 0.00334 \times 16 = 0.05344 \\ 62 \text{ new } ,, \ 10\% \text{ larger} = 0.0036 \times 62 = 0.2232. \end{cases}$$

Total $= \dfrac{0.2766}{2} = 0.1383$ mfd.

This would give period of the system:

$$T_p = \frac{2\pi}{10^3} \sqrt{0.1383 \times \frac{85}{10^6}} = \frac{21.54}{10^6} \text{ and } n = 46,425 \text{ per. sec.}$$

It is of interest to find what value for the inductance of the secondary will be obtained by substituting in the equation for the secondary $T_s = \dfrac{2\pi}{10^3} \sqrt{L_s\, C_s}$, the value for C_s found by measurement before recorded. Since T_s must be $= T_p$ we have

$$T_p = \frac{21.54}{10^6} = \frac{2\pi}{10^3} \sqrt{L_s \times \frac{3600}{9 \times 10^5}} \qquad C_s = 3600 \text{ cm.}$$

or $\dfrac{2\pi}{10^3} \sqrt{0.1383 \times \dfrac{85}{10^6}} = \dfrac{2\pi}{10^3} \sqrt{L_s \times \dfrac{3600}{9 \times 10^5}}$ and from this:

$$L_s = \frac{9 \times 10^5}{3600} \times 0.1383 \times \frac{85}{10^6} = \frac{0.1383 \times 765}{36,000} = \frac{0.1383 \times 85}{4000} \text{ henry or}$$

$$L_s = \frac{10^9 \times 0.1383 \times 85}{4 \times 10^3} = \frac{138,300 \times 85}{4} = \mathbf{2,939,000} \text{ cm. only!}$$

While this estimate is not correct in principle it shows, nevertheless, that the capacity measured in a state of rest is not that which enters as an element of the vibration. It was thought from this result that the secondary might have responded to the first octave and the two primary cables were joined in series, but results proved inferior.

It is possible that when the primary cables were connected in series, owing to the less satisfactory working of the spark gap, the e.m.f. on the secondary was smaller than it ought to have been if both the primary and secondary vibrated at the same rate. Furthermore, it should be borne in mind that when the cables were in series and the vibration in primary reduced to half the number per second, the induced e.m.f. in the secondary turns could have been only about one half of that in the first experiment. If in the latter experiment with the primaries in series the true note of the secondary was struck then, assuming the capacity to be 3600 cm. or thereabouts, the inductance of the secondary as modified by the primary would have been still only $4 \times 2,939,000 = 11,765,000$ cm. It is, therefore, probable that the capacity which enters as an element of the secondary vibration is much smaller than that which is found by measurement, which is as might be expected, since the cable can not be fully charged and discharged at each alternation, as is evident from the constants.

Observations during the experiments today. When the secondary worked very well the spark was very noisy and nearly an inch thick, judging by the eye, and about 3 feet long. The sparks passed all the time over the lightning arresters as the secondary discharge was playing and, at times, for a short interval they were extremely vivid and thick. This seemed to occur chiefly when the secondary arc became louder and roaring, this indicating a better working of the arc and a higher e.m.f. for a given length of the path through the ground. The sparks on the arrester's arc, as is now established beyond any doubt, are due to the propagation to the ground through the earth wire, and it is now plain that although they take place when the oscillation is slow, they are more easily produced with a quicker oscillation. Perhaps the higher harmonics enter prominently into their formation. The secondary arc was adjusted to a length of 31″, then the sparks on the arresters were very violent. It was thought that, if the vibration was propagated through the earth wire and caused the sparks on the arresters in this way, by adding capacity to the earth wire the action on the arresters would be increased. Accordingly, a sphere of 12″ diam. *s* was connected to the wire as shown in diagram and, indeed, the play on the *arresters was intensified.* By now reducing the gap still a few inches the display on the arresters seemed to increase further. When the secondary discharge was permitted to pass continuously for about 5 minutes the fuse on the supply circuit primary gave way showing that energy was taken at a rate of about 20 H.P. This also indicated that the connection between the primary and secondary of the oscillator was fairly close and that the secondary was capable of taking up considerable energy. When two external gaps were used in series, with gaps in box, less energy was taken from the supply circuit, this indicating that the arc in primary short-circuited the secondary of W.T. to some extent.

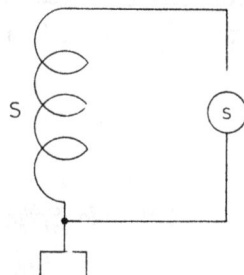

Colorado Springs

July 26, 1899

Investigating vibrations of "additional" or "extra coil": from observations before made it would appear, as I believe it has been stated already, that when the impressed electromotive force was increased, in other words, when the movement in the secondary was made greater, the free vibration of the extra coil did not readily assert itself. At least this has been noted in a number of experiments with the object of ascertaining this. A complete analogy is afforded in mechanics. In order that free vibration may take place, and readily, there must be a loose connection with the part impressing the movement. This truth is obvious. Considerations of this kind led to experimentation with an arrangement, as illustrated in the sketch below: In this the connection with the secondary S impressing

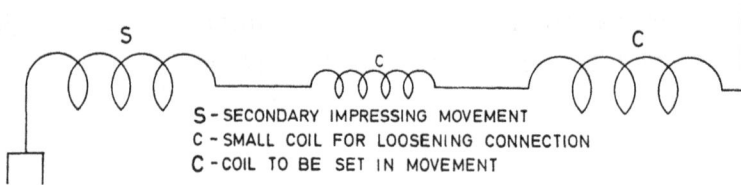

S – SECONDARY IMPRESSING MOVEMENT
C – SMALL COIL FOR LOOSENING CONNECTION
C – COIL TO BE SET IN MOVEMENT

the movement was loosened, so to speak, by the insertion between it and the coil C to be excited another coil c which was generally adjusted to suit the conditions. This amounted to the same as increasing the momentum of coil C and rendering it more preponderating and capable of freely asserting itself. These experiences lead to a rule long recognized, that development of the oscillator must be in two directions: either in the direction of obtaining a high impressed electromotive force by transformation ratio, when the connection between the secondary and primary is *rigid*; or obtaining a high e.m.f. by an excited extra coil in loose connection not reacting inductively.

Colorado Springs

July 27, 1899

Experiments were made today with spark gaps constituted as indicated in the sketch below: The idea in this scheme was to make gaps in the box, which varied from

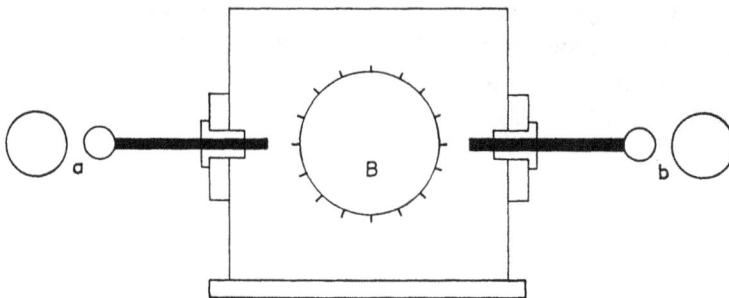

a very small to a great length owing to the movement of the break wheel B, very small and the gaps a and b very large, so large that the discharge could break through only when the gaps in the box were at their minimum length. Thus the loss in the box itself was greatly reduced and owing to the great velocity of separation of the electrodes, a greater suddenness of disruption was obtained. This was, of course, certain since the gaps in the box could not be bridged except for a short interval since gaps a and b took up the e.m.f. Thus the velocity of separation was the greater the smaller (in length) the arcs in the box were made. The adjustment of their length was effected by merely varying the length of gaps a and b. In these experiments spheres of various sizes were used to constitute the additional gaps *a* and *b* and it was observed that unless the induction coil be capable of giving a large current, spheres of considerable size were not the best to employ in the usual arrangement of apparatus. The reason was that the arc was formed with difficulty, hence it had to be made shorter, but when the current broke through, the resistance of the small arc was very low and the secondary of the induction coil was short circuited too much. This may not be always true. By using additional gaps the primary was closed during a shorter interval of time and also the secondary was less short-circuited. The latter was an advantage but the former was decidedly a disadvantage because the primary circuit could not vibrate very long. The energy taken from the supply transformer or coil was, of course, smaller. All the results obtained in these experiments seem to indicate, contrary to former opinions, that a higher economy is obtainable with one gap that with a greater number. This observation was, however, made before in the New York apparatus. Finally, two gaps in series were adopted as convenient and giving greater velocity of separation. But the best results were obtained with two electrodes in the form of toothed disks rotated in opposite directions. The apparatus in this form was more troublesome to run but worked decidedly better. In this form also an improvement was practicable, which I have since adopted in some form of mercury breaks, and that was to make the number of teeth on each disk such that the total number of the makes and breaks was the product of both the numbers. In this form a small number of teeth was found sufficient for a great number of breaks and the arc could not follow from one to an adjacent tooth.

Colorado Springs

July 28, 1899

The following arrangement was found particularly efficient in applying the method of magnifying the effects of feeble disturbances by means of a condenser. Two instruments were fixed up in the manner indicated in diagram. A similar plan of connections was used before only the sensitive device *A'* was differently placed, as it was found to disadvantage. The sensitive devices *A* and *A'* were prepared as described on a previous occasion and showed, unexcited, a resistance of over 1,000,000 ohms, but when excited the resistance fell in both almost exactly to fifty ohms. Later, instead of the device *A'* another was emplo-

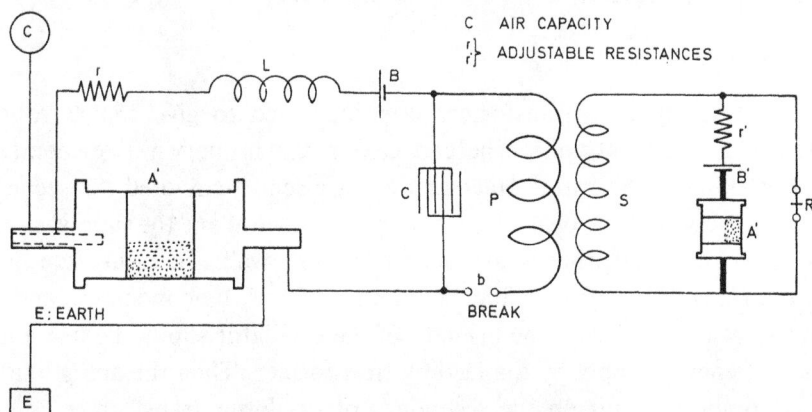

yed of a higher resistance when in the excited state and the coil PS was replaced by one with more turns in the secondary. As finally adopted the secondary S had 160 turns in each layer and 32 layers making 5120 turns. The relay R had a resistance of 998 ohms, wire No. 36. The primary P had 50 turns of lamp cord No. 20. The self-induction coil L had 1900 turns of wire No. 20. All coils were wound on spools 4″ diam., 4″ long with wooden core 1 1/4″ in center. The condenser C and 1/2 mfd. Battery B: 8 cells 11.1 volts; Battery B': 4 cells about 5.7 volts. Speed of rotation of the devices A and A' was about 24 per minute. The break b was 72 per second. The break wheel and arbors of devices $A\,A'$ being driven by clockwork. The break whell has 180 teeth, a small very thin platinium brush bearing on it. The devices readily responded when four persons joining hands would shunt the device A'. In one instance the devices recorded effects of lightning discharges fully 500 *miles* away, judging from the periodical action of the discharges as the storm moved away.

Colorado Springs

July 29, 1899

As has been observed before, in order thal the free vibration of an excited coil may predominate it is necessary to make the momentum of the coil very large relatively to the impressed vibration. With the object of bettering the conditions favorable for the free vibration, a new coil was wound on same drum 2 feet in diam. and 6 feet long. This coil had, instead of 260 turns as before, about 500 turns of cord No. 20. Its inductance was therefore nearly 4 times that of the old coil or about 40 million centimeters roughly. The coil was connected to the free end of the secondary and resonance was observed with 32 jars on each side, there being on each side two tanks in series, so that the total capacity was only 4 jars in the primary. Taking the capacity of one jar at 0.00334 mfd. the total primary capacity was $4 \times 0.00334 = 0.01336$ mfd. The primary turns were in series, so that the primary inductance was $L_{\mathrm{p}} = \dfrac{4 \times 7 \times 10^4}{10^9} = \dfrac{28}{10^5}$ henry. This would give the period of the system approximately:

$$T_{\mathrm{p}} = \frac{2\pi}{10^3} \sqrt{0.01336 \times \frac{28}{10^5}} = \frac{2\pi}{10^5} \sqrt{0.037408} = \frac{2\pi}{10^5} \times 0.1934 = \frac{1.215}{10^5}$$

or $n = \mathbf{82{,}300}$ approx.

First the Westinghouse transformer was connected to give 15,000 volts, but later it was made to give 22,500 volts. The capacity in the primary was evidently too small for the best working of the transformer and the arc schort-circuited the secondary considerably, this causing a great deal of energy to be drawn from the supply circuit. This is always the case when the primary arc does not work well. To insure the best working conditions the transformer should first be able to charge the condensers and the rate of energy delivery of the latter into the primary of the oscillator should be just a little greater than the rate of energy supply by the feeding transformer. Then the arc is loud and sharp and there is no short circuit on the secondary of the latter transformer as the currents over the gap are of very high frequency and the low frequency current of supply — or if it be a direct current — can not follow. The system then works economically and the economy is much greater than might be supposed judging from the unavoidable losses in the arc. As the capacity was too small a *flaming* arc often formed in the box, a sure sign of bad working, the curious feature being that the arc was of a decidedly red color. This may be due to the alumina which was formed as the break wheel was of aluminium. As was expected, the use of two additional gaps improved the working of the apparatus, reducing the trouble due to the short-circuiting of the Westinghouse transformer. It was observed that when the gaps were made so large that the arc did not break through, a lamp on the supply circuit near the condensers — about 6 feet from the same — would brighten up. I am not quite sure that this was due to resonant rise in the Westinghouse transformer, for it may have been due simply to electrostatic action from the jars, as I have observed a similar effect before. When the electrostatic influence is strong the gas in the bulb is excited, the discharge passing through the same though, of course, it is not visible on account of the intense light of the filament. Particles are thrown off and against the carbon and the same is, on the one hand heated to a higher temperature while on the other hand, owing to the hotter environing medium it can not give the heat away so fast as normally — hence it brightens up. Possibly also a small part of the current of supply passes through the excited gas and slightly more energy is drawn from the mains. It was evident that, as was expected, the free vibration of the coil took place more readily than before when the coil with 260 turns was used, owing to the larger momentum as before explained. The streamers were larger than with the old coil but not quite so large as it was surmised they would be. Partially because of this fact, and partially also because not enough energy could be supplied from the Westinghouse transformer to the primary, owing to the small primary capacity, it was decided to change the connection so as to get the next lower or fundamental tone in the primary, this being in all probability the true note of the coil. The capacity in the primary was made 32 jars on each side in multiple, making the total capacity $16 \times 0.00334 = 0.05284$ mfd.

The primary vibration was now just an octave lower than before but the results proved inferior to those first obtained. There was now only one thing possible and that is, that the tone was right after all, in the first experiment, but the results were not quite satisfactory because the primary capacity was too small, thus unfavorable for the best working of the Westinghouse transformer. Accordingly, the same vibration was again secured in the primary but this time by using a capacity four times larger and reducing the inductance to one fourth, which was done by putting the two primaries in multiple. Now, indeed, the results were satisfactory, for the Westinghouse transformer could supply much more energy, practically four times as much as before. The streamers were now much stronger, extending to a distance of 6 1/2 feet from the top of the coil and they were abundant and thick. I can not understand why they should be of such a *deeply red color*. Those in New York never were such. Perhaps it is due to the smaller atmospheric pressure in this locality. Their movement, and darting about is also much quicker and more explosion like. At times a big cluster of them would form and spatter irregularly in all directions. Sometimes it apeared as if a ball would form above the coil, but this may have been only an optical effect caused by many streamers passing from various points in different directions. Many times sparks passed from the top of the coil to the point where the lower end of the coil was connected to the secondary "free" terminal. These sparks *were 8'—9' long.*

New condensers proposed: old ones being inadequate to stand the strain beyond 15,000 volts on two dielectrics, it would be necessary to resort to four sets when using higher pressure and this would make condenser boxes too bulky. It is now proposed to use new bottles of lead glass (Bethesda Mineral Water). These are, as nearly as can be ascertained, twice or rather more than twice thicker than the old bottles. The comparison of capacities was made today for this purpose. The new bottles were filled up to 10″ from the bottom and immersed in a tin tank. The old bottles were filled up to 9″ from the bottom and immersed in a tank. A solution was prepared from rock salt as concentrated as practicable and care was taken that the liquid was at equal height outside and inside. The readings were:

New bottle:				*Old bottle:*		
Volt	Defl.	Average:		Volt	Defl.	Average:
180	6.5	Defl. Volt		180	9 trifle less	Defl. Volt
181	6.5+trifle	6.5 181		182	9 much less	9 182
179	6.5			180	9 trifle less	

Now:

$$\frac{\text{Weight of old bottle}}{\text{Weight of new bottle}} \text{ measured} = \frac{17}{21}$$

From this result the thickness of the walls was first estimated. This not affording a sure test, some bottles were broken and the average showed that new bottles were twice as thick as old ones.

$$\frac{\text{Capacity of new bottle}}{\text{Capacity of old bottle}} \text{ from measurement before made was as } \frac{65}{90}.$$

Taking for comparison specific inductive capacity of old bottle glass 1, we have $65 : 90 = 0.722$. Now, for the same glass, the capacity of a new jar would have been $0.722 \times 2 = 1.444$ times that of old, hence the new glass has 44.4% higher specific inductive capacity. It is still not certain, in view of unequal size, whether *25* or *20* or possibly only 18 will go in tank. Taking 18 to be the lowest figure for new bottles in one tank then $\frac{\text{New tank}}{\text{Old tank}}$ will be as $\frac{65}{90} \times \frac{18}{16}$ or about $\frac{8}{10}$. For *20* bottles in the new tank this ratio will be nearly $\frac{9}{10}$, for 25 jars nearly 1.11. In the *least* favorable case, the first, we shall be able to transform at least 60% more energy than with the old jars; in the best case 2.22 times as much as before. *This is good.*

Further observations in experiments with *coil 500 turns* before described. In these experiments two external gaps were used in addition to the gaps in the box, all being in series so that the total length of the spark gap varied from 2 1/8″ minimum to about 5″ maximum. The two outside gaps were of a fixed length, 1″ each, while the gaps in the spark box varied rythmically with the rotation of the disk. The coil was connected as shown in Diagram 1. to the free end of the secondary, the lowest points of the coil being about 6 feet from the ground. Resonance was obtained with 7 tanks of capacity on each side of the primary. As it was thought that the tanks might perceptibly differ in capacity Diagram 2. is added showing their position. The primary capacity total was from approx. 0.04336 — 0.0498 mfd., according to connections. The effects observed were in many ways interesting. The streamers produced on top of the coil were generally seven feet and sometimes eight feet long, thick and violently darting about. They did not seem so red as those

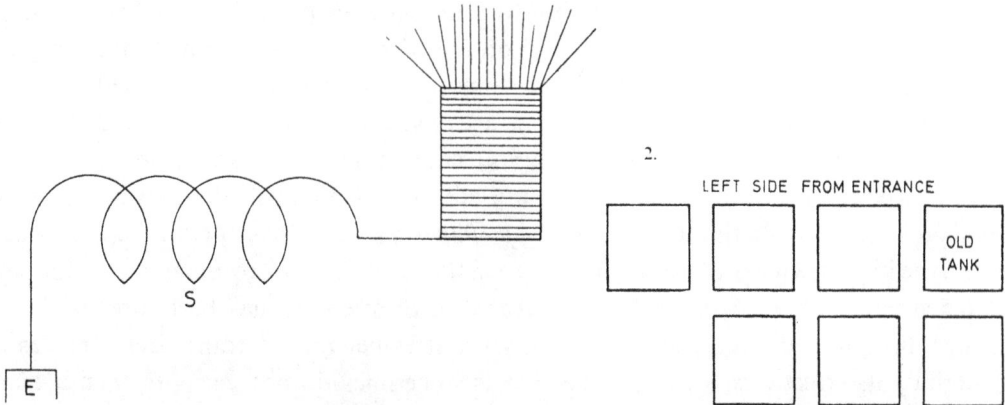

2.

LEFT SIDE FROM ENTRANCE

OLD TANK

S

E

produced before under similar conditions. Very often strong brilliant sparks would pass from top to bottom of the coil. The remarkable feature of the sparks was that they would go in a curve, almost a semicircle, as if they would start out originally in another direction and then be deflected to the lower end of the coil. Certainly they could have reached the lower end by a route shorter by 40 or 50%. During the display it was observed that *no sparks* passed over the *lightning arresters*. This was an indication that only a comparatively small electromotive force per unit length of ground was set up, too small to bridge the gaps on the arresters. In order to see whether the coil would respond to the next fundamental tone the primary cables were joined in series, this making the frequency of the primary oscillations just one half. The coil did respond but the effect was, as anticipated, small, about one quarter. Presently one of the large balls, 30″ diam, with a wire of some 20 feet was connected to the top end of the coil. The self-induction coil being adjusted, very strong streamers were now obtained showing that the ball had reduced the period of the coil considerably. But the vibration of the coil with the ball was still too fast for the primary and another ball, 30″ diam. was connected in multiple with the first. By again adjusting the regulating coil good resonating action was obtained. The sparks and streamers were now stronger, the former passing sometimes to the top of the secondary, a distance of about 8 1/2 feet in a straight line. But owing to their curved path the sparks were actually much longer. The sparks passed sometimes also to the ground from a kink in the wire connecting the top of the coil with the balls, the distance of these sparks in a straight line was 103″. The sparks were much fuller, thicker and louder with the balls then without them, they were particularly strong and bright when passing from top to bottom of the coil. It was plain that much longer sparks could be obtained with *more* turns in the extra coil as then the capacity on the end could be reduced. But the experiments also showed that the amount of electrical movement in the coil was not very great owing to the small section of the wire, for when an arc was established between two large balls it ceased to pass as soon as they were separated a distance of about 1 foot. The streamers were visible and sometimes strong on the wires leading to the balls and particularly on the wire leading from the top of the coil but still the sparks failed to bridge the gap. This showed that there was not enough energy available to charge the balls to a sufficiently high potential while, of course, the passage of the sparks was rendered more difficult owing to the large radius of the curvature. The density ought to be inversely as the radius of the curvature, hence the density on the wire leading to the ball is much greater than on the ball itself, in other words, which really means the same, the ball is charged to a *lower potential* than the wire. This seems to me a somewhat novel view to take. Without much thought I would at once assume that the pressure on the ball and on the wire is the same, but it must be greater on the wire since it can leak out from the same while it does not from the ball.

The thin wire, or any projection or surface of small curvature becomes thus equivalent to a small hole or leak in a pipe or reservoir containing a fluid under pressure and it is plain that such surfaces of small curvature will greatly diminish the maximum pressure obtainable in an oscillating circuit. It is very *important*, as I have often noted, in order to insure the high efficiency of the apparatus, to make provisions for overcoming the formation of the streamers and to this subject a great deal of attention has been already devoted. In signalling to a distance, the formation of a streamer on the transmitter impairs very materially its effectiveness so that the signals sometimes do not go more than a quarter of the distance or even less just on this account. By using a body of considerable surface, which should be spherical or a cylinder with hemispherical ends better results are obtained than with a wire leading to a height alone, not so much because of the increased capacity, but generally only because there is less opportunity for a leak and the system is more economical in producing an electrical vibration in the ground. A large sphere or surface, provided it is not too large as to interfere with the vibration of the transmitting system is better than a small one for the same reasons.

I have, however, observed long ago in this connection that when the transmitting system is formed by conductors of a considerable *mass* of *metal* a greater suddenness, or a greater rate of variation per unit time is obtained and the transmitter is more effective in producing disturbances at a distance. One obvious cause of this is that usually in such a case $\frac{pL}{R}$ is larger than if the conductors are not of great mass, but as far as I have been able to judge, the chief reason is that an electromotive force, acting upon a circuit so constituted, must give rise to a much greater current, in the first moment when in any manner, as by the passage of a spark, a great and very sudden variation in the electromotive force acting in the system is produced. I make a distinction between these two effects. One raises the pressure *gradually*, the other is responsible for the great suddenness. Thus, a mass of metal of minute electrical resistance behaves towards a sudden manifestation of electrical pressure much in the same manner as a mass of metal of *great inertia* behaves towards a sudden pressure caused by a blow. In both cases there is an increase of the pressure or force. When in the experiments, presently described, a small wire was attached to one of the balls, the other ball being left as before, the sparks passed readily between them at a distance of 6 1/2—7 feet.

Nevertheless although the sparks were very brilliant and to all appearance highly effective, *no sparks* very visible on the arresters when the discharge was playing. Evidently, the electromotive force developed per unit length of ground was small despite the power of the sparks. Upon thinking over the causes of the absence of the sparks on the arresters, it was soon recognized that in the connection as used only a comparatively slow vibration was transmitted upon the ground, the secondary effectively preventing the upper harmonics, which would have been competent to produce the sparks, from passing through the ground. The connection as first used, which is illustrated in Diagram 3. was now changed into the one shown in Diagram 4. Although the coil was now excited by a small impressed e.m.f.

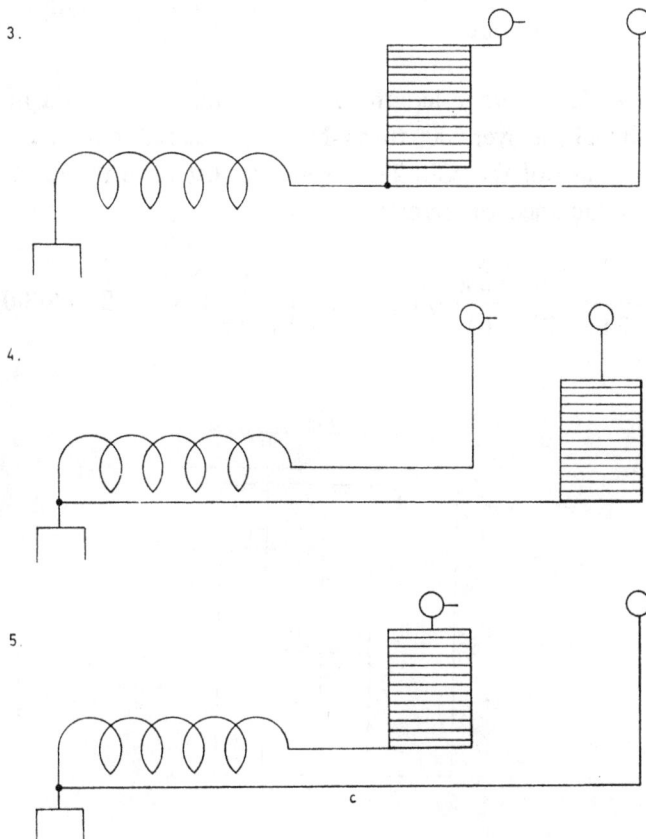

the sparks between the balls were nevertheless quite strong reaching more than 2 1/2 feet. This showed that there was also an induced e.m.f. cooperating with the impressed e.m.f. in the coil to bring about the great rise. Diagram 4. suggested the connecting of the lower end of the excited coil to any other point of the secondary thus regulating the impressed e.m.f. at will. Again, in the connection as shown in Diagram 4., there were *no sparks on* the arresters, for the reason pointed out. But when the connection illustrated in Diagram 5. was made, they appeared and became stronger when conductor *c* was constituted by a very heavy cable. This was to be expected from the above. It demonstrated the obvious fact that short waves are more effective giving higher e.m.f. per unit of length in the ground.

New induction coil for apparatus involving *method* of *magnifying* the effects by means of a *condenser* designed for the purpose of investigating: the *propagation of waves* through the ground and telegraphy.

A quick vibrating system was constituted comprising a ball of 30″ diam. and a stout cable. The period of vibration of this system was found to be, by resonance method, 240,000 per second. It was excited by sparks passing through a gap of about 7 feet, or less, from a wire connected to the top end of a coil excited by the secondary as last described. It was hoped that stationary waves might be produced by this apparatus as it seemed powerful enough. It was desirable to have an induction coil the secondary system of which would vibrate with the same period if possible, and a coil was wound on spool the dimensions of which are indicated in diagram. Now the system being 240,000=n, this gives the wave length

$$\lambda = \frac{186,000}{240,000} = 0.775 \text{ miles} \quad \text{or} \quad \frac{\lambda}{2} = \frac{0.775}{2} \text{ miles}$$

or 2000 feet approx. The average length of one secondary turn is a little over 4.5″, this will therefore require a little over 5000 turns. Now it is desired to use the 1/2 mfd. condenser on hand for the primary of the coil. The primary must have the same period, hence we have to find the inductance of primary

$$\frac{1}{n} = \frac{1}{24 \times 10^4} = \frac{2\,\pi}{10^3}\sqrt{L_p \times 0.5}, \quad \left(\frac{1}{480\,\pi}\right)^2 \times 2 = L_p = \textbf{9000 cm.}$$

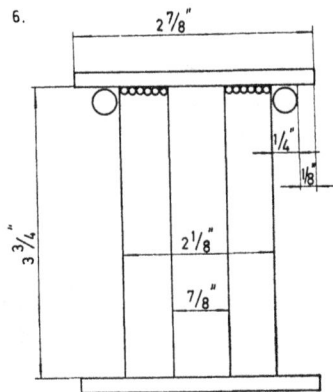

The coil actualy wound for the above object has 32 layers, 160 turns per layer in the secondary. This makes the length of the secondary very nearly$=\dfrac{\lambda}{2}$. The primary has 50 turns=n. Calculating inductance we have: length of spool: 3.75″=9.525 cm. Area= =28.3 sq.cm., from this $L_p = \dfrac{28.3 \times 2500}{9.525} \times 1.257 = L_p = \textbf{9358 cm.}$ Very nearly the value required.

Note: It will be better, of course, to adopt plan used in New York and design coil with a lot of copper.

In order to observe the rise, the spark gap ordinarily used was made so large that the secondary discharge could not break through and the voltage on the primary, when throwing in the switch, was observed. Several values of primary capacity were experimented with. The diagram below indicates the position of the set of condensers on the right side looked at from the center of the building towards the entrance. On each side of the primary there were then 2 sets in series, the connections in one instance being indicated in diagram. In each old tank there were 16 jars, the capacity of each jar being 0.00334 mfd. In each new tank there were 16 jars, capacity of each jar 10% more, being 0.00367 mfd. With tanks a and b off we have, calling C_1 total capacity:

$$\frac{1}{C_1} = \frac{2}{(0.00334 + 0.00367)\,16} + \frac{2}{0.00367 \times 16 \times 3}$$

or

$$C_1 = \frac{0.00493}{0.144} = 0.03424 \text{ mfd.}$$

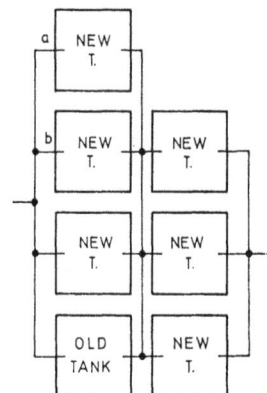

With this capacity on the secondary of the Westinghouse transformer the rise on the primary, as observed by Weston voltmeter, was from 102 to 122 volts. With tank b on each side added, the total capacity being C_2, we have:

$$\frac{1}{C_2} = \frac{2}{(0.00334 + 2 \times 0.00367)\,16} + \frac{2}{0.00367 \times 48} = \frac{1}{0.08544} + \frac{1}{0.088}$$

and from this $C_2 = 0.04336$ mfd. In this case the rise was from 102 to 126 volts. With tank a still added on each side to the preceding, the value C_3 of capacity in the secondary was 0.0498 mfd. and the rise in this instance was from 102 to 130 volts.

Note: In all cases when the switch was thrown in, the pressure rose higher at first and then settled down to the values recorded which are once more given in the results summed up:

Results:

Capacity in mfd. in secondary W.T.	on primary Westing. Trans.	
	Voltage initial	Voltage res.
0.03424	102	122
0.04336	102	126
0.0498	102	130

Proposed condenser from *Mantion Water* quart bottles

Comparative test with sample bottles showed as follows:

Defl. 7 bottles Maniton average three readings 20.66 } degrees on
„ two old bottles   „   „   „   19.7 } scale

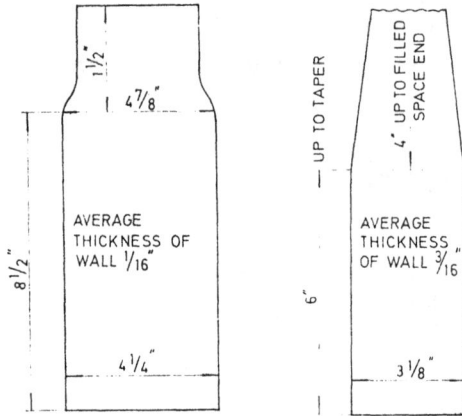

In the first case the e.m.f. was   164 volts } average values
„ „ second „ „   „   „   167 „   } of three readings

One new bottle defl. $\dfrac{20.66}{7}$. One old bottle defl. $\dfrac{19.7}{2}$. Reduced to the same voltage one of the new bottles would have given $\dfrac{20.66}{7} \times \dfrac{167}{164} = \dfrac{21}{7} = 3$ approx. This gives a ratio of capacity of one old bottle to new $\dfrac{\frac{19.7}{2}}{3} = \dfrac{19.7}{6} = 3.30$ or approximately

$$\frac{\text{new bottle}}{\text{old bottle}} \text{ capacity} = 0.3.$$

Now we may have in one tank 39 new bottles whereas only 16 of the old bottles could go in. This will give the capacity of the new tank to that of the old as

$$\frac{39 \times 0.3}{16} = \frac{11.7}{16} = 0.731.$$

Now mean diam. of cylindrical part of old jar outside $= 4.5625''$. Now mean diam. of cylindrical new bottle jar outside $= 3.125''$. Allowance for upper part on old jar 1 1/2'' taken of same diam. as cylindrical or nearly cylindrical part.

Similar allowance for new bottle 8/3''.

These figures would give:

$$\frac{\text{Surface of old bottle}}{\text{Surface of new bottle}} = \frac{\dfrac{\pi}{4} \times 4.5625^2 + \pi \times 4.5625\,(8.5 + 1.5)}{\dfrac{\pi}{4} \times 3.125^2 + \pi \times 3.125\,(6 + 8/3)} = \frac{50.829}{29.525}.$$

Now a fair idea of the thicknesses of the walls in the two bottles may be obtained by taking their weight. Measured repeatedly and changing the bottles many times it was found that $\dfrac{\text{Weight of new bottle}}{\text{Weight of old bottle}}$ was $= \dfrac{24}{32} = \dfrac{3}{4}$. Taking into consideration the surfaces as calculated before this would give

$$\frac{\text{Thickness of new bottle}}{\text{Thickness of old bottle}} = \frac{50.829 \times 24}{29.525 \times 32} = \frac{152.487}{118.1} = 1.3 \text{ only!}$$

Since the thickness ratio is much greater as found in this way the determination of the thickness by weight as above is not practicable without making allowances. The glass is evidently uneven, much more so in the old bottles than in the new. In the former particularly the bottom is heavy which vitiates the result inferred from the weight of the bottles. Many bottles were broken and it was ascertained that the average thickness of new bottles was three times that of the old. It was quite certain at any rate, that the *weakest* spot on the new bottle was fully three times the thickness of the weakest spot on the old. This was the most important thing to ascertain for the bottles give way at the weakest place. Now since the capacity of the old bottle in relation to that of the new is found by measurement to be 1 : 0.3 approx. and the surfaces are as $\dfrac{50.829}{29.525}$ we can get an idea of the specific inductive capacity of the latter with respect to that of the former. The new bottle would have for the same thickness, that is one third of the actual, 0.9 instead of 0.3 and for the same surface it would have $\dfrac{50.829}{29.525} \times 0.9$ or 1.55 times the capacity of the old, both things considered so that the *specific inductive capacity* of the glass in the new bottle must be something like **55%** greater than that of the glass in the old bottle.

Vichy water syphon bottles tested with the object of using them in the proposed new condensers. Dimensions: 3.8″ outside diam.
The glass is from 1/4″ to 1/4″+1/64″ thick, **very uniform.**

Height available 6 1/2″

Compared with Mantion water bottles: $\dfrac{\text{Mean diam. Vichy}}{\text{Mean diam. Maniton}} = \dfrac{3.5″}{3″}$

$\dfrac{\text{Thickness of Vichy}}{\text{Thickness of Maniton}} = \dfrac{17}{7}$. Now $\dfrac{\text{Vichy surface}}{\text{Maniton surface}} = \dfrac{3.5\,\pi \times 6.5}{3.125\,\pi \times 8} = \dfrac{23}{25}$

approx. Now the deflection — average of three readings was with same e.m.f. $\dfrac{\text{Deflection Vichy}}{\text{Deflec. Maniton}} = \dfrac{4.3}{9}$. The capacities are in this ratio and the test shows that, while the Vichy bottles would make excellent condensers, the capacity for two sets in series as desired would be too small. The reason is that the wall is unnecessarily thick. If it were convenient to use only one set of condensers nothing better could be desired. It having been practically decided to adopt the Maniton bottles, tests were made to see how much pressure these bottles would stand safely. Accordingly 7 of these bottles with wall rather weaker than normal, hardly 1/8″ thick, were placed in a tank with the other bottles. The concentrated salt solution reached to a point about 3″ from the top. Paraffin oil was poured to nearly the top. First 7500 volts were turned on, then the tension was raised up to 15,000 volts and the bottles withstood. The e.m.f. was then raised to 22,500 volts when the bottles began to give way, three being broken after some time. The conclusion was that two sets of bottles in series would withstand quite safely at least 30,000 volts. The glass is *really excellent*. These tests were made with 144 cycles per second. A curious observation was that when one bottle gave way others followed, this being due to the violent oscillations caused or else to the concussion upon explosion.

Approximate determination of secondary modified in construction for the purpose of *overcoming the drawbacks of distributed capacity.*

As before remarked, on a number of occasions, one of the chief difficulties encountered in the operation of a large oscillator is the distributed capacity, owing to which the efficiency of the machine — in transmitting an electrical movement to the environing media — is greatly impaired. The distributed capacity becomes particularly hurtful when the turns are of large diameter or when, generally stated, there is a great difference of potential between portions of the wire not far apart. The adjacent portion of the wire acts like a condenser in which energy is stored at each alternation, and the amount of this stored energy is proportionate to the square of the difference of potential existing between the portions of the wire constituting the condenser. Now most of this movement of electricity, occasioned by this distributed capacity, takes place within the coil and does not, unless in a very small part, appear in the external circuit. Since the movement in the latter circuit is the chief object, the charging and discharging of the condensers formed by the turns of the coil is mostly lost for the purpose for which the machine is designed. In a properly designed oscillator of this kind all the movement produced in the coil should be propagated to the external circuit, but this condition can never be rigorously realized.

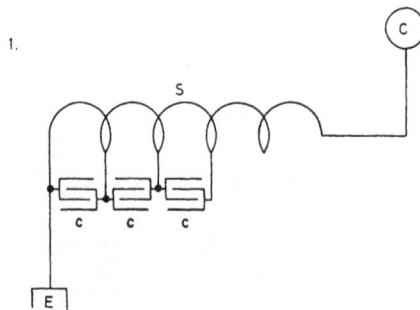

The object of the design which will be presently considered is to approach this degree of perfection as nearly as practicable. Referring to Diagram 1. illustrating a secondary as here used, the object of which is to produce the greatest possible movement of "electricity" from E into C and vice versa, it will be seen that the succeeding turns form small condensers c, c, c, in which a local movement takes place which is not transmitted through the entire wire. Such a coil S with considerable distributed capacity will, therefore, not be efficient in producing disturbances such as I contemplate using for a number of purposes. The distributed capacity is particularly hurtful when it is desired to produce a very high potential at C and, as this is generally the case, it is important to adopt a construction and observe working conditions such as will reduce the evil to the minimum. I have in similar instances attained the object more or less by constructing a secondary in parts connected in series through condensers, and a secondary to suit the present apparatus designed on this plan is now to be considered. In order to explain better how such condensers in series act in reducing the effect of distributed capacity, as pointed out before, reference is made to Diagram 2. which shows turns or portions of the secondary connected through condensers C_1, C_2, C_3. Suppose there would be n such parts as t joined in series and let C_1 be the total capacity of the coil or secondary S, then the capacity of the turns or portion t would be $\dfrac{C_1}{n}$. But now, if there be on the ends of this portion t condensers C_1 and C_2 it is plain that the distribution of electricity along the portion t will be greatly modified by the presence of these condensers and, if their capacity be very much greater than that of the wire t, almost all of the electricity will reside on the coatings of the condensers, hence there will be very little local action in the portion t, and most of the electrical movement created in the wire will be transmitted along the entire length of the same from E to C and vice versa. The machine will then act more efficiently and will be much more suitable for the transmission of energy through the terminals E and C for whatever porpose the energy be intended and in whatever manner it be used. It should be stated that in the particular form of oscillator as here illustrated, the lower turns of portions which are closer to the earth, and therefore at a lower potential, are not nearly as hurtful as the

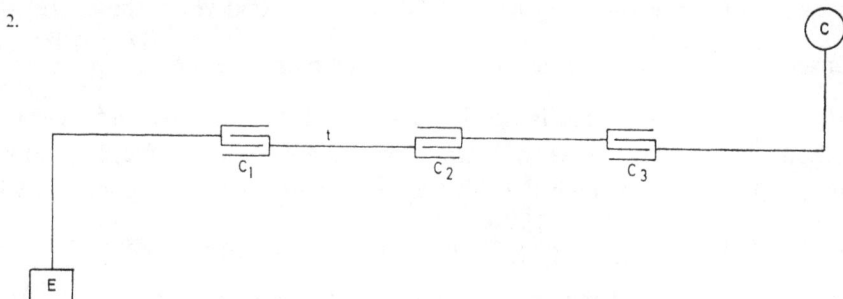

upper ones, or to put it more generally, those which are farther away from the earth connection and closer to the free terminal. The latter are, namely, at a much higher potential and there exists generally a greater difference of potential between portions of wire of the same length when the resonant rise is considerable; as the energy stored in the distributed condensers is proportionate, as before stated, to the square of the difference of the pressure between adjacent portions, it is, as a rule, of advantage to make the upper turns of smaller diameter or to put them farther apart. But in certain instances this very drawback can be turned into an advantage and by placing a few of the turns near terminal C as close together as practicable a greater electrical movement may be produced in the lower turns or those nearer to the earth connection and the effect exerted at distance may thus be increased, though I prefer not to resort to this means as a better result is obtainable in other ways. Coming now to the consideration of the secondary to be modified in the manner proposed let Diagram 3. illustrate the arrangement as contemplated in the present instance, a condenser C being placed between each succeeding turn of secondary S. I shall

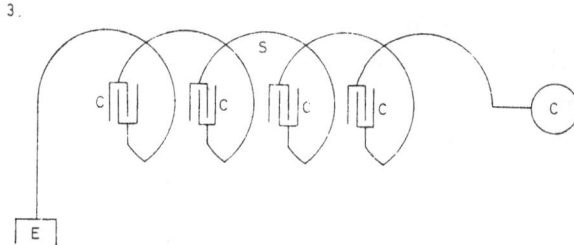

make a rough estimate on the basis of *one* primary turn and the maximum primary capacity available, which is 80 jars on each side giving a total capacity of 40 jars or $40 \times 0.0036 = 0.144$ mfd. approx., there being two sets in series. This would give $n = 50,000$ per second roughly. Assume we work with 30,000 volts from the Westinghouse transformer and retain, as at present, 35 turns of the secondary. In this case, considering the distance of the secondary turns and their decreasing diameter, the induced e.m.f. in the secondary will be much smaller than might be inferred from the ratio of transformation which will be 35 :1; with reference to measurements of mutual inductance made before, with a secondary wound on the same frame, it may be estimated at 18 times the primary e.m.f., that is, the induced e.m.f. in the secondary may be assumed to be $18 \times 30,000 = 500,000$ volts *nearly*. Taking the number of breaks at 1200 per second as used in some of the preceding experiments with the apparatus, and assuming the charge of condensers always effected at an average pressure, say 6/10 of 30,000 or 18,000 volts, the energy stored in the condensers per second and delivered in the primary will be $\dfrac{18^2 \times 10^6}{2} \times \dfrac{0.144}{10^6} \times$ $\times 1200$ watts or 28,000 watts, that is nearly 38 H.P. will be taken by the apparatus under these conditions. This is, of course, only an approximate estimate. We may now, on this basis, estimate the current through the Westinghouse transformer's secondary, which will be roughly estimated $\dfrac{750 \times 38}{30,000} = \dfrac{285}{300}$ of an ampere, or with losses etc., say *one ampere*. The performance will thus be still much below the maximum output of the transformer as it can deliver 1 amp. at 60,000 volts pressure and continuously and, I have no doubt, much more, as so far not the slightest increase of temperature has been noted after prolonged working. Assume then the current through the supply secondary to be *one ampere*, the current through the primary of the oscillator will be increased by the factor $\dfrac{Lp}{R}$ approx. Here $L = \dfrac{7}{10^5}$ and $p = 6\,n$, roughly$= 3 \times 10^5$. R is difficult to estimate since the primary circuit includes the arc over the spark gap. I have adopted a method for determining the resistance of the arc with fair accuracy and this will be the subject of a later consideration. The resistance of the primary cables in multiple is entirely negligible, there being in each 37 wires, No. 9. Taking 1265 feet per ohm, the resistance of about 160 feet of such two cables would be $\dfrac{160}{1265 \times 74}=0.0017$ ohm only. Not considering the resistance of the arc the factor $\dfrac{Lp}{R}$ would be enormous,

$$\text{being}=\frac{7 \times 10^4}{10^9} \times \frac{3 \times 10^5}{\dfrac{17}{10^4}}=\frac{21 \times 10^4}{17}=\mathbf{12{,}353!}$$

While, of course, such a condition can not be realized in practice we may approach this value more or less by doing away with the arc in the primary as, for instance, in the form of an oscillator with mercury breaks, which I have devised to work with low tension so that the arc practically does not occur. Or a condenser may be placed in shunt to the primary as has been already considered on a previous occasion. By experience I know that the initial currents in the primary reach certainly several thousand amperes showing by this that the resistance of the arc can not be great. For the present I shall assume that it is 18 ohms so that if the condensers are, as supposed above, charged to 18,000 volts during ordinary performance, the initial current would be about 1000 amperes in the primary decreasing logarithmically. With this maximum in the primary, the loss in this circuit will not be unduly great. Now the secondary condensers should be of a capacity to carry the secondary current at the frequency used. Calling now e the e.m.f. induced in the secondary per turn, c the capacity of one of the secondary condensers as before, we will have the current through the turn $i=e\,c\,\omega$, ω being here 3×10^5 as before assumed. Now the e.m.f., taking it on the average, will be per turn of secondary $e=\dfrac{500{,}000}{35}=15{,}000$ volts approximately, and on the preceding assumption of 1000 amp. maximum in the primary turn, the largest value we may assign to i would be $i=\dfrac{1000}{35}=30$ amperes, roughly, and this would give

$$c=\frac{30}{15000 \times 3 \times 10^5}=\frac{30}{45 \times 10^7}\ \text{farad or}\ \frac{3 \times 9 \times 10^{11}}{45 \times 10^7}=\frac{30{,}000}{5}=\mathbf{6000}\ \textbf{cm.}$$

We would thus require a capacity of about 8000 cm. in each of the condensers to carry the secondary current in the oscillator. But this is really too high an estimate and it is quite certain that a smaller capacity would do. Since a jar has a capacity of $0.0036 \times \times 9 \times 10^5=3240$ cm., two jars would be amply sufficient and possibly also one jar between each turn of the secondary. Taking it on this basis, the total capacity of the secondary would be $\dfrac{2 \times 3240}{35}=185$ cm. approx., while the measured capacity was 3600 cm. The effects of distributed capacity would thus be reduced by the use of secondary condensers to about 5%. These secondary condensers will, of course, have to be so constructed as to withstand not so much the strain on the dielectric — for this they will support easily — but the sparking over the condenser coating. Let the spark length on the secondary be, say, 12 feet and suppose we had 36 secondary condensers, then on the average they ought to be able to prevent sparking when the pressure on each is such as to cause a spark of a length of $\dfrac{12 \times 12}{36}=4''$. There will not be much difficulty encountered in this respect, if the condensers are properly designed. The virtue of condensers used in such a manner is well established and I think it resides in the fact that when they are used the charge does not distribute itself along the wire, but accumulates on the coatings of the condensers thus reducing the effect of mutual electrostatic induction of the adjacent or near positions of the wire, to a large extent, and reducing in this manner the amount of energy stored in the coil itself. It now remains to consider the capacity on the end of the secondary which is "free". This capacity ought to be so large that it can take up all the current of the secondary at the frequency used. Or, to put it otherwise, it should be able to store all the energy the secondary is able to give. There is, however, another consideration which must be made in case the secondary is capable of free vibration, and that is that the capacity on the end should be so determined as to secure resonance with the primary. As regards this capacity there are then three relations to be borne in mind when deciding upon this and they are:

$$i = E c_s \omega \qquad c_s = \frac{i}{E \omega} \quad \cdots \cdots \cdots \cdots \cdots \cdots 1)$$

Watts in secondary $W = \dfrac{c_s E^2 \omega}{2}$ or $c_s = \dfrac{2W}{E^2 \omega} \cdot \cdot 2)$

$$L_p C_p = c_s L_s \quad \text{or} \quad c_s = \frac{L_p C_p}{L_s} \quad \cdots \cdots \cdots \cdots 3)$$

E secondary e.m.f.

ω ,,   frequency

L_p inductance $\Big\}$ in primary
C_p capacity

L_s inductance secondary

This to follow up.

Colorado Springs Notes

August 1—31, 1899

For want of time following items, partly worked out, have been omitted:

Aug. 4. General observations of electrical phenomena here with particular reference to stationary waves.

Aug. 6. Experiments with stationary waves on water pipe. *Note:* Distance from groundplate to End of main *exactly* 550 *feet*.

Aug. 31. Patent matter worked on:

　　　1) Production of stationary waves and use of same in general
　　　2) Distribution of universal time observatories etc.
　　　3) Indication of direction for ships etc.

Colorado Springs Notes:

Aug 1 — 31 1899.

For want of time following items,
partly written out, have been omitted:

Aug. 4. General observations of electrical
phenomena here and particular
~~reference~~ to stationary waves.

Aug. 6. Experiments with stationary waves
on waterpipe. Note: distance from
groundplate to end of main exactly 550 feet
~~experiment~~ carried on:

1) Production of stationary waves and
use of same in general

2) Distribution of universal time observations etc.

3) Indication of direction — for ships etc.

Colorado Springs

Aug. 1, 1899

Various observations. In the course of these experiments and particularly during the past month a number of highly interesting observations have been made which will be presently dwelt upon.

First of all one is struck by noting the extraordinary purity of the atmosphere which is best evident from the clearness and sharpness of outlines of objects at great distances. In low regions, especialy where moisture is in excess, the outlines of objects become more or less indistinct and confuse at distances of but a very few miles while here at many times such distances the outlines appear perfectly clear and sharp. When a train is moving up Pike's Peak it is very often quite easy to distinguish not only the engine and cars but even the windows and wheels of the same perfectly, although the distance from the experimental station is from 10—12 miles. Quite frequently also the house on top of Pike's Peak can be clearly seen with the naked eye. The ranges of mountains 100—150 miles away or more can be perceived perfectly. A range at a distance of about 50 miles can be seen plainly even at night when the sky is clear. It is wonderful how at times immense objects appear dwarfed, while small objects as horses, carriages or men assume unnatural gigantic dimensions.

Pike's Peak Range appears at times so close and so ridiculously small, that anyone not knowing the reality would be apt to fire a modern rifle at some object on the mountainside believing it to be within shot. Nor is this statement exaggerated much as it seems so. At other times again Pike's Peak appears far remote and its height much beyond what would seem natural. The arc lamps at the foot of the mountains five to seven miles away or more shine with a brilliancy as though they were only as many blocks from the observer and under certain conditions an ordinary incandescent lamp of 16 c.p. seems to give out as much light, judging from a distance, as ordinarily an arc light does. It appears also as big as the latter. This penetration of the light is due to the wonderful purity and extreme dryness of the atmosphere.

The moonlight is of a power baffling description. I have been told that the best photographs of the mountains have been obtained by moonlight and I do not doubt it. Exposures of half an hour ought to give clear photographs revealing all details although the exposures are as I am told from 1 1/2 to 2 hours. I have nowhere seen such a light. Italy is famous for moonlight nights but in my estimation that country can not even compare with Colorado. I think this extraordinary brightness of the moonlight is chiefly due to the absence of moisture, for there are many places, as in Central America, which are located much higher and yet the moonlight, I am told, is not so intense and I can see no other reason for this except the presence of more vapour in those places. It is not a mere saying, but literally true, that during full moon in these parts it is "as light as day". Objects can be clearly perceived at distances of many miles and one can easily recognize a friend or familiar object at a distance of something like a quarter mile if not more. The shadows cast by the moonlight are extraordinarily black and sharp. They suggest the Crockes' shadows noted in vacuum bulbs and on this account the moonlight is particularly interesting and suggesting thought and stimulating the imaginative powers. The shadows of the clouds on the plains and mountains are quite dark and clearly defined and it is interesting to behold the patches as they speed over the ground. When the moon is absent and the nights clear the number of stars visible and their brilliancy is amazing and the sky presents

127

a truly wonderful sight. The twinkling of the stars is very pronounced, they seem to move in orbits of as much as ten or fifteen of their own diameters across. At times one observes a star burst out into great brilliancy. This is probably due to the removal of an invisible cloud or of a layer of air at a great altitude containing some kind of particles which cut off a large portion of the light. One sees shooting stars quite frequently, also colored rings around the moon, generally in the advanced hours of the night, at times when the air is slightly misty. As this happens generally during very cold nights I believe the colored rings are due to minute crystals of ice.

Owing to the extraordinary purity and dryness of the atmosphere the sounds penetrate to astonishing distances. This is particularly true of *high notes* as nearly as I can judge. Certain conditions, entirely exceptional, concur at times and produce effects of this kind which are startling. A bell will ring in the city several miles away, and it would seem as though the bell would be before the very door of the laboratory. During certain nights when sleepless I have been astonished to hear the talk of people in the streets and sounds of this kind in a large radius around the dwelling not to speak of the grinding of the wheels, the rolling of wagons, the puffing of the engines etc. which are perceptible in such a case, and with painful loundness though coming from distances increadibly great. These phenomena are so striking that they can not be satisfactorily explained by any plausible hypothesis and I am led to believe that possibly the strong electrification of the air, which is often noted, and to an extraordinary degree, may be more or less responsible for their occurence.

The dryness of the atmosphere, which is still further enhanced by the low pressure, is such that wood or other material is made what is called kiln-dry inside of a few hours, and is rendered an insulator far more perfect than wood is ordinarily. The nails on the hands and toes dry out to such an extent that they break off very easily, in fact one has to be careful in trimming them. I found the claws of a cat as brittle as glass. The skin on the hands dries out and cracks up and is apt to form deep sores particularly if, as often in experimentation, one has to wash the hands frequently. The hair gets perceptibly thinner owing to the drying out. Colorado is not a good country for hair. This may be of interest to people with a tendency towards baldness. People even very sick do not cough and expectorate evidently owing to the dryness of the atmosphere. One does not perspire as the sweat is immediately evaporated. It is curious how quick the body gets dry when a bath is taken. Still more this is noted when the body is rubbed with alcohol. These observations are not often made, unfortunately, as the opportunities for comfort are not such as one might desire.

In many respects one is disappointed with the aspect of the country itself although it is far famed. I think it very uninteresting and even the celebrated Pike's Peak is insignificant. Most of the country is barren, practically a desert, with little vegetable and animal life in places. Prairie dogs are about the only animals one can see on the plains. One rarely sees a bird and the country must be a tedious one to live in for any one with tastes for hunting and fishing. But as much as the country is devoid of interest and beauty, so much and far more, is the sky beautiful. The sights one sees here in the heavens are such that no pen can ever describe. The cloud formations are the most marvelous sights that one can see anywhere. The iridescent colors are to my judgement incomparably more vivid and intense than in the Alps. Every possible shade of color may be seen the red and white preponderating. The phenomena accompanying the sunrise and sunset are often such that one is at the point of not believing his own eyes. At times large portions of the sky assume a deep red almost blood-red color, so intense that superstitious people might will be frightened when first seeing it as by some other altogether unusual manifestation in the heavens. Sometimes, particularly in the forenoon, huge masses of what appears to be snow are seen floating in the air and they are so real and tangible, so sharply defined, that it is difficult to believe them to be composed merely of vapor. The purity and dryness of the atmosphere explains to a degree the sharpness of definition of the boundaries of these formations of mist, but it is quite possible that some other causes as electrification of the particles cooperate in rendering them so compact as they appear to be. Of course, the purer the air, the greater is the difference between the region filled by cloud and that surrounding it as regards the passage of light rays, and the boundaries of the cloud appear sharper and quasi-solid much on this account. The whiteness and purity of these masses of cloud is such that one has the idea that nothing, not even an angel, could come in contact with it without soiling it. Very often when the sun is setting, a considerable portion of the sky above the mountain range presents the sight of an immense furnace with white-hot molten metal. It is absolutely impossible to look at the melting away clouds without being blinded, so vivid is the light. On a few occasions I have seen the mountains covered with a white silvery veil most beautiful to see, an unusual occurence and caused by a fine mistlike rain in the mountain region. The intensity of the light on these occasions was really wonderful. What was remarked before of the shadows of the moon is, and to a much greater degree, true of those thrown by the sun. They are inkblack and sharply outlined. The shadows on the plain and mountains thrown by the clouds appear like big patches of inking blackness hurrying along the ground. Particularly interesting are shadows thrown across the sky resembling often large dark streamers, or those which under certain conditions are formed and are visible like dark columns extended from the ground to the sky. These shadows seem to be best visible in the middle of the afternoon or a little later when the sun in fairly down and on days when it has been extremely hot and sultry in the forenoon and the clouds are formed quickly and are of greater density than usual.

A very curious phenomenon is the rapid formation and disappearance of the clouds. One can watch them continuously forming and disappearing rapidly and one merely needs to turn away for a few moments when he may see that the aspect has changed, new clouds having replaced those he saw before. On many occasions, just after sunset, I have seen seemingly dense, white clouds appear as by enchantment below the mountain peaks. So quickly did these clouds or mist form that their appearance was much like the projection of an image on the screen. The wonderful beauty of the cloud formations as seen here is, however, enhanced not only by the incredible sharpness of the outlines and vividness of color but also by their accidental arrangement and forms they assume. Not unfrequently one can see clouds resembling all kinds of known objects, this adding much to the enjoyment one finds in observing them. In fact I have scarcely ever watched the clouds here without noting among the shapes resembling some or other familiar object. It is probably owing to the peculiar character of the clouds here that phenomena of this kind may be almost daily observed whereas in other parts they are very rare. Very often I have seen low on the horizon what appeared to be immense fields of ice as a sea frozen in the midst of a storm but so wonderfully real that it would be impossible to give an idea of it by a description however vivid. At other times there appeared ranges of mountains which one could not distinguish from the actual, on the horizon or the wide ocean, with its deep green, or dark blue, or black waters stretching out as far as the eye could reach.

Nor was this an ordinary resemblance which one could banish from the mind by a small effort of will, but was rather of nature of those visions or hallucinations which make it necessary for one to pinch himself to fully realize that his senses have been deceiving him. More than once I have seen this ocean dotted with green islands or populated with glittering icebergs or sailing vessels or even steamers not less real to the eye because they were formations of mere mist or cloud.

Almost every evening, after sunset, and when the sky is clear, the horizon towards the plains becomes peculiarly tinged with colors of surprising vividness, all the colors of the rainbow being represented, the strata higher above the horizon beginning with red and passing through all nuances, the lowest strata finishing with blue, violet and black. As it grows darker the black line rises continuously above the horizon. This phenomenon illustrates in an interesting manner how the sun's rays are deviating from the straight course and are being continuously deflected downwards to the more dense strata of the atmosphere. Among the seemingly infinite variety of clouds there are four typical forms regularly observable which are of surpassing beauty. They are:

1) *Red clouds,* which are seen very frequently in the early morning hours at sunrise and, though less frequently, in the evening when owing to a greater percentage of moisture the clouds are denser, more like rain clouds. They reach an intensity of color equal to that of a ruby of the "pigeonblood" species. They are particularly beautiful when appearing in detached masses.

2) *White clouds* which are seen chiefly in the forenoon or in the early part of the afternoon though not so often. The whiteness and purity of these clouds and their sharpness of contours which has been already referred to makes them a unique sight. It would be difficult to offer to the eye a greater treat than it finds in the contemplation of these masses of mist, generally floating in big detached lumps in the blue sky. I note that these clouds are seen generally after a short rain when the wind, springing up suddenly, clears the sky, leaving only a few large and separate masses of vapor.

3) *Clouds* presenting the appearance of immense lumps of gold. These are iridescent clouds witnessed chiefly at sunset. They present a striking sight, particularly when they are small and detached from each other and the sun's rays can penetrate them more freely thus heightening at times to a degree really incredible the intensity of the iridescence. Their color is absolutely like that of gold and the similarity is rendered complete by the forms they assume which are those of gold nuggets found in nature, but generally they pass from pure yellow to a reddish yellow of the kind peculiar to gold found in certain countries or generally gold containing a small percentage of copper. A feature of these most beautiful clouds is that they persist in their iridescence but a very short while. Usually they last only from five to ten minutes and often even not so long, although the yellow color may generally persist on the edges for as much as half an hour, more so in the morning than in the evening hours.

4) Clouds resembling lumps of incandescent metal. These clouds are most wonderful to behold and the intensity of the light emitted by them is such that it baffles description. I have never before seen anything of this kind in the Alps or elsewhere. One can see all nuances of color exhibited by heated metal or coal, from dull red to blinding white incandescence such as is seen in silver furnaces known in German as the "Silberblick". But most generally these clouds present the appearance of lumps of glowing coal surpassing, if anything, the latter in brilliancy and intensity of color and the sense of sight is still more

completely deceived by the gradual burning away of the glowing mass offering to the eye the spectacle of a mass of charcoal which is being quickly consumed in a furnace with a very strong draught. How can the intensity of the light emitted by these clouds be explained? They throw out at times a light which to the eye is as intense and blinding as that of the sun's disk itself, yet they present a surface many hundred times greater than that of the sun's disk. Is it not possible that in this intense iridescence, not to say incandescence, we see not only a phenomenon of reflection and refraction of the rays of light but also, at least partially, of conversion of dark radiations of the sun into such which cause in our eye the sensation of light? Or, if not exactly this, might it not be possible that the dark rays being absorbed in the mist in some way or other reduce the absorption of the light rays and render the process of reflection and refraction of the latter more economical? I can not recollect any experiments carried on with the object of ascertaining the influence of temperature on these processes. A hot glass lens ought to be more efficient in letting the light rays through than a cold one. But, reasoning in the same strain, it would appear that reflection from a surface ought to be impaired by heating the latter. Furthermore I should think that it can not be indifferent for these two processes at what temperature the body reflecting or refracting the rays is maintained, at least one must infer so from the accepted theories according to which the dark and luminous radiations merely differ in their wave lengths but are otherwise identical. The most plausible view on the above phenomenon still seems to me that first expressed, according to which invisible radiations are partially converted into luminous rays or radiations thus supplying the additional light which it is difficult to account for otherwise. It is not impossible that a phenomenon similar to fluorescence might be produced by heat rays falling upon the particles of mist thus heightening the light effect or there may be caused, by the dark rays, a decomposition or falling apart of the vapor particles (as Tyndall demonstrated) — and this process may be accompanied by some evolution of light. Certainly the particles capable of producing such vivid iridescence must be very minute, much smaller than ordinary particles composing the clouds and their form can not be but a passing one as is evidenced by the rapid disappearance and reappearance of clouds already mentioned. These four types of cloud, which can be observed here almost *daily* and which in purity, brilliancy and depth of color and sharpness of outlines surpass by far such clouds noted in other parts, constitute the chief attractions of the incomparable beauty of this sky. These phenomena would be more appreciated if they were more rare, but the fact is that for most people they loose a large portion of their charm by forcing themselves upon the eye too frequently. We are used to speak of "Sunny Italy" but compared with Colorado that country might be almost likened to foggy England. They tell me that there are scarcely 10—20 days in a year, on the average, when the sun does not shine and even this estimate is rather exaggerated. Since my arrival here about the middle of May, with the exception of a few passing thunderstorms, the days were clear with just enough clouds in the sky to break the monotony of the blue.

No wonder that consumptives and generally people in feeble health are getting on here so well. The purity of the air, the altitude, which compels exercise of the lungs to be continuously and unconsciously practised owing to the lesser density of the air and smaller percentage of oxygen (about 20% less than at sea level), the dryness of the air which is altogether exceptional, all these causes may cooperate more or less efficiently in improving the condition of the patients, but I believe that the chief cause of betterment is to be found in the profuse and cheering sunlight. Whether the light produces a specific germicidal effect is a matter of conjecture as yet, as far as I know. I learned here that experiments had been carried on to ascertain whether there are any Roentgen rays emitted by the sun or produced in other ways by the sun's rays but the results were negative. Similar experiments, I am told, were conducted for a long period on Pike's Peak but no action on a photographic film, which was the means of these investigations, was noted, at least not such as might be attributed to Roentgen rays. I think though that rays of this kind must be ultimately demonstrated to exist in the radiations of the Sun as well as of most other sources of intense light and heat. It is possible that such rays are, in a measure, active in arresting the process of decay caused by the bacillus. I conclude that, since the bacillus of tuberculosis is an organism developed under *exclusion* of light, such rays of short wave length, made by any means to penetrate the tissues and reach the affected parts of the same, must needs be inimical to the development of the microbes not used to such rays. Though this conclusion might not prove true, still there is a good foundation for it, and I am hopeful that with the apparatus I am now perfecting for other purposes as well as this, it will be possible to produce Roentgen rays of great intensity which will furnish the long sought for means of successfully combating these dreaded deseases of the internal organs. Whatever be the cause of the marvelous improvement noted in patients it is a fact that most people afflicted with these ailments, and often pronounced beyond medical help, recover and get soon seemingly quite well here. A short while ago I was induced by a friend to go to a dinner he gave in my honor where I met a number of more or less interesting people. The conversation during the entire evening was an animated one and the entertainment highly enjoyable. Everybody seemed to be in high spirits and excellent health. But my pleasure was spoiled in the end when I learned before parting, with painful astonishment, from a friend who is a very skilled and competent physician, that of the two dozen people I met scarcely one individual had more then one whole lung left, the majority of them being in fact "much farther gone" as he said, so that they would infallibly die in a very short time if they would leave here. I soon learned that there were thousands of consumptives in the place, about the only healthful people being coachman, and I concluded that while this climate is certainly in a wonderful degree healthful and invigorating, only two kinds of people should come here: Those *who have* the consumption and those *who want to get it*. That the sun's light and heat exercise a highly beneficial effect on these sick people may be inferred with certainty from its effect upon people who are quite well. It is curious to note how agreeable and indispensable the sunshine becomes here after a while. Even healthful people become sad and unstrung when the sky gets clouded and dark. I have however, observed such an effect before but it is quite natural that it should be so here where the sun shines constantly day after day. I do not suppose that in London or even in New York, where the weather is comparatively fair much attention is paid as to whether the sky is clear or clouded, but here every laborer laments when the sun does not shine. Despite the beautiful spectacle offered by the parting sun one feels sad when its disk sinks behind the mountains and one is thoroughly glad to see it rise again. These feelings are experienced, of course, everywhere, but somehow they are of greater intensity here than elsewhere. Considering the elevation, the small density and exceptional purity and extreme dryness of the air, the scantiness of the vegetation and particularly the scarcity of protecting timber, the vastness of the practically desert prairies over which the wind can sweep unimpeded, the geographical position of the country and other causes and conditions determining the character of the climate it is not difficult to guess the general nature of the weather in Colorado. Nevertheless it is a surprise to learn that the climate is mild in an extraordinary degree, the storms coming but seldom and lasting one or two days at the most, the snow remaining scarcely over more than thirty--six hours on the ground.

In fact Colorado people seem to be particularly proud of their winter climate. I expressed to a friend my delight at the wonderfully fine and bracing weather we had so far, but he astonished me by saying: "This is not a fair opportunity to judge. To form a correct opinion of the qualities of this climate you must come here in wintertime". I could scarcely conceive how it could be possibly finer and more agreeable then so far experienced. I expect to get data as to the pressure, temperature, moisture etc. The pressure at present is about $24''$ average, considerably less than at sea level but, owing to the bracing air, one does not feel much the effect of the rarefaction of the atmosphere except when performing some physical work, when one gets quickly out of breath. The humidity must be extremely small otherwise one would feel both the heat and cold much more. The mean temperature presently at noon is about $80°$ in the shade but in sunshine it is different. I believe the good people here are more or less inclined to find the days in summer cooler than they are in reality, and they seem also to prefer to be silent about cold snaps which occasionally come in wintertime. But from some indiscreet persons I have learned that the thermometer was at times very near $40°$ below zero and in the plain sunshine of summer it is apt to be "way up" as my informants told me. I feel sure it can not be far from $150°$. The power of the sun's rays on certain days when the atmosphere is particularly calm, dry and pure, is such as to positively surpass belief. The waterpipe passing for some distance across the field to the laboratory being partly uncovered the heat was as a rule so fierce that the water came out boiling and steaming like in a Russian bath. It would be impossible to hold the hand in it, even for a few moments, for it would at once cause a severe pain. One day, about five o'clock in the afternoon, the rays fell through the open door on a high tension transformer which I had brought from New York and, before anybody could notice it, melted out all the insulation, rendering the apparatus completely useless. I observed the danger a few days before and warned the assistants to watch the machine, but unfortunately on that day the usual precautions were omitted. Several barrels filled with concentrated salt solution were placed outside of the laboratory, and the pressure in them rose every day as in a steam boiler, and a few of them were damaged! When the cock was opened the water squirted out to a great distance across the field and it was thought advisable in order to avoid bursting and damage, to leave a small opening in the barrels for the escape of the steam. The most astonishing experience of this kind was, however, the heating of a wooden ball covered with tinfoil, which was supported above the roof, to a point it was deemed unsafe to expose it to the sun's rays. It emitted a dense vapor actually like smoke, and the tinfoil crumbled away! This excessive heating seemed to take place suddenly. I believe that it occurs when, owing to the removal of a layer of impure air, a particularly clear path is opened for the sun's rays, which then pass through the pure medium without much loss. Often I have felt a scorching pain on the cheek or neck to come on *suddenly* when working in sunshine, and I can only explain it with the above assumption. But the most interesting of all are the electrical observations which will be described presently.

Aug. 2, 1899

New induction coil for portable apparatus, designed for investigation outside contained in box. *Condenser method*. Particulars: Secondary wound with wire No. 30, 32 layers plus one layer with thick cord. Turns 180 per layer. Total number of turns 5670. Least turn $1.25 \times \pi = ?$ longest turn $3.5\pi = ?$ average $2.125 \times \pi = 6.675''$ or 17 *cm.* approx. for average length of one turn. Resistance of secondary 375 ohm. The dimensions of spool are as in sketch below.

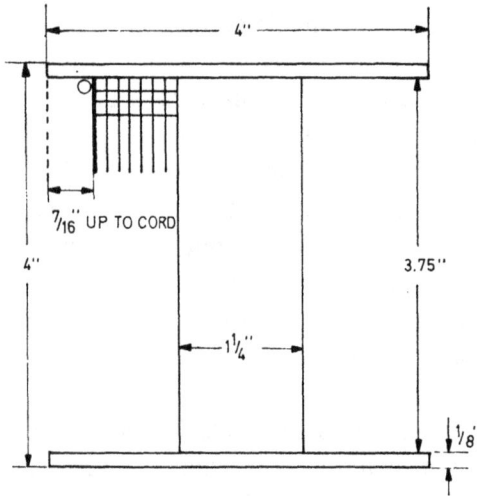

The available length of coil $4'' - 1/4$ for two fibre flanges $l = 3.75''$. The primary 50 turns as a coil before wound cord No. 20, Res$=0.51$ ohm. The length of average turn $10.6'' = 26.93$ cm. Total length of wire in primary $530'' = = 1346.2$ cm.

In connection with this there is to be used: a condenser to be adjusted, a charging coil experimental like wire to be adjusted to the condenser and break, a Thomas clockwork with wheel 180 teeth for break, the arbor carrying also a sensitive device so that same clockwork will serve for break and first sensitive device. A second sensitive device in secondary of oscillating transformer driven by another Thomas clockwork. This latter need not be used when self-exciting process with condenser employed. Relay brought from New York 996 ohms or thereabouts. *This final.*

Mantion bottles to be used compared with *Champagne bottles*. The latter would seem to be better suited. The tests showed as follows: Comparing 2 bottles of each kind filled as far as practicable and placed in tank with rock salt solution as in previous instances charged to *same* potential 356 *volts* approximately, the average of four readings was for: Maniton bottles 13° defl. for Champagne bottles 9.5° defl.

This gives: $\dfrac{\text{Capacity of Maniton}}{\text{Capacity of Champagne}} = \dfrac{130}{95}$

How by taking weight: $\dfrac{\text{Weight of Maniton}}{\text{Weight of Champagne}} = \dfrac{23}{36}$

These figures give a slight advantage to glass in Champagne bottle, but the latter was larger than the Maniton about $1/4''$ outside and contained a trifle *over* one *quart*. Furthermore, there is the usual hollow bottom (to deceive customers) and this increased the surface in the Champagne bottle. In view of this the conclusion is that the glass is not greatly different in respect to dielectric qualities in both the bottles. The Champagne bottle would unquestionably break down first because of hollow bottom, as it would be difficult to exclude the air. It would also be difficult to get the required number of such bottles in this quiet town. This compels use of *Maniton*.

Consider the following case: A condenser is connected in series with another to a generator of high tension. A circuit making and breaking device is arranged in a bridge between the condensers as illustrated in diagram. When the circuit is closed through this

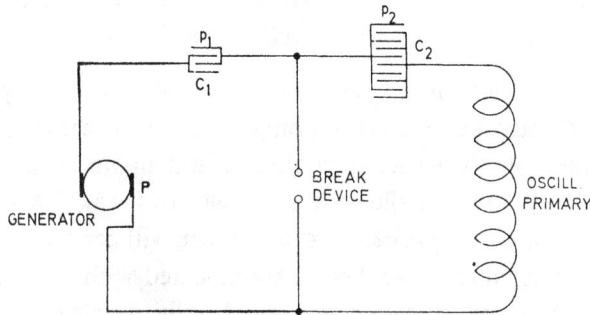

device the condenser included in the circuit of the generator is charged to the full potential, but when the device breaks the current path, the charge is distributed over the two condensers. Such an arrangement with two condensers has certain valuable features in connection with oscillators, particularly when they are worked from a generator of very high and constant e.m.f. The condenser included in the generator circuit prevents short circuiting of the generator in case of defective action of the make and break device and the amount of energy drawn from the source is limited to a quantity which can be exactly determined beforehand. The arrangement is sometimes of value also with alternate current generators. The condenser C_1 then performs the function of a reducing valve on a reservoir such as is used in connection with a distribution system of some gas under great pressure. By means of such an arrangement an oscillator may be worked safely from a generator of any e.m.f. and at any desired pressure. In the case as illustrated the total capacity is

$$C = \frac{C_1 C_2}{C_1 + C_2}$$

and the energy stored by one charge in the system is

$$\frac{1}{2} P^2 \frac{C_1 C_2}{C_1 + C_2}$$

Now
$$P = p_1 + p_2 \qquad \dots 1)$$

and
$$\frac{1}{2} p_1^2 C_1 + \frac{1}{2} p_2^2 C_2 = \frac{1}{2} P^2 \frac{C_1 C_2}{C_1 + C_2}$$

and from this
$$p_1 C_1 = p_2 C_2 \qquad \dots 2)$$

and
$$p_1 = \frac{P C_2}{C_1 + C_2}; \qquad p_2 = \frac{P C_1}{C_1 + C_2} \qquad \dots 3)$$

Suppose the process be such that C_1 first be charged to full pressure P and then disconnected and charge distributed over the two condensers, C_2 being the condenser belonging to the oscillator, whereas C_1 is the regulating condenser, then since

$$\frac{1}{2} P^2 C_1 = \frac{1}{2} p_1^2 C_1 + \frac{1}{2} p_2^2 C_2 C_1 (P^2 - p_1^2) = C_2 p_2^2 \quad \text{and} \quad p_2^2 = \frac{C_1}{C_2} (P^2 - p_1^2)$$

is the pressure on condenser C_2 which it was the object to find.

Modifications of apparatus, involving *condenser method* of magnifying effects, experimented with "*Self-exciting*" process.

In this special mcdification of condenser method, an effect prcduced upon the sensitive device is rendered accumulative not only as in some other modifications, but more so by a process comparable to the self-excitation of a dynamo. Thus much feebler initial effects are made sufficient to cause the sensitive device to break down and the receiver to be operated. This process accomplished in this manner will certainly have many valuable uses. A few arrangements which have been experimented with are recorded below. They are self-explanatory. The lettering makes Diagram 1. fully clear with reference to previous

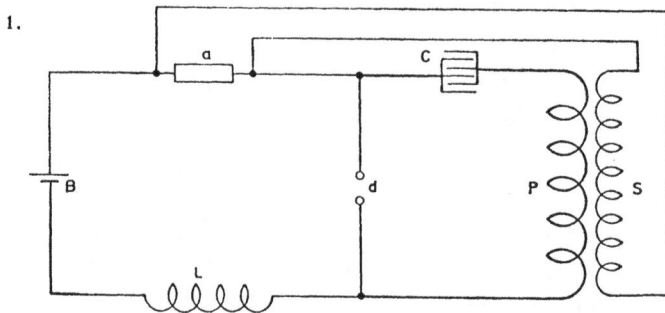

diagrams of this kind. *a* is the sensitive device, *d* a make and break device, *C* a condenser *P* the primary and *S* the secondary of oscillatory transformer, *B* a battery and *L* an inductance suitably adjusted. The sensitive relay usually employed may be in the circuit containing battery *B*, inductance *L*, and the devices *d* and *a*; or, it may be in the circuit of secondary *S* and device *a* in which case this circuit will contain an additional battery. There will also be in the circuits the usual adjustable resistances to adjust the instruments and insure the best action. From the diagram it will be easily seen that when the device *a* is at first very slightly affected, the condenser is charged more and the secondary currents become more strongly excited in their turn, the device and so on, until the device breaks down and diminishes sufficiently in resistance so that the relay is operated. This method has been

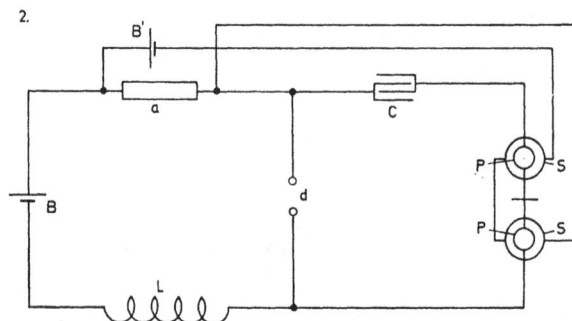

found excellent and will have besides telegraphy many valuable uses since by its means effects, too feeble to be recorded in other ways, may be rendered sufficiently strong to cause the operation of any suitable device. A number of modified arrangements as have been experimented will now be recorded. Referring to diagrams which follow in Fig. 2.

an arrangement identical with that in Fig. 1. is shown, only the relay itself is made to be the transformer by being suitably proportioned to the break and condenser and having on top of primary windings PP a secondary SS wound with fine wire and containing a battery B' as shown. The secondary excites the sensitive device a until it breaks down when the relay is operated. This arrangement is not the most preferable to employ as better results will be obtained with an independent transformer and relay, but it has the features of compactness and simplicity. It can be, however, still further simplified by doing away with inductance L.

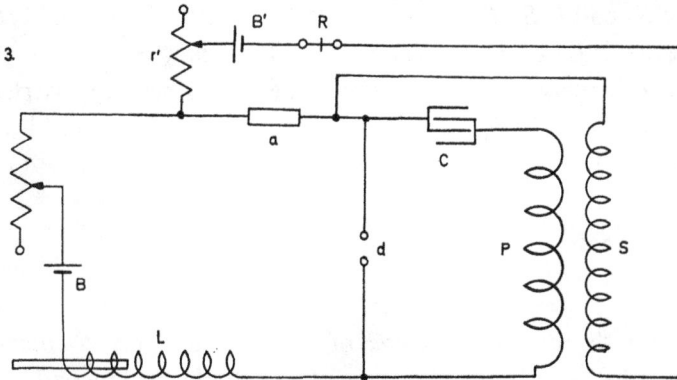

The batteries B and B' may be connected to cooperate in their acts upon sensitive device a or to oppose each other. The former seems preferable. The condenser should be

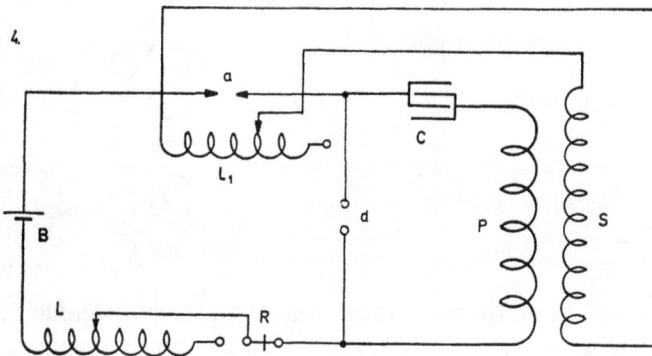

of *large capacity*. In Diagram 3. the form of connections is illustrated which was found most convenient for experimentation. An independent sensitive relay is used and adjustable dead resistances r and r' in primary and secondary circuits. The inductance L is also made adjustable and so is also break device d though this is not indicated in the diagram. In

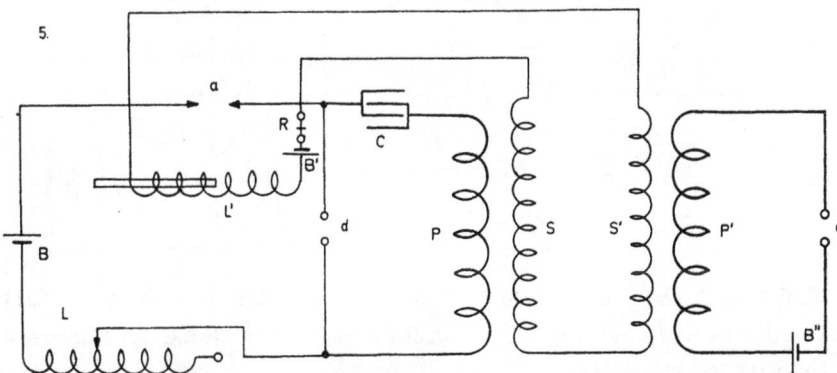

Fig. 4. again plan is shown which is suitable when, instead of a sensitive device as has been described before and which is based on the properties of minute conducting grains, a minute gap *a* is employed. This special device comprises two points almost in contact and in an atmosphere or medium the insulating properties of which are impaired to such an extent that it breaks down readily upon a slight increase of the electrical pressure. The additional adjustable inductance L_1 serves to bridge the gap and allow normally a small current to pass and to charge condenser *C*, to strain sufficiently device *a*. The relay *R* may be otherwise placed. Finally, in Diagr. 5. a modification is shown with an additional induction coil $P'S'$, B'' and d', the latter device making and breaking the circuit and straining device *a* by currents generated in secondary S'. The diagram is otherwise self-explanatory. The relay may be, as stated before, otherwise inserted.

Colorado Springs

Aug. 5, 1899

Experiments with *condenser method* of *magnifying effects continued.* More of the modifications described: In Diagram 1. a resistance, preferably inductive, is placed around

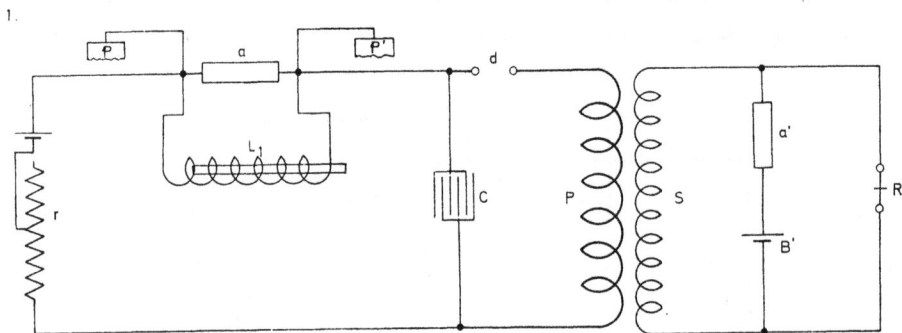

sensitive device *a* for the purpose of regulating the charge of condenser *C* and thereby determining the degree of excitation of device a' in secondary circuit of transformer. The adjustable resistance *r* serves to regulate strain exerted upon device *a* by battery *B*. The terminals or plates $p\,p'$ are placed in suitable locations of medium or media, one in the air, the other in the ground generally. Otherwise the diagram explains itself.

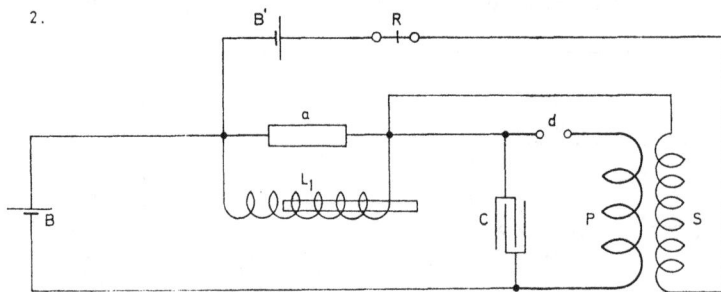

In Diagram 2. the second sensitive device is omitted and the secondary *S* is connected around sensitive device *a*. Other accessories, as adjustable resistances, are likewise omitted for sake of clearness.

In Diagram 3. again the coil *L* is made the secondary of another coil which is supplied from primary oscillating transformer *P S*, the latter being controlled in its performance by sensitive device *a*.

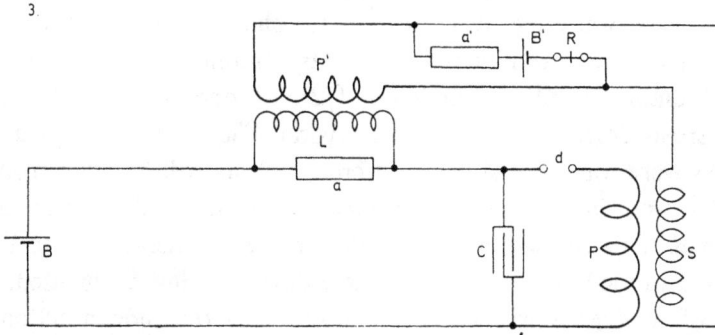

In Diagram 4. a similar arrangement is illustrated as shown on a previous occasion, the secondary *S* being connected around sensitive device *a* and containing another sensitive device, relay and battery, other accessories being omitted for reasons above given.

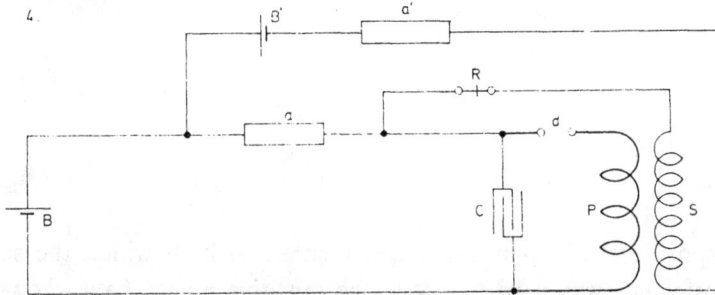

Colorado Springs

Aug. 6, 1899

Experiments with *Condenser method* of magnifying effects. More of the arrangements experimented with described:

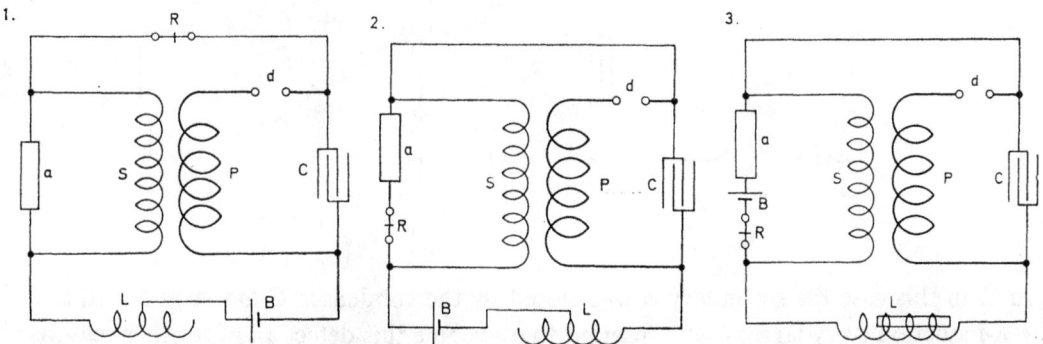

The three diagrams shown illustrate arrangements as variously carried out in a form of portable apparatus referred to before. Referring to Fig. 1. the sensitive device *a* was one consisting of a small glass tube and two metalic plugs, the tube being rotated by a Thomas Clockwork. Coarse nickel chips prepared as before described were used in the tube. An improvement was effected by cleaning the chips first with dilute acid and alkaline solution and destilled water and alcohol at the last. In one apparatus the plugs were 1/8″ apart, tube 1/4″ diam. *half* filled. Condenser 1/2 mfd. one of the small ones before used. *L, S, P* and *d* were as described on another occasion. The results *were good.* In Diagram 2. the same devices were used in a slightly different way as will be plain from the diagram. The receiver *R* was put in series with device *a* so as not to be affected much by operation of break device *d*. This seemed better, results are very satisfactory. In Diagram 3. the improvement was carried still further the same devices being again used. The delicately balanced lever of receiver *R* was very little affected by the sudden action of the break, the pull was steady in consequence and a better adjustment was possible. The results were now most satisfactory. One cell of Leclenché dry being quite sufficient.

Evidence of stationary waves water pipe

Colorado Springs

Aug. 7, 1899

Some dispositions of apparatus experimented with in which the secondary of an oscillating transformer was used to excite the sensitive device have shown results vastly better and it seems that the increased sensitiveness is due partially at least, to the fact that one terminal of the secondary connected to a sensitive device (which unexcited has a resistance of about 100,000 ohms or more) is practically *open* and that therefore by the slightest disturbance a high pressure can freely manifest itself and break down the insulation of the device. To investigate further the capacity of such arrangements a great number were tried of which some follow: One of the earlier arrangements was as illustrated

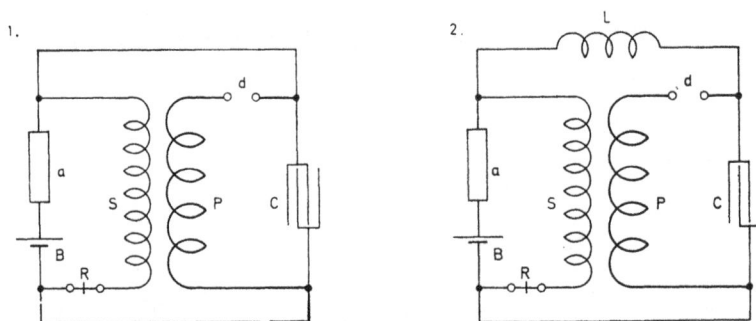

in 1. In this case the secondary *S* was closed by the condenser *C too* large for it, and the self-induction (very large) *L* was inserted to overcome this defect and for other reasons (2).

The next arrangement adopted was as shown in 3. In this case the terminal of secondary, or respectively, its continuation up to *t*, was open ended by 'a' or at least practically so and the apparatus was more sensitive when the secondary had a great number of turns, it was better to place battery *B* and receiver *R* between the other end of secondary and the corresponding terminal of the condenser, but when the secondary did not have as many turns the apparatus generally worked better with battery and relay placed as shown in 3. as

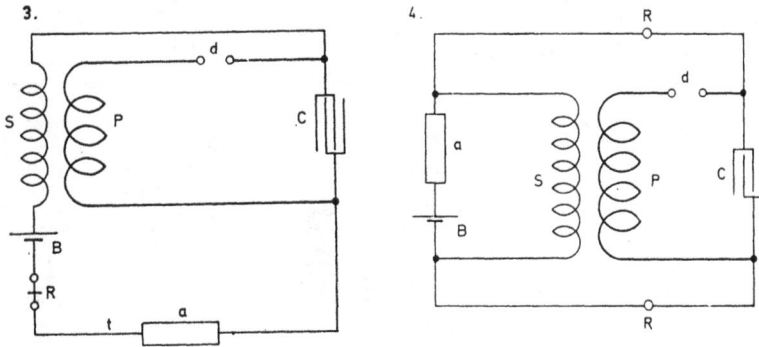

they caused a certain rise — by their capacity and self-induction — of the electrical pressure on terminal *t* of sensitive device.

A modification of Fig. 2. is shown in Fig. 4. In this case the self-induction *L* was replaced by relay *R R* which had one of its legs or coils inserted in each of the two branches of the circuit leading from the condenser as illustrated. As the relay had a very high winding and high resistance nearly 1000 ohms, this left the secondary practically opened and free to work on sensitive device *a*.

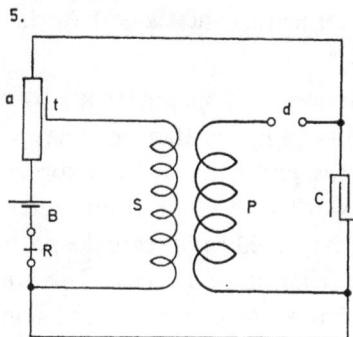

It was observed in some experiments that a sensitive device becomes more responsive to feeble disturbances when, instead of being excited by direct connection to a source, it is strained from a source from some distance. In some instances the apparatus was affected and the relay responded to a small bell from a great distance. It is probable that the increased sensitiveness is due to a certain freedom or looseness of the grains of nickel which were used and which does not exist to such extent by direct connection to a battery. These observations led to investigating the capacities of some such arrangements typically illustrated in Diagram 5 in which the device is excited by induction.

Other ways of connecting apparatus when using open acting secondary for exciting sensitive device.

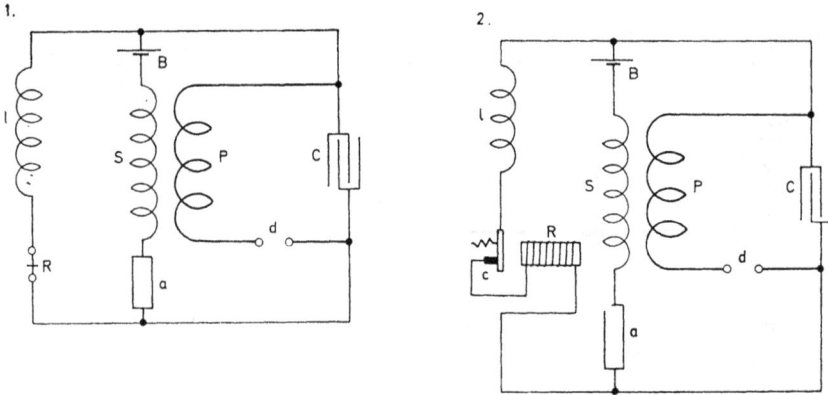

As in some previous experiments the fine relay R was affected by the break d, the relay with an adjusted high ohmic or inductive resistance l was placed in a special branch circuit (1.) It being found, furthermore, that when the sensitive device is very delicately adjusted, it often would not loose the excitation quickly enough by rotation (when a rotating cylinder as often experimented with before was used), but it would always loose the excitation by breaking the battery circuit — the disposition illustrated in 2. was adopted in which the relay R was made to break the battery circuit by opening contact c which was fixed similarly to that of an ordinary bell or buzzer with a fine spring so that the relay could complete the contact underneath (not shown) working the printing apparatus or other appliance.

The under contact was, however, dispensed with by connecting around contact c a circuit of *very high* resistance including another fine relay which was brought into action whenever the lever of relay R was pulled toward the core.

Instead of using the clockwork with break before referred to, an ordinary magnetic circuit breaker with contact was employed to operate the primary P and to generate thereby the currents in secondary S for excitation of device a. This simple arrangement is shown in Diagram 3. To provide for excitation of condenser and consequently of sensitive device up to the point desired various plans were investigated of which some follow here:

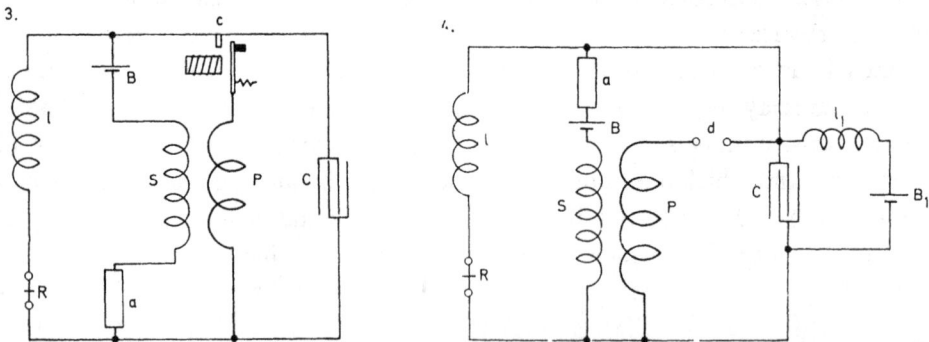

In Fig. 4. for instance, the resistance of device a unexcited being practically infinite an other battery B_1 with self-induction coil l_1 was placed around the condenser to excite the same and by the action of secondary S also sensitive device a.

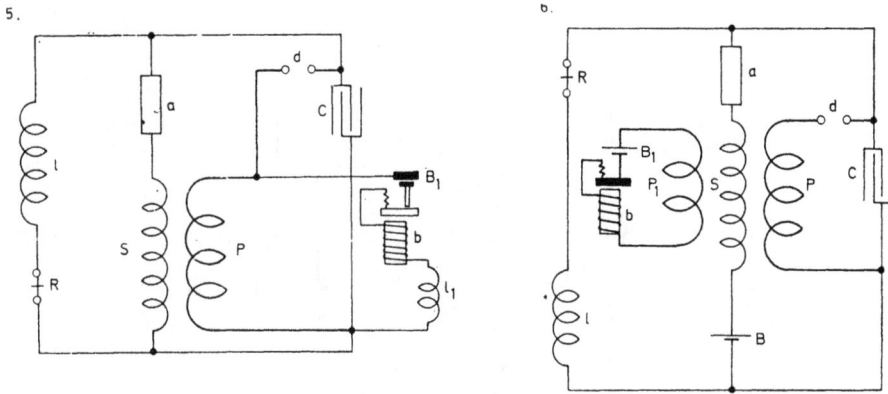

Again in Fig. 5. around the ends of the primary P was connected a Battery B_1 in series with buzzer b and adjusted self-induction l_1. The latter was so graduated that the induced currents in S would strain the device a to the point of breaking down. Still another such plan was tried by placing a separate circuit comprising coil P_1 Battery B_1 and buzzer b at a suitable distance of secondary S, such that the preparatory excitation of device a was effected.

Colorado Springs

Aug. 9, 1899

Other dispositions of apparatus experimented with. One of the plans before described was modified in the manner illustrated in 1., the battery B_1 which effected the preparatory excitation being placed as indicated so as to work through the primary P and *conjointly* with main battery B.

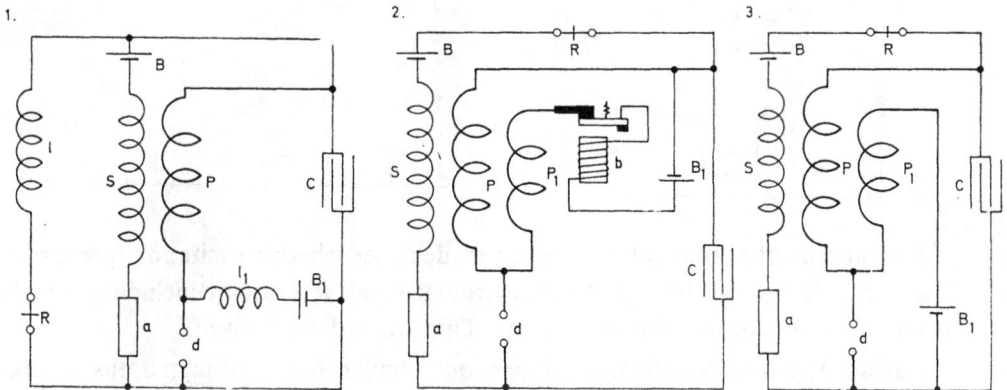

Fig. 2. illustrates a disposition similar to one experimented with before, only in series with buzzer b and auxiliary exciting battery B_1, a primary which was adjustable

(P_1) was used. This arrangement was modified to the one illustrated in Diagram 3. The intention being to use the same break for both primaries P and P_1. Neither of the plans (2 and 3) seemed capable of such results as were readily obtained in some previous dispositions. These experiments showed that the proper way was to modify plan 3 into one illustrated in 4.

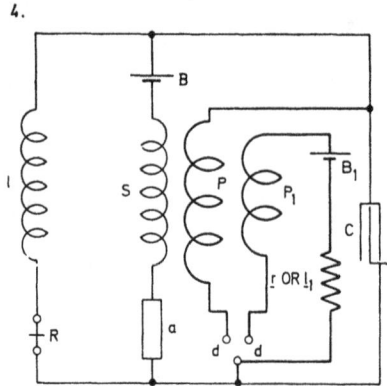

In this case a break device with *two contacts* is provided which make and break *simultaneously* both the main primary P and auxiliary primary coil P_1. In auxiliary circuit P_1, for purpose of adjustment, a resistance r or inductance l is included.

Colorado Springs

Aug. 10, 1899

Further modifications of apparatus experimented with.

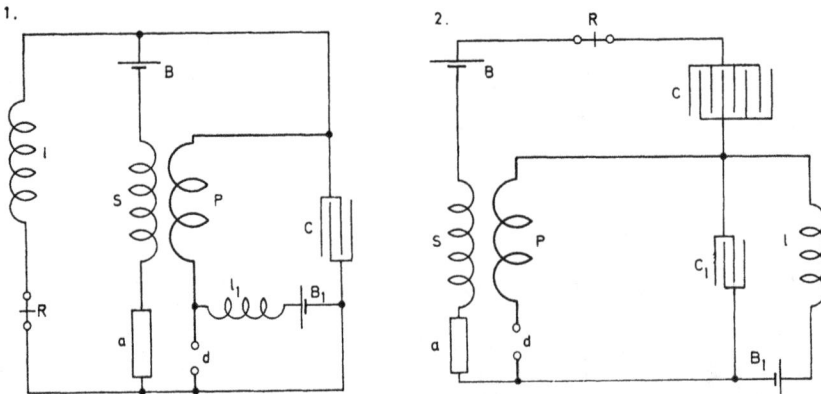

An arrangement as illustrated in 1. was made to see whether excitation (preparatory) could be conveniently effected by shunting around break d a circuit including very high graduated self-induction l_1 with battery B_1. This worked fairly well.

To avoid certain disadvantages in previous similar forms of apparatus — like a permanent charge of the condenser — a number of modifications based on the use of two condenser was resorted to and experimented with, some of which are illustrated in diagrams following.

In Diagram 2. two condensers C and C_1 are placed in series, one of them being shunted by graduated self-induction l and battery B_1. The other condenser C, being larger, allows the current from battery B to pass through break and coil l when device a is excited. Not very good. A modified plan is illustrated in 3. In this case the auxiliary battery B_1 charges the two condensers C and C_1 in series, whereupon one of them is discharged through the break d. This is fair.

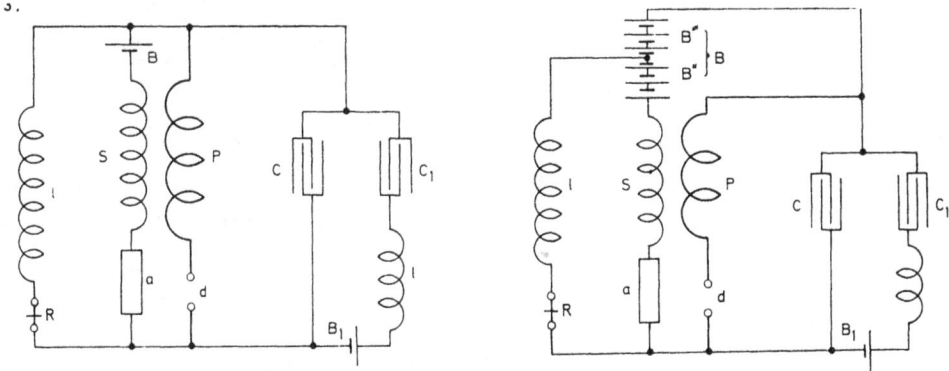

In 4. the battery (main) is so placed that a high e.m.f. is charging the condenser yet current through the sensitive device is small.

Colorado Springs

Aug. 11, 1899

Measurement of capacity of the *new condensers* prepared from *Maniton Water bottles.*

With the exception of three tanks in which bottles with green glass were used all the remaining bottles were of *dark glass.* The test did not show much difference between the two kinds of glass and all the tanks separately measured were found to be, after proper filling in of the solution, of the same capacity or very nearly so. All the bottles were first connected in quantity and the capacity compared with that of 1/2 mfd. standard condenser. The deflections were almost exactly as 36 : 44, 36° for the 1/2 mfd. and 44° for the bottles. The capacity of all the latter was therefore $\frac{11}{18}=0.611$ mfd. This would give for one bottle, since there were 576 bottles, the average value of $\frac{11}{18 \times 576}=0.00106$ mfd. or $0.00106 \times \times 9 \times 10^5 = $ **955 cm.** As there are 36 bottles in each of the tanks we get for one tank $36 \times 0.00106 = 0.03816$ mfd. or $36 \times 955 = $ **34,380 cm.** As each side has 8 tanks the capacity of each side would be $0.03816 \times 8 = 0.30528$ mfd. or **275,040 cm.** and when, as mostly the case, the two sides are in series the total primary capacity will be $\frac{0.611}{4} = $**0.15275 mfd.** or $0.15275 \times 9 \times 10^5 = $ **137,475 cm.** When working with one primary turn, taking the inductance of primary as $\frac{7}{10^5}$ henry we get

$$T = \frac{2\pi}{10^3} \sqrt{0.15275 \times \frac{7}{10^5}} = \frac{2.054}{10^5} \quad \text{and} \quad n = 48,700 \text{ per sec.}$$

Other modifications of signalling apparatus experimented with.

Several ways of providing for initial excitation when using oscillatory transformer principle are illustrated in diagrams which follow:

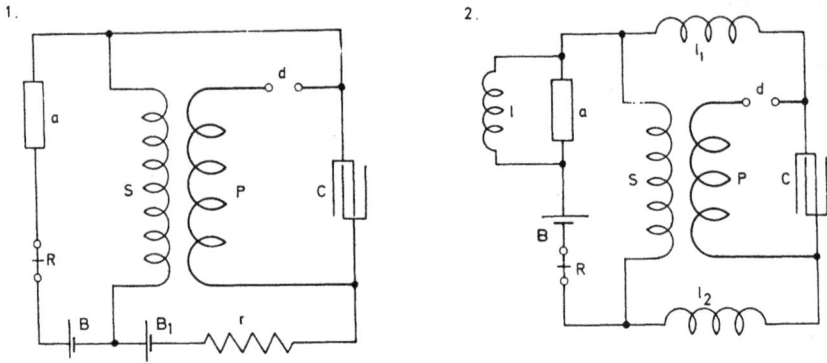

In the first diagram the excitation is provided through auxiliary battery B_1 the strength of which is regulated by adjustable resistance (inductive or ohmic) r. The two batteries B, B_1 are so connected that they join in straining device a.

In Diagram 2. an inductive resistance *very high* is connected around device a and coils $l_1 l_2$ are also employed to prevent the secondary S being closed and potential diminished through comparatively large condenser C. The inductance l and also $l_1 l_2$ are adjusted.

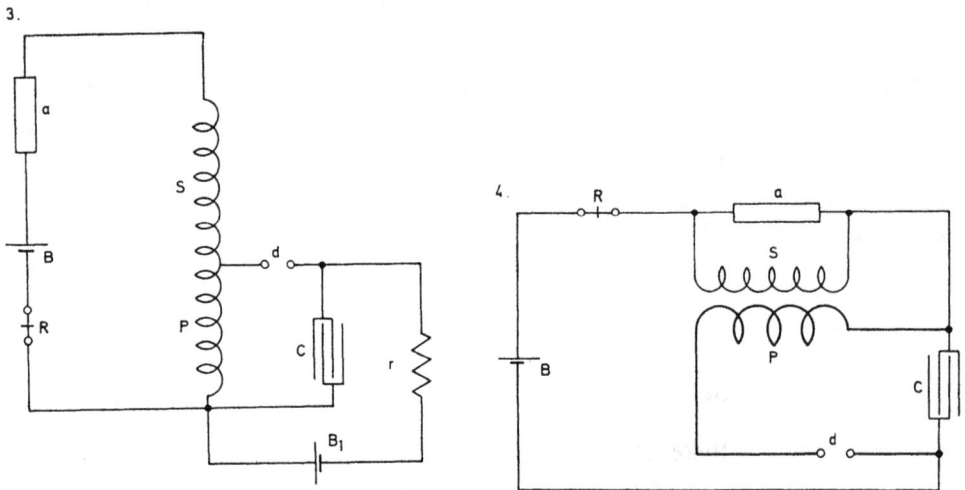

Again in Sketch 3. is shown a plan convenient to use when the secondary is wound in one single layer on a drum. A part of the secondary is made to serve as a primary through which condenser C, loaded to the desired point by battery B_1 (graduated by resistance r) is made to discharge. In this manner it is easy to adjust the action of the secondary on device a.

In Diagram 4. is illustrated a simple way experimented with before which secures good results. This to follow up.

Further modificatons in signalling apparatus.

Figs. 1. and 2. show ways of securing initial excitation by means of very high inductance $\mathfrak{L}$ connected as shown.

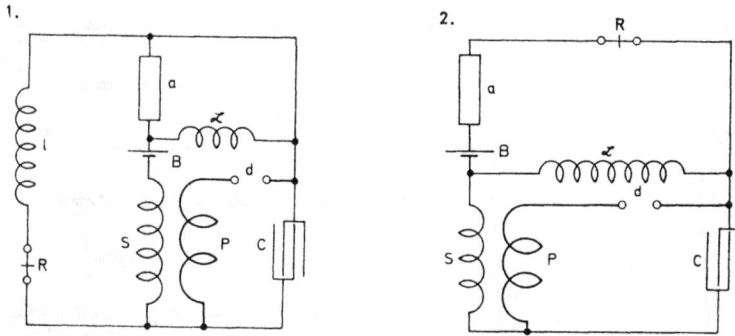

Diagrams 3. and 4. illustrate similar plans of connection. In the former the battery *B* is in the main circuit, in the latter in a shunt to device *a*.

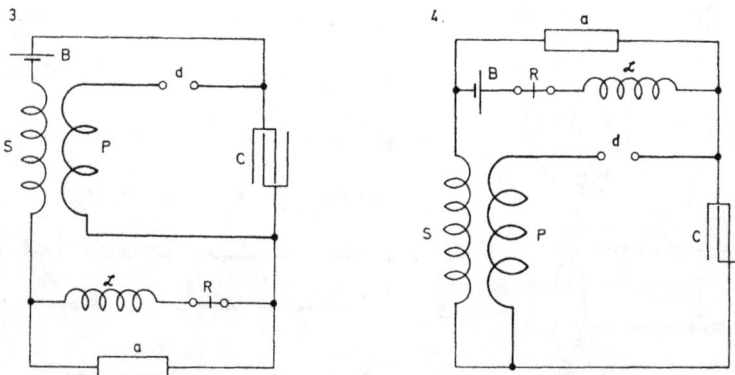

In Figs. 5. and 6. other modified connections are shown. In 5. an auxiliary secondary S_1 with battery and relay is connected around device *a* which is excited by main secondary *S*. In 6. a similar connection is used with a buzzer *b* to excite device *a* through secondary *S*.

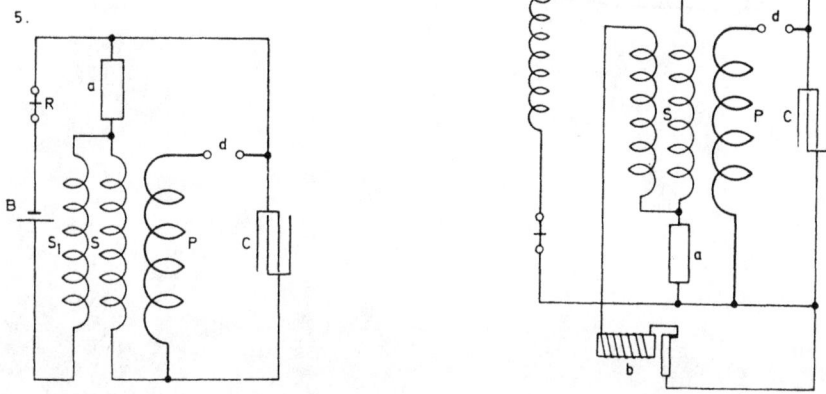

Colorado Springs May. 12. 1899.

Further modifications in signalling apparatus.

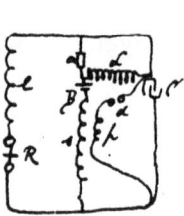

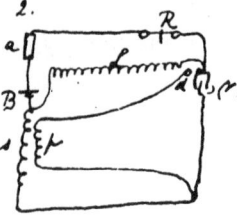

Fig. 1. and 2. show ways of securing initial excitation by means of very high inductance L connected as shown.

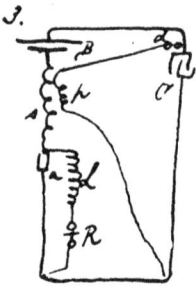

The diagrams 3 and 4. illustrate similar plans of connection. In the former the battery B is in the main circuit, in the latter is a shunt to device a.

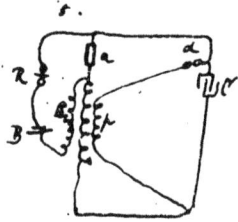

In fig 5. and 6. other modified connections are shown. In 5 an auxiliary secondary s, with battery and relay in connected actuates device a which is excited by main secondary S. In 6. a similar connection is used with a battery E to excite device a through secondary s.

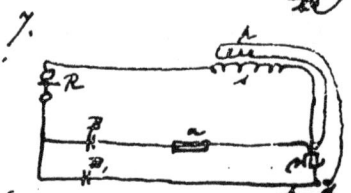

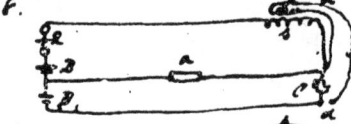

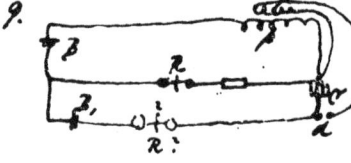

Fig. 7,8, and 9 again illustrate other arrangements in which two batteries are employed one generally to secure initial excitation

Figs. 7., 8. and 9. again illustrate other arrangements in which two batteries were employed, one generally to secure initial excitation.

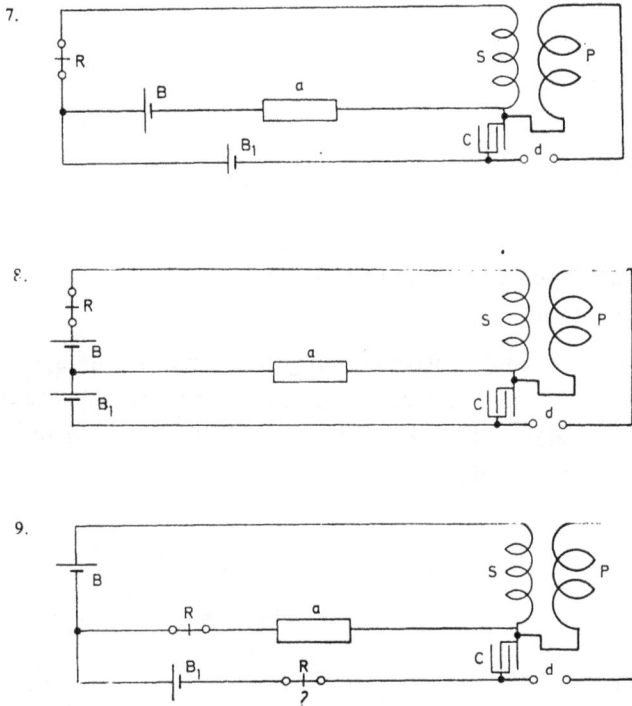

Colorado Springs

Aug. 13, 1899

Experiments with oscillator 35—35 1/2 turns. Tension on Westinghouse Transformer 15,000—22,500 volts. Supply transformers connected 100 volts.

1 primary turn, all jars tension 15,000 volts, effects would indicate capacity too small.

To ascertain this the connection was changed to 2 primary turns in series and 1/4 capacity (2 tanks). The capacity was now varied but resonance effects moderate. All experiments show clearly too much capacity and comparatively little self-induction in secondary. There is a large movement in the wire, but pressure can not appear on end as it would in the absence of capacity.

By adding capacity on one end better results indicate that this view is true.

One of the balls 38 cm. on end results much better, sparks on arresters much stronger.

Two balls connected — effects still stronger, sparks livelier on arresters but the tension still too small. Needs much more capacity on the end to overcome internal capacity distributed along cable.

Now again changed to one turn as oscillation much better. The extra coil was added and adjustment of capacity made. Best results with 3 2/3 tanks capacity on each side. An empty tank placed on top of the coil for capacity. The effects with 22,500 volts on W.T. remarkable. The streamers very rich red, quickly darting up to 9 feet long. Many brilliant sparks would jump up to a 10 foot distance.

Colorado Springs

Aug. 14, 1899

The following arrangements with two sensitive devices were the subject of consideration and experiment today:

This disposition (1.) though it worked fairly had the disadvantage that a diminishing of resistance of device *a* was not very effective in increasing the charge of the condenser, but by making the secondary and relay circuit of very high inductance and resistance this defect was to a degree remedied.

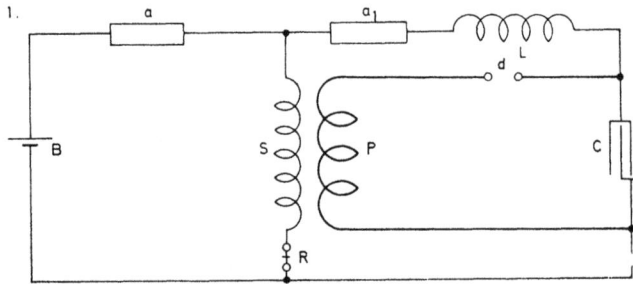

By changing connection to the one indicated in the second diagram the condenser was stronger and more effectively charged upon the falling of the resistance of either of the devices *a* a_1.

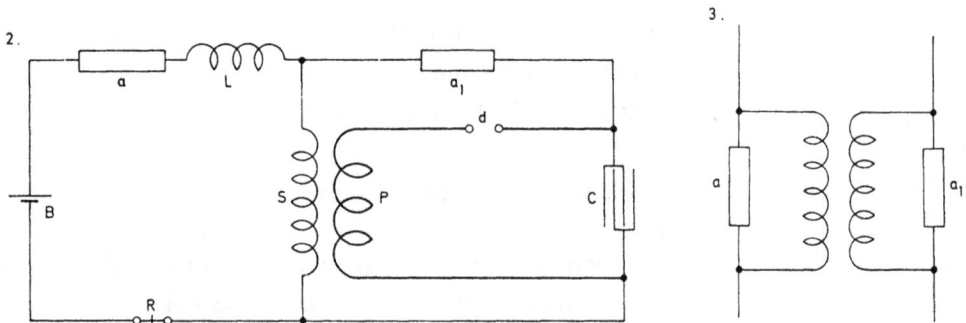

The conclusion arrived at from many experiments with two devices, which seem to indicate that two such sensitive devices are better than a single one as regards sensitiveness, was that the devices should be arranged as in Sketch 3 so that a change in one will produce a change in the other which in return should react upon the first and so on. This general scheme is to be further considered.

Other arrangements of apparatus with open secondary for exciting sensitive device.

In this plan (4.) the secondary S is connected to the terminals of sensitive device a through a small condenser CC_1. A very small condenser is sufficient to cause the excitation.

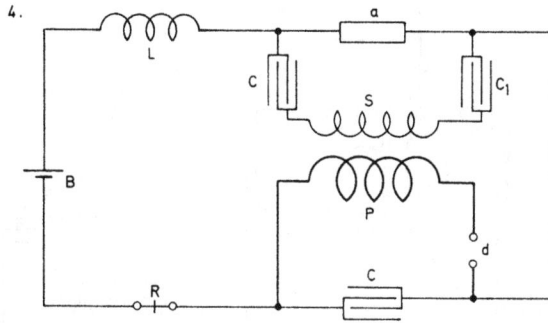

This is a modified arrangement (5.) there being only one condenser and besides a *very high* self-induction around device a to provide for initial excitation when device a is originally of practically infinite resistance. The relay R may be placed around device a instead of self-induction l.

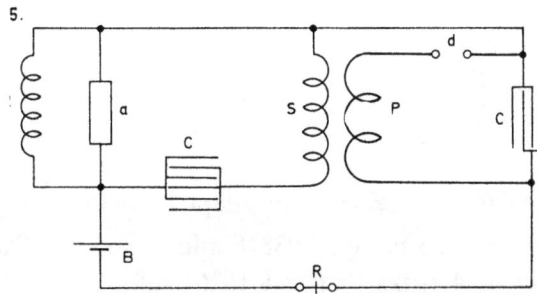

In the diagram below (6.) is shown a manner of connecting apparatus to a circuit $L_1 C_1$ which is adjusted to be in synchronism with the primary vibrations of the oscillator and excites device a.

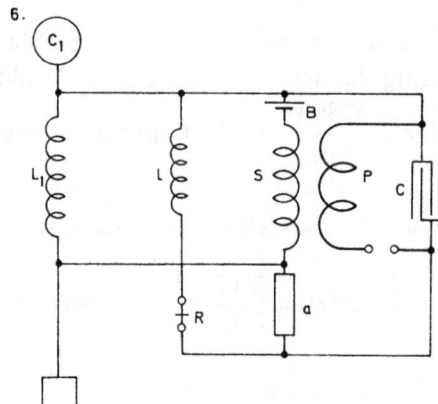

Again in Diagram 7. a special synchronised circuit is done away with, the secondary itself being adjusted to the primary vibrations. The plan adopted in New York apparatus of winding secondary and primary on a large drum (8.) serving at the same time as table is best. Tuning is easy, apparatus cheap, a large amount of copper may be easily placed in the synchronized circuit.

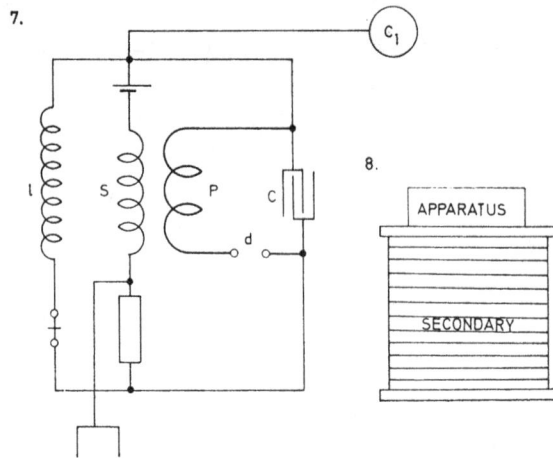

Colorado Springs

Aug. 15, 1899

Change of secondary of oscillator to adapt it to the jars.

Capacity for one tank 36 bottles 0.03816 mfd. Two sets of tanks, 8 in each, give in series a total capacity of 4 tanks that is 0.15264 mfd.

From this $$T = \frac{2.054}{10^5}$$ approx. and $n = 48{,}700$ or nearly 49,000.

This gives $$\lambda = 3.8 \text{ miles, or for } \frac{\lambda}{4} = 5016 \text{ feet.}$$

Now resonance of present secondary 5280 feet was obtained with total capacity of 6 tanks instead of 4. Reducing the figures for length we should have for smaller length a capacity larger in proportion $\left(\dfrac{5280}{5016}\right)^2$, or 1.11 that is instead of 4 tanks we should have had 4.44 for 5280 feet.

The required length for 6 tanks capacity. This length would be

$$\sqrt{\frac{4.44}{6}} \times 5280 = \frac{2.17}{2.45} \times 5280 = 4677 \text{ feet.}$$

5280 present length
4677 required   „ } to take off 603 feet.

With this length the oscillator will require same capacity as extra coil and good results may be expected.

Owing to high self-ind. of secondary of W.T. and large ratio of transformation and also great inductive drop in supply transformers (which are poor) of inadequate capacity —desirable to work with two circuits as adopted in small size oscillators with mercury break. Various advantages are thereby secured chief of which: double break number, smaller resistance in gaps and increased capacity of W.T. for charging condensers. Connections may be as illustrated in I and II.

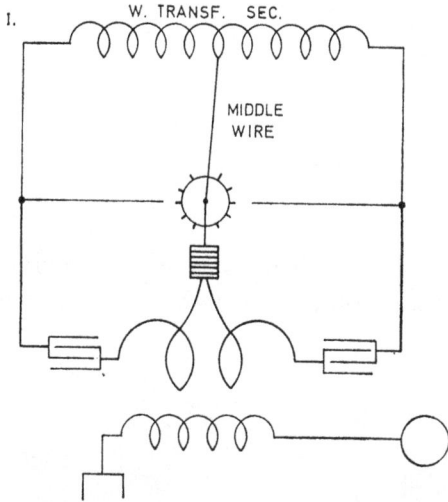

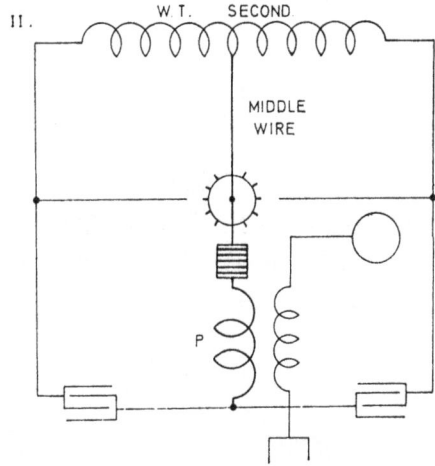

This connection shows well with small oscillators provided the *short circuit* of *secondary* of supply transformer avoided.

Colorado Springs

Aug. 17, 1899

In the working of transformer as before illustrated I. or II. the short circuit of secondary is an inconvenience which is overcome by having a few teeth at a good distance, but this diminishes the number of breaks which it is practicable to secure. Another way is to adopt a process also successful with small oscillators — of charging the condenser, next disconnecting the same and finally discharging. But this has also the disadvantage of reducing the number of breaks.

Plan here illustrated seems free of these objections:

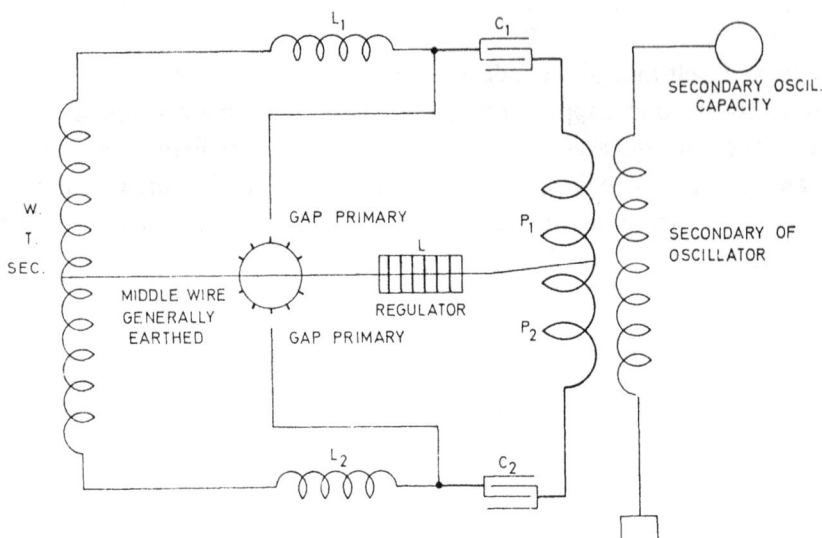

Here the short-circuit is avoided by the use of self-inductions L_1 and L_2 which must be well insulated to stand high tensions. A *single* self-induction inserted in the middle wire may also be used with like effect.

Colorado Springs

Aug. 18, 1899

The best result with apparatus at command, that is, 3 supply transformers, West. Tr. and condensers newly constructed is obtained by employing two dielectrics giving total capacity of condensers equal to that of four tanks. This allows working with 22,500 volts safely.

The results will probably be the same with 45,000 volts connection and 4 dielectrics but then capacity is only one tank and sparks in primary are longer and more difficult to control.

With connection illustrated before (Diagr. I or II) it is also practicable to work with 45,000 volts total by connecting the tanks on each side in a series so that capacity of each of the two alternately working circuits is equal to that of *two* tanks. That is $2 \times 0.03816 = 0.07632$ mfd.

With *one* primary turn this gives

$$T = \frac{2\pi}{10^3} \sqrt{0.07632 \times \frac{7}{10^5}} = \frac{2\pi}{10^5} \sqrt{0.053424} = \frac{1.4553}{10^5}$$

and $n = 68,710$ adding 10% for mutual induction, now smaller because smaller number of turns in secondary (29), we have

$$n = \begin{Bmatrix} 68,700 \\ 6900 \end{Bmatrix} = 75,600 \text{ say } n = \textbf{76,000} \text{ and } \lambda = 2.45 \text{ miles} \quad \frac{\lambda}{4} = \textbf{3234 feet.}$$

Aug. 19, 1899

Previous experiments showed in a number of cases good results with connection and quantities as indicated:

Capacity in primary circuit 6 tanks on each side; 1 turn primary about 3/4 of self--ind regulator.

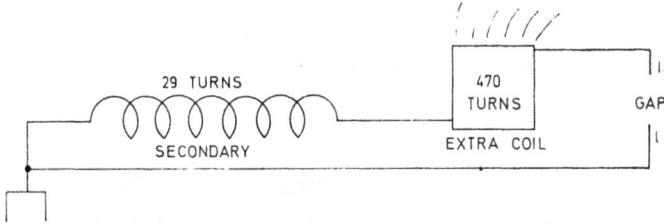

From reaction of capacity on secondary of W. Transformer it was probable that more capacity was required, but the transformer was overloaded when more tanks were joined.

With 22,500 volt connection the overload was very marked and lamps would go down 50%. When the connection was changed to 15,000 volts the lamps instead of falling would go up some 35% — 40%. No other change in capacity or otherwise was made and this showed that effect not merely due to an interaction and self-induction and capacity but that the e.m.f. was also a determining factor.

With the first connection effects were brilliant, sparks in gap 8—11 feet according to charge and adjustment. Above considerations led to changing to *two* turns primary. Capacity first 1 1/2 tanks on each side. e.m.f. on transformer 22,500 V would go up possibly 25%. Capacity was gradually increased to 3 2/3 tanks on each side when with Regulator all out the effects were best. The rise of e.m.f. about 35—40%. The sparks were curiously fierce, no direction seemingly, darting pass terminal *l*.

(Here fire started on coil).

Aug. 20, 1899

Exp. with oscillator secondary 29 turns continued to ascertain free vibration more exactly. Connection 2 series on each side, 4 dielectrics, total capacity 1 tank. Tension on Westinghouse Transformer 30,000 volts approx. Spark gaps outside about 3″ each, inside *one turn*. Results on the whole less satisfactory showing clearly that difficulties increase as tension becomes greater. Middle box on one side broke down, sparks following through the mahogony frame to a screw and jumping from this a distance of 4″. This can be only due to rapid vibration and suddenness as tension on that box only $\dfrac{15,000}{2}$ volts. There are some doubts as to the distribution of e.m.f. in condensers in *series* when vibration takes place. Strong (probably inductive) drop on supply circuit (exceptionally so).

Observation: When lamps increase strongest on supply circuit then spark will not jump over the gaps, showing that then e.m.f. on secondary of W.T. smallest.

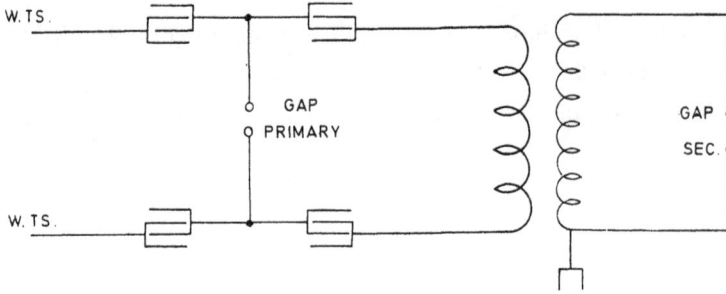

Connection was changed to that indicated in sketch for the purpose of avoiding effect of short circuit of secondary of W. T. through primary arc. Absorbed energy was great. Sparks on switch serious. Lamps would go up very much when arc would break through. But general results not satisfactory. The condensers directly on W.TS. take strong current.

Colorado Springs

Aug. 21, 1899

Other experiments with oscillator secondary 29 turns. Simply spark gap. Connectios used:

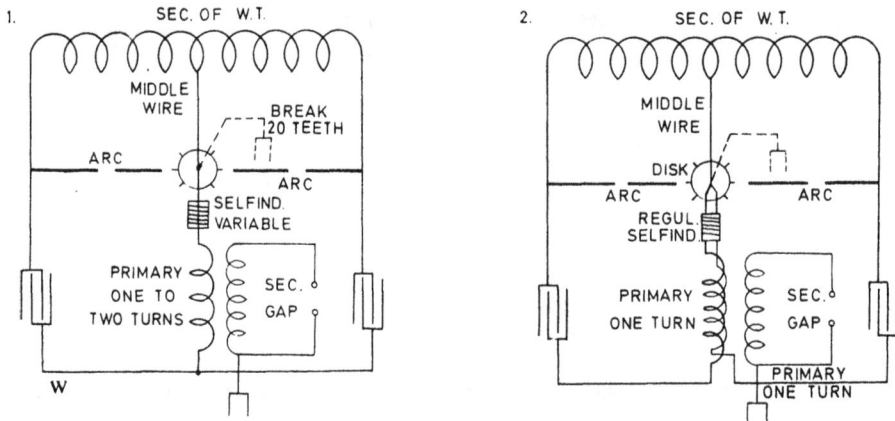

Plan illustrated in first diagram was adopted to obtain double number of breaks with same disk and securing other advantages. Also to better utilize W. Transformer. It was found that when one side on it worked remarkably well, sparks about 4 feet. The tension on each half of transformer being 11,000 volts approx. when both parts on interaction hurtful. The chief drawback being short circuiting of secondary. The arc was *snappy* and *loud* indicating short circuit and rapid vibration through wire *W*. The secondary discharge was thick but spark not long about 3 feet. All tanks were in on either side and the transformer charged them full when separate. When both parts on evidently the secondary of W. T. was overloaded.

In arrangement illustrated in 2 short circuiting was largely overcome but the short circuit of secondary of W. T. remained the same. The results were similar no matter in what direction both primaries were connected. The necessity of overcoming short circuit in both arrangements became soon more and more important.

To improve — arrangement illustrated in Diag. 3. was used. It proved itself more economical but the amount of energy was limited. The hurtful short circuit was entirely obviated and the lamps were less affected.

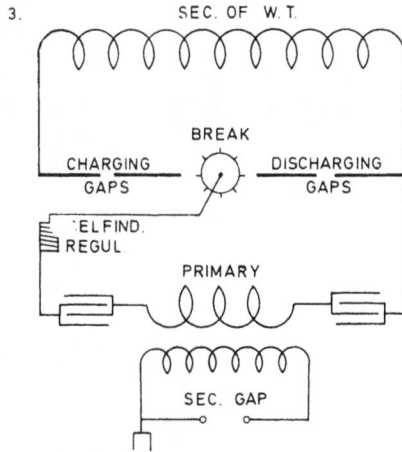

Colorado Springs

Aug. 22, 1899

Arrangements for telegraphy tried. In these the chief point was to keep one end of secondary spool open so to allow full rise of pressure on this end. The sensitive device, one of the before described, excited fully resistance 12 ohms approx. Not excited over 100,000 ohms.

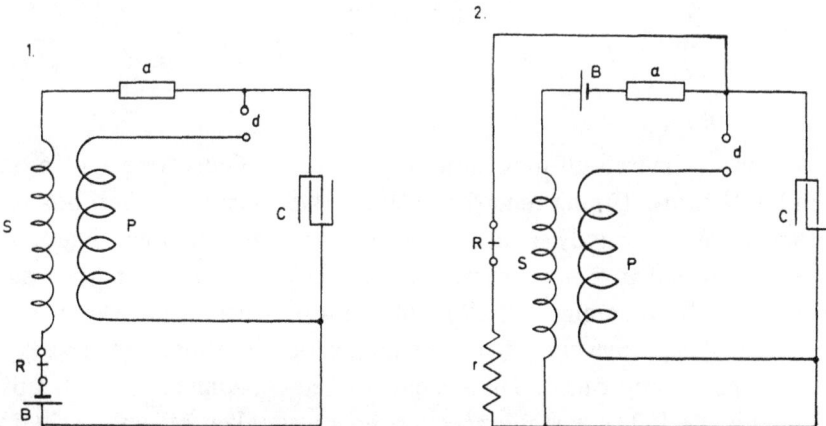

In Diagram 1. first experimented with, a disadvantage was found to exist: namely, the receiver R was operated through the break device d. This inconvenience was done away with in arrangement illustrated in Fig. 2. which allowed more sensitive adjustment of Relay R and the apparatus worked better. Capacity of condenser was varied up to 20 mfd. with changing success. Best results seemingly with small capacities up to 1 mfd. Secondary about 4,000 turns, primary Lamp cord No. 10 turns 28. The apparatus responded freely to small pocket coil at a distance of several feet with *no capacity attached* and *no adjusted circuit*. Consequently will go at great distance.

In these arrangements, as in the previous ones involving the same principle, the effect on the sensitive device is accumulative and a difficulty arises that namely the sensitive device will not readily de-sensitive. By inserting large resistance r in circuit with receiver R this effect upon the latter is largely reduced as the current through receiver and device is kept to a minimum. By adjusting speed of rotation of sensitive device the inconvenience is also overcome.

Colorado Springs

Aug. 23, 1899

Experiments with new extra coil placed in center of primary. The spool 75″ diam., 12 feet high, 160 turns in all. 120 turns wound close together in the adjacent grooves and 40 turns the upper ones at three times that distance, that is, two empty grooves between each two turns. Breaks on two sides alternately approximatelly 2,400 breaks per sec. On top (free end) ball of 38 cm. capacity. Resonance was obtained with 5 2/3 tanks on

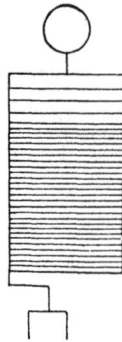

each side, one turn primary, self-induction in box 4 turns. Gaps were 1 1/16 on each side plus gaps in box 2 turns. Tuning remarkably exact, 1/8 turn of self-ind. box reducing the effect very much. When exactly 4 turns in box, sometimes streamer 8 foot long would shoot out from a defective spot on wire. The ball on top reduced streamer capacity and prevented streamers from coming out all along the top turn as usual. The spark gaps work extremely well, loud explosive character indicating good vibration. Such sparks are always noted when secondary well tuned. The system worked economically, the lamps in supply circuit not falling at all. The earth connection now was taken off and oscillator of same period — (the secondary 29 turns connected). Both had now same period, the secondary and the extra coil. On first throw of switch a spark darted to roof above from the ball and the cord caught fire. Fortunately, it was extinguished before doing damage.

This accident showed that better provisions against such an accident have to be made. The roof to be fixed with a guard of wire gauze which would prevent the wood from catching fire through sparks darting up. As it was dangerous to work further without guard against such an accident another ball 38 cm. supported on high was connected to earth and placed at varying distances from the ball on the end of the extra coil. The sparks jumped from the upper turn of the coil to the Earthed ball and sparks of seven feet were easily obtained. It was evident that the distance could be much increased but this was deemed hazardous. As it was the sparks of seven feet were probably the longest obtained from such large balls or surfaces of such small curvature.

The connection of the primary circuit was now changed, two turns being used in series. This reduced the period to one half and it was thought that this would respond to the fundamental note of both secondary and extra coil. Experiments were disappointing for the display was not remarkable, the sparks were up to four feet long but much *thicker* and whiter. I believe that the true vibration was not struck but skipped. As time pressed, further experiments with the view of ascertaining the fundamental note were postponed and the first connection with one primary turn again made. Both balls were now connected in multiple to top of the coil and to the upper rod of a spark gap, the lower rod being earthed. There was no danger of setting fire in this way. The display was remarkably noisy. The sparks were up to 14 feet long, snapping quick, explosive and very white. Sometimes streamers would shoot out fully 11 feet. Often several sparks at once. No particular direction in striking. The capacity in the primary circuit was varied up to 8 tanks on each side. Always striking effects. The ground wire had no capacity and no sparks were seen on arresters *but before*, with only one ball and *no streamers*, sparks of 5/16″ were *drawn from water pipe in distant room.*

Colorado Springs

Aug. 24, 1899

Experiments with new extra coil and oscill. secondary 29 turns continued.

The ball on top was disconnected and a bare copper wire run around the upper rim of the coil to produce streamers. Capacity in primary circuit on each side was 5 2/3 tanks with 4 turns self-induction box in. It was not advisable to work because by the throw of switch some streamers would dart up to the roof a distance of 12—13 feet. The other ball used in previous experiments was placed at a distance of 11 feet from coil and also it was unconnected — except that it had a wire of about 8 feet hanging from it — the sparks would fly to it from the rim of the coil.

A curious feature is that the streamers are very sudden, explosive. This is due probably to the suddenness of the break. Occasionally an unusually long streamer would shoot out. This probably owing to resonance of break or temporarily short circuit over break, probably the former cause responsible. Desirable either synchronous break as worked in New York, or a *very rapid one*. The speed of motor is to be increased to double for this purpose.

Coil was disconnected from the oscillator and connected to the ground. The period corresponded to that of the primary with *7 tanks* on each side, no self-induction. 5 2/3 tanks, 4 turns, and 4 tanks and *9* turns. Thus 3 tanks made only a difference of 5 turns on self. box. With four tanks tuning wonderfully close, twice it was missed before finally located.

(The roof of building was fixed today, cords done away with)

Colorado Springs

Aug. 25, 1899

Experiments continued with extra coil on wooden frame 12 feet high, 6 feet diam., 160 turns No. 10 wire. A bare copper wire was supported on top, the wire forming a circle not closed of about 8 feet diam. Another copper wire was supported 4 feet below and at a distance of about 13 feet, all around the diameter of circle being approximately 34 feet. This circle (also not closed) was connected to ground. Very powerful streamers were produced sometimes extending the full distance between the wire circles, but still they showed tendency upward in spite of presence of ground circle. Often sparks would pass in curved paths between the two circles. During the display *no sparks* on arresters, small sparks in adjacent room from water pipe. Capacity on each side from 5 2/3—7 tanks. Longes streamers with former value. The circle of 8 feet diam. was then taken down and another one about 10 feet placed on top of coil. Streamers now showed some tendency to pass to grounded circle. Sparks to the latter more frequent and brilliant. *No play* on arresters and small sparks in adjacent room as before.

One of the balls was now connected to the ground but although sparks of eleven feet jumped to same *no* sparks on arresters. The vibration was evidently slow, that pertaining to extra coil and *harmonics* in earth wire from ball *did not* preponderatingly appear.

Colorado Springs

Aug. 26, 1899

Experiments with oscillator secondary 29 turns and extra coil last described continued.

The alternate motor was put on 200 V with self-ind. coil in series, latter regulated so that motor could drive disk of break with twice the speed, that is 4200, the speed of motor being approximately 2100. This gave, since disk had 20 teeth and two alternately working breaks, $\dfrac{4200 \times 20 \times 2}{60} = 2800$ breaks per second.

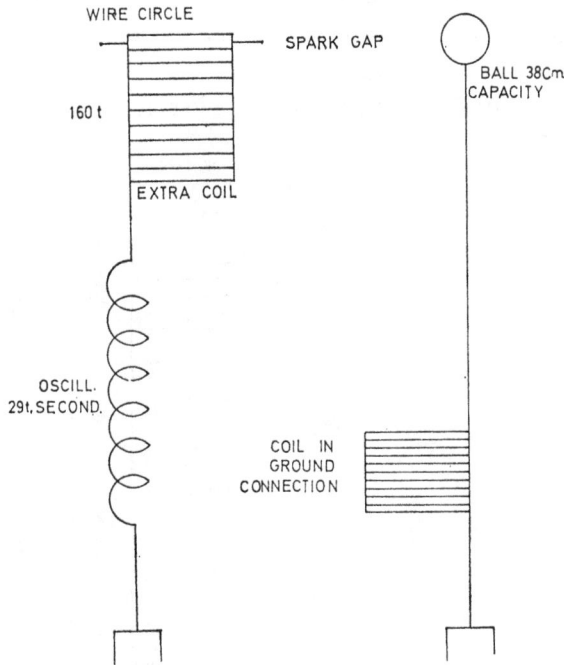

In the first trials connections were made as in sketch. The spark gap between wire circle on top of extra coil and ball supported was 8 feet.

Sparks passed readily and the display on arresters was remarkable. Thick arcs joined the arrester contacts on *both lines* and jumped also through one of the choking coils. This was the strongest effect so far on arresters.

A choking coil was now inserted in ground connection to see whether by lengthening the period of the earth wire the sparks on the arresters would be diminished. This coil was 34″ diam. wound with one layer wire No. 16, thick rubber insulation (layer 10″ high) 50 turns. This coil did not weaken the effect much probably because frequently sparks would jump between the turns. Otherwise it was surmised that the vibration of the secondary itself with the extra coil might be responsible for most of the e.m.f. generated between the ground and line. *Singularly*, despite this strong effect as evident from arresters but very small sparks were drawn from water pipe in adjacent room, this seemingly indicating that in this experiment the earth acted as a nodal region.

The conclusion from these first experiments as to the efficiency of the break was that double number of breaks decidedly better. Nor did it short circuit the transformer more because of the increased number, but on the contrary less as far as could be judged from the lamps on the supply circuit which went *up* as the switch was thrown in.

It was evident, furthermore, that the large circle of wire which was before supported above the secondary and grounded, strongly interfered with the action on arresters because it allowed local vibration which was not effectively transferred to the ground and the air.

To decide surely whether, and to what extent the *long waves* were responsible for the difference of pressure evident on the plates of arresters, the secondary and extra coil were connected as in sketch. In this connection only *long* waves could be effectively transmitted upon the ground. It would have been desirable in this and previous experiment as well to take off the wire circle and substitute a ball on top of the extra coil but this being inconvenient the circle was left. As the extra coil had now only a small initial pressure the e.m.f. obtainable in the spark gap was much smaller and the gap was reduced therefore

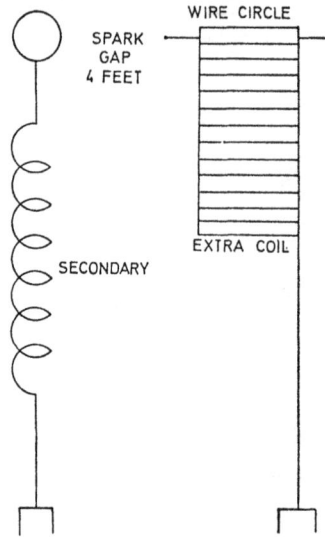

to 4 feet at which distance sparks readily jumped. The play on arresters — though weaker — *took place* nevertheless, this important result showing that waves 3—4 miles long *could* produce these e.m.f. sufficient to cause the sparks to pass between the plates of the arresters. Now it is important to consider: is the earth a nodal region or the crest of a wave (that is, the region immediately adjacent to point of attachment of secondary to ground). If a nodal region then the e.m.f. set up at the small distance of 60 feet separating the point of attachment and the ground of lightning arresters was only a *small part* of the total e.m.f. But if a crest then the e.m.f. set up and causing sparks was nearly the total e.m.f. produced by the apparatus. If a nodal region near the point of attachment of the secondary, then at a distance of about 4000 feet there must be a strong effect, but if a crest, then at that distance there would be no effect. This is to be decided by further observations. The connection was now changed to that indicated. It was thought that both vibrations would cooperate and produce a stronger effect, but it was at once evident that so long as streamers (which were about 10 feet) formed on top of the extra coil the effect must be smaller, since *all* energy came from the secondary and the streamers caused loss. A condenser ought to be used instead of a gap to make such an arrangement most economical. Although owing to *nodal* point the length of spark in adjustable gap was small, the display was strong on arresters, but not nearly as strong as when the extra coil was entirely left off. In the latter case the action was very rigorous so that often flames would form on arresters showing short circuit of dynamo. Also the other choking coil would break through. Evidently then the extra coil did not in this instance prove useful in intensifying vibration contemplated.

Experiments continued: extra coil was now lowered 2 feet nearer to ground, distance now being about 4 feet from floor and 5 feet from ground.

Capacity 5 2/3 tanks on each side in primary. The transformer (W. Co.) works very well (22,500 volts). The lamps go up 35—40% when the arc does not break through, the gap being made large for this purpose, and when the arc breaks through they still rise

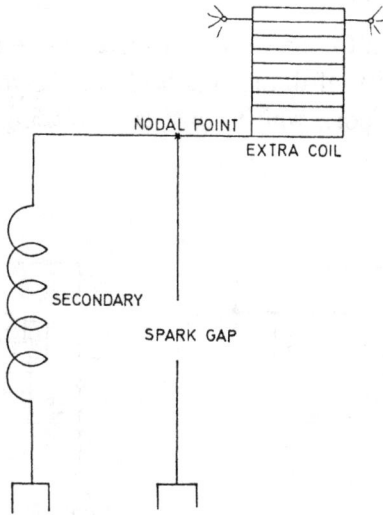

slightly above normal. The gaps outside 1 1/4" each approx. Inside 1 1/2—2 turns. Streamers produced were still more powerful being made so owing to approach of secondary. They would dart out to a distance of 12 feet sometimes.

Important. Strong arcing on arresters, although no spark would pass to the ball used before, which was placed at a distance of about 9 feet. Could the sparks be produced by static induction upon wire through the air and not chiefly by conduction through earth? To test this a coil 50 turns referred to before was inserted in the ground wire of the lightning arresters. It was expected that it would weaken discharge across, but did not probably because the current was small and the choking action likewise for this reason.

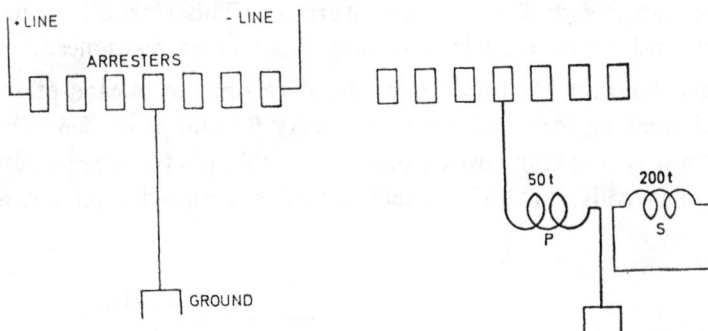

To see whether there is some current passing through the earth wire to the line, another coil was placed in inductive relation to the ground wire coil and strong sparks 3/8" were obtained on former. Sparks, lively 3/8" approx., were also obtained from coil *P*. Note: Sparks to ball sometimes, at other times streamers would dart past the ball. The streamers horizontal when sudden, when switch was held longer they would waver. In last experiments only half of wire circle on top of spool was used.

Older plans experimented with and modified arrangements of apparatus for wireless telegraphy further considered.

These connections used to relieve the sensitive device from the strain of the battery after excitation. The necessity of doing this leads to the reconsideration of an old plan experimented with in New York which consists of placing the sensitive device between

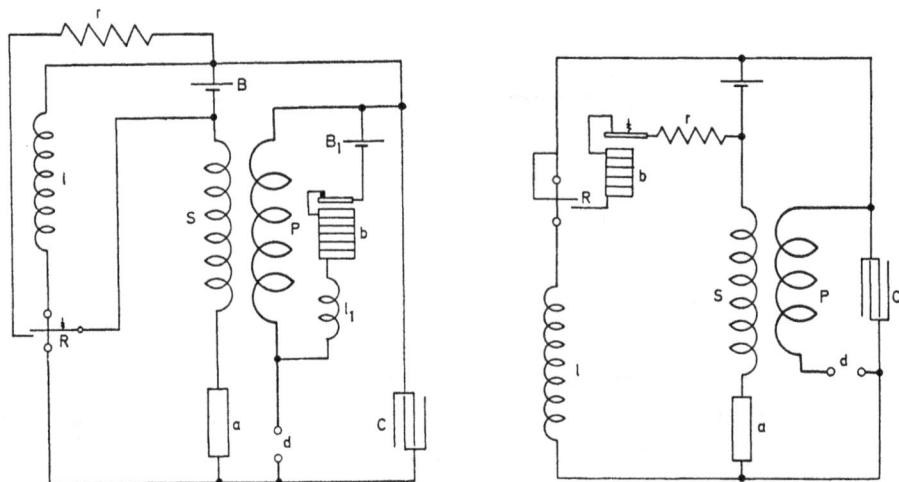

condensers in circuit so that each time only *one* current impulse can pass through the device. This is illustrated in a general way in the little diagram below. The battery strains the device *a* through the condensers $C C_1$ but when, upon the device *a* becoming excited, the condensers are suddenly charged the current impulse caused by the charging automatically stops. It is then necessary to reverse the mains, or discharge the condensers to make the apparatus ready for a second operation. This plan allows use of very high pressure on the sensitive device which should be of great resistance.

Plan in last diagram illustrated consists of raising, by means of inductances $l l_1$, condenser *C* and break device *d*, the e.m.f. of battery *B* so far as needed to bring the device *a* to the point of nearly breaking down. The quantities should for a better result be adjusted as usual. Both relay coils *R R* and inductances $l l_1$ are placed symmetrically.

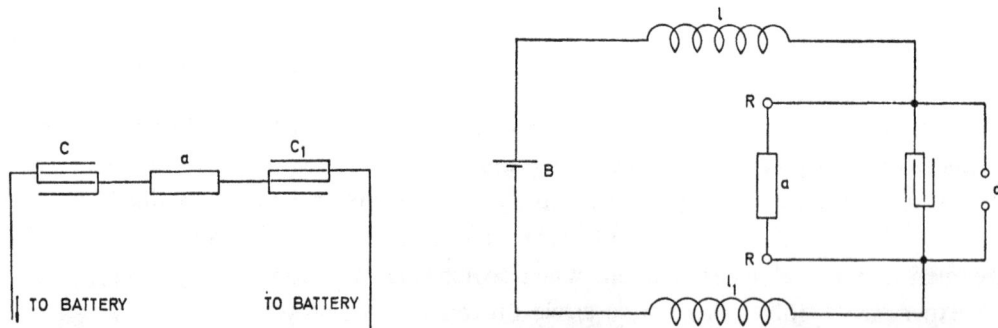

Experiments with oscillator, secondary 29 t. in series with extra coil before used (160 t) were continued tcday and showed the following:

Capacity in primary being from 5 2/3—8 tanks on each side, varied to observe shifting of nodal point, play on arresters and behaviour of streamers and spark discharges.

A half circle of bare wire on top of extra coil was left and in addition a larger half circle of bare copper wire (No. 14) was supported on wooden strips 4 feet below the former half circle. Both the bare wires were connected to the free end of an extra spool. The lower half circle was 9 1/2 feet away from a circle of the same bare wire which was supported on oscillator secondary frame and formed the terminal of the secondary. Abundant sparks and streamers were produced. The play on arresters was also observed at each throw of the switch. The rain and lightning were just beginning. Magnificent intense white light witnessed below Pike's Peak, something very unusual. It resembled a white hot silver furnace. The lightning on the mountains was very frequent and the discharges of unusual brilliancy. Twice a curious phenomenon was noted. Lightning striking in one part of the mountains from *cloud* to *earth*, there was seen in another part a few miles away from a high *peak* a lightning discharge which to all appearences came from the *peak* to the *cloud*. The discharge was much thicker at the root and branched out towards the sky spattering itself in many branches and disappearing in fine streams. The astonishing phenomenon was witnessed a second time and subsequently, though there was much uncertainty about the direction in the latter cases; a few times a similar discharge took place from other peaks. Is it possible for a discharge to go from Earth to cloud? As far as the visual impression is concerned there can be no doubt. The discharge in all cases followed a preceding lightning discharge in another region, and apparently from cloud to earth. Perhaps it can be the effect of an intense vibration started by the first discharge which results in another discharge towards an oppositely charged cloud. The clouds were unusual in configuration and grouping. A large portion of the sky was quite clear. The wind at times was very strong. An instrument by its constant play indicated strong electrical disturbances through the earth, even when there was no display of lightning as far as could be seen or heaid.

After some time the experiments were continued and presently it was observed that the usual sparking on the arresters was *no longer to be seen* when the switch was thrown in. The only change made was to take the upper half circle off leaving only the lower one. This gave a smaller streamer surface and consequently longer streamers. The display was fine. In order to see whether the upper half circle of bare wire was responsible for sparks on arresters the wire was replaced but still no result. Then it was thought that other causes for the sparks not appearing were responsible and everything that could have the slightest bearing upon this was investigated. Still nothing was arrived at. The sparks did not appear no matter what change was made in the adjustment of the vibratory circuits. What could be the cause? The only explanation at present is that the roof was rendered slightly conducting (although there was little rain in this locality) and that this produced the change. *Important to find out*. Observation: The lightning lighted two houses about two miles away.

Experiments were made with receiving apparatus comprising an oscillator with mercury break and two devices of the kind before described. The oscillator was of a later pattern, mercury break by 2000 rev. per minute gave $\dfrac{2000}{60} \times 24 = 800$ breaks per second, there being 24 teeth in the pulley. The condenser in the instrument was 1 mfd. approximately. The instrument was used as a sender and the experiments were intended to test its efficiency as a receiving apparatus. Accordingly, the connections were made as in sketch, the method of magnifying by oscillating transformer being made use of to increase sensitiveness. As far as practicable all connections and parts of instrument were used

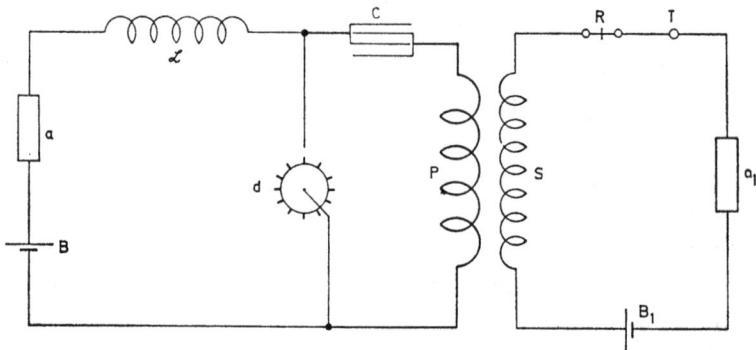

The connections of primary circuit including break remaining the same, only a battery B. (1—4 cells dry O.K.) and sensitive device a being inserted instead of a generator. In the high tension secondary were connected a receiver R (relay), telephone T, battery B, and another similar sensitive device a_1. The motor was driven from a small direct current generator which in turn was driven by the alternate current motor usually employed to drive the break disk of the large oscillator. This apparatus was extremely effective, merely the addition of small capacity on a was sufficient to make the receiver respond. Evidently this effectiveness is due to the efficiency of the oscillating transformer and excellent working and high frequency of the mercury break.

Experiments were continued for a short while with oscillator and extra coil. The frame of the secondary was repaired and a board for connections of the transformers put in place and other work took most of the day, it being late when the investigation was resumed. A netting of wire gauze (iron) had been placed around the opening of the roof to diminish danger of inflaming the building. But on the first throw of the switch the streamers and sparks darted against the netting a distance of about 12 feet and sparks were seen to go from netting on to the wooden structure of the roof. It was advisable to stop work and the roof was removed. Now the ball on top of the extra coil was connected to the latter by a wire No. *10,* 40 feet long, very heavily insulated with tape over the rubber covering. One turn on the outside and nearly another complete one in the inside were made and the end of the wire connected to the ball. The latter could not be lifted up and the experiment was tried with the ball in place. The streamers now appeared on the ball copiously when the current was turned on, their tendency being to go straight up into the air. The longest were only about 4 feet as it was deemed unsafe to strain the apparatus higher until further provisious for safe working were made. The lightning arresters were observed but *no sparking*. This showed that the absence of sparks was not due to rain or moisture as was concluded yesterday, since the weather was very warm and dry.

Colorado Springs

Aug. 30, 1899

Experiments were resumed with resonating coil to be used in connection with receiving apparatus. The coil was wound a week before on a drum 25 1/4″ diam. of bicycle hoops and a thin board, the idea carried out before in New York being followed to make the drum with coil serve, at the same time, as a table for instruments. The drum was 3 1/2 feet high, only partially wound on upper part. The wire was ordinary magnet wire No. 20, 516 turns. The self-induction was approximately calculated from the following data: diameter of drum 25 1/4″ or 64 cm.; length of wound part 20″ or 50.8 cm.

$$L = \frac{4\pi}{10} \times \frac{\pi}{4} \times \frac{d^2 n^2}{l} = \frac{d^2 n^2}{l} = \frac{64^2 \times 516^2}{50.8} = 0.02 \text{ henry approx.}$$

Taking n approximately 50,000 per second it was close enough for the purpose to assume $p = 300,000$.

The resistance of wire being 34 ohms we had

$$\frac{Lp}{R} = \frac{300,000 \times 2}{3400} = 177 \text{ fairly good.}$$

The coil was now tuned with oscillator in response to a somewhat higher note with small capacity on free terminal, the other being connected to the water pipe. Sparks of 3/4″ were obtained while from the water pipe alone a very minute spark, scarcely perceptible, could be obtained. Induction from primary being carefully eliminated, the sparks were still 3/8″ long and white.

Colorado Springs

Aug. 31, 1899

Experiments were continued with the extra coil and secondary conditions as before. The ball in the center was connected again to the top of coil and elevated a little above the roof, the latter being opened as wide as possible. The experiments were begun in the afternoon while the Sun was very bright. Scarcely any streamers from the ball could be seen but occasionally sparks would go to the roof from the center wire leading to the ball. The distance was 12 feet. There was a pronounced tendency in the sparks to fly to the roof which might have been due to dampness of the latter owing to rain the day before. During the few trials which were cut short because of the danger threatening from the sparks, the lightning arresters were observed but no spark was noted. In the forenoon the mains were tested and it was found that one of them was fairly grounded which to some extent also made the other defective. This probably was the reason why the sparks no longer appeared on the arresters.

A number of curious observations were made during the trials with the elevated ball. A fly was seen to light on the top of the ball and when the switch was thrown in the insect disappeared evidently thrown off with great force. Another such insect alighted on the under part of the ball, and the current being thrown in just about at the moment when the fly started off, the fly was seen to fall from a distance of about one foot from the ball straight down to the floor, evidently killed in the flight. Still more curious it was to see a moth at a distance of fully eleven feet from ball, near to the wooden frame fall straight down as the switch was thrown in. The strongly electrified ball evidently exercises a strong attraction on a small insect which is drawn towards it every time the ball is electrified. This was repeatedly tried.

An observation less amusing but more useful was that when the ball with its circuit were well tuned and *no streamers* appeared, owing to good insulation of leading cable — — there was a decided tendency to break the jars in the primary. Evidently, when there are no streamers the vibration is effected with lesser loss and hence there is a great rise of e.m.f. reacting upon the primary. This at least appears the most plausible reason for the phenomenon observed.

Light seems to interfere decidedly with the streamers from ball and wire and it is also unmistakably noted that the noise of the discharge is lessened when the sunlight falls upon the apparatus.

Spark gaps were established in a number of ways as by connecting both coil and secondary to ground and each to one of the balls and establishing a spark gap between the latter.

Finally the ball was again connected as before and elevated, a point being first placed on top to facilitate formation of streamers. It was curious to observe that the streamers were carried away horizontally, and eventually blown out by the wind. The resonating action was strong but the length of the streamers could not be estimated. From the leading cable the discharge would sometimes leap to a distance of at least 10 feet. The action of the wind suggests the idea of preventing the formation of wasteful streamers by a current of air.

Colorado Springs Notes

Sept. 1—30, 1899

The following items, partly worked out, omitted for want of time:

Sept. 9. Experiments to be made with st. waves. Exact distance measured to point from ground plate 1938 ft.

Sept. 10. Completed text on ways of producing electric oscillations for wireless telephoning etc. by a) insulation impairment b) change of pressure c) condenser tuning.

Sept. 30. Completed text on a) gass battery b) voltameter as detector

From 1—30, Sept. Method of increasing magnifying factor of res. circuits by cooling.

Colorado Springs Sept. 24 . 1899.

One of the difficulties in telegraphy and in
transmission in practical introduction is the elevated
terminals of capacities as proposed by me. It is difficult
to elevate a structure to a desired height and keep
the same insulated from the ground as will very
powerful oscillations in here produced even an insulator
will leading up offers difficulties because of streamers
which reduce the force and the effect of distance.
I propose to overcome this by the following
plan. A structure is to be elevated from the
ground up to the desired height and an insulated
wire from oscillator is to be brought up to a
point of the structure and connected to the same or
else brought into proximity as for instance when
a spark is used. This point should be so
located that there is at the same moment say,
a position maximum on top and a negative on
bottom or ground, that is the top and bottom should be
one half the wave space or a multiple thereof.

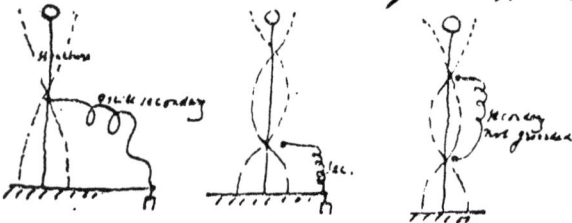

When winding is grounded as
usual it will probably be
advantageous to make the
wavelength in both systems so that
they will be maximum on the ground.

Various ways of connecting instruments on receiving station experimented with and considered as to their merits:

Diagram 1. illustrates one of the earlier dispositions involving the principle, before described, of exciting by means of energy stored in the condenser. This principle has proved itself highly effective as it secures self-excitation and great magnification of an initial feeble effect. In Fig. 1 the defect is that no initial excitation of the condenser is provided for, which makes it difficult to employ a sensitive device of *very high* resistance which, for other reasons, is desirable. This fault is overcome in Fig. 2. by providing an additional battery B_1 for charging initially the condenser and thereby exciting device *a* to the point of breaking down. Still in the latter diagram there is the inconvenience that the relay is traversed by a pulsating current during the time when device *a* is not excited.

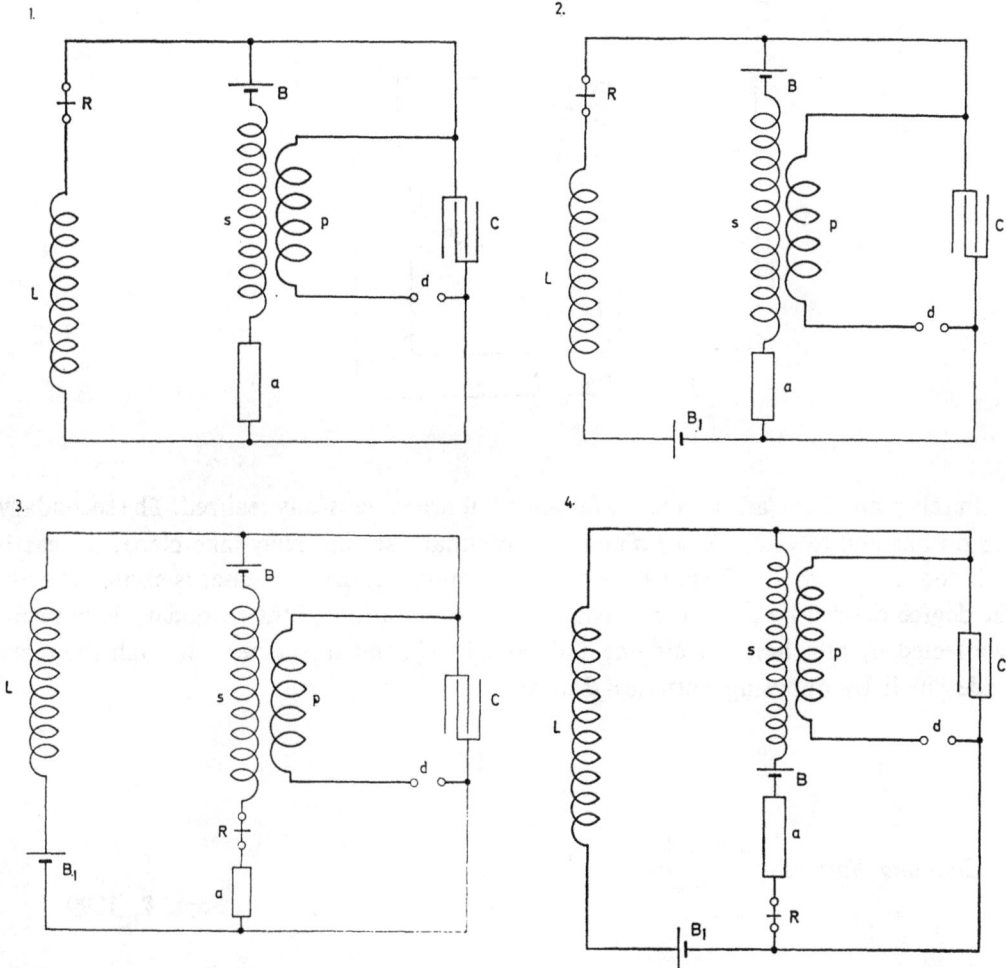

The improvement illustrated in Diagram 3. does away with this drawback and this makes it possible to adjust the relay much better. Still the relay by its self-induction is apt to interfere with the vibration of the tuned secondary *s*. This consideration led to the modification illustrated in Diagram 4.

In this case the battery was placed either near sensitive device *a*, as shown, or in series with the other end of secondary *s* and the rest of the apparatus.

To work best, however, it was recognized, that: there should be no capacity to speak of on the free end of the secondary which is connected to the sensitive device, and on the other end of the sensitive device there should be as much capacity as practicable. Various other considerations finally led to the adoption of the connection shown in Fig. 5 as the best suited so far.

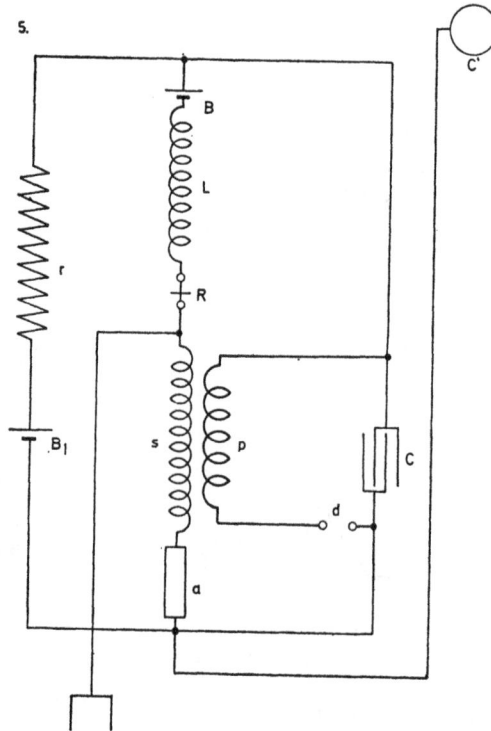

In this plan all the advantages so far aimed at are successfully realized. The secondary is *free* on one end towards device *a* and the potential rise can freely take place; the earth and air connections are both very advantageously situated; the condenser is excited exactly to the degree desired by adjusting resistance *r*. The vibration of the secondary is not sensibly affected by attaching the air line and capacity C', and the current through the relay is made small by opposing batteries B and B_1.

Colorado Springs

Sept. 2, 1899

The plan of connections of the receiving apparatus, which was last described, was modified as shown in the present diagram. The battery B_1 instead of being in branch including resistance *r* was included in the other branch circuit containing the condenser. Furthermore the batteries B and B_1 were disposed in a number of ways and graduated with reference to each other to study the best conditions of working with the plan. The device *a* was here chiefly strained through the induced currents in *s* the strength of these being graduated by adjusting resistance *r*. Therefore the strain by the batteries themselves was insignificant. Now these batteries were connected either so as

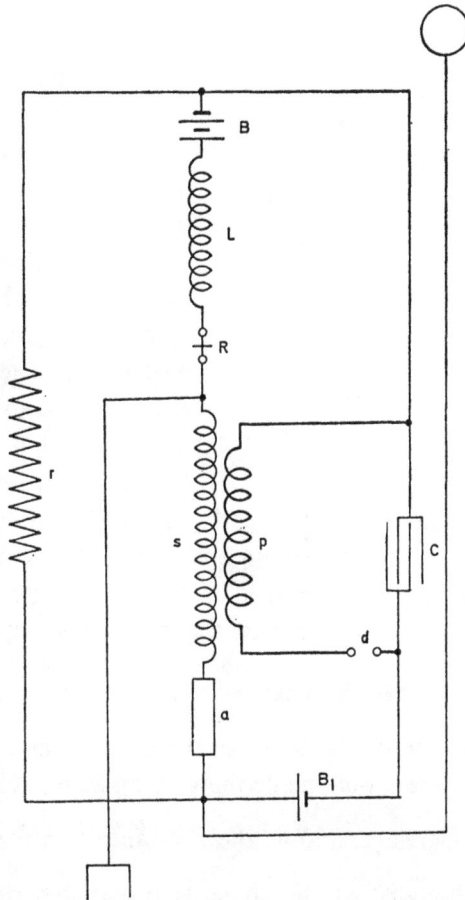

to add together in charging the condenser when device *a* was diminished in resistance, or they were made to oppose each other. In the former case a small diminution of the resistance of *a* tended to produce a change in the same sense and the apparatus possessed the feature of self-excitation, while in the latter instance when device fell in resistance, the condenser charge was diminished and the excitation ceased automatically. This secured small current through the sensitive device. Any condition could however be readily secured graduating the batteries.

Colorado Springs

Sept. 3, 1899

Experiments were resumed with oscillator the connection being as illustrated in diagram.

The extra coil and secondary were both connected to ground and on top of each a ball was placed of 38 cm. capacity. On extra coil, to facilitate the pumping of the spark and thereby enable the balls to be placed at great distance, a wire was fastened to the ball. The spark gap being about 8 feet. As both oscillator secondary and extra coil vibrated the same period but were displaced in phase sparks passed readily and the vibration was that due to each separately, the harmonics being practically prevented to pass to earth.

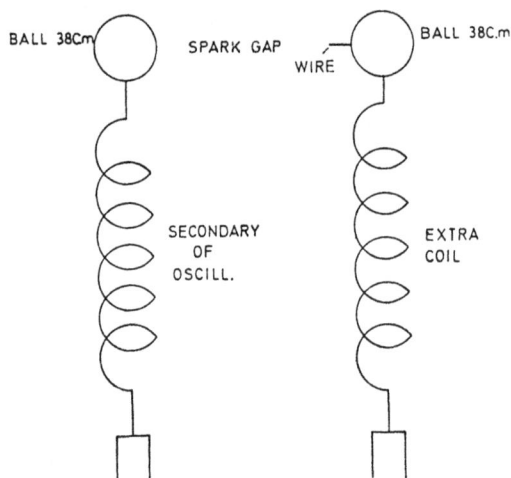

An experimental coil was then fastened to the water pipe with one end and the adjustment for the same period was made. The coil was so placed as to exclude any inductive effect from the vibrating system so that the vibration in the coil was due only to that transmitted through the water pipe. The wire on the coil was previously wound upon a drum approximately 25 1/4 inches in diameter, there being 516 turns of wire No. 21, res. 45 ohms all wound in a single layer. This coil gave on the free end — with induction from vibrating system aiding the vibration — a spark of 3/4″; with the induction eliminated the spark was fully 3/8″ long. The spark on the water pipe itself was scarcely visible, say 1/64″ long, so that the coil increased the pressure many times.

Now it was of importance to increase the magnifying factor $\dfrac{Lp}{R}$ and for the purpose of investigating the best conditions the same wire was wound on a form 22 1/2″ diam., 1″ wide, 1 1/2″ deep. 18 layers were made there being 28 turns in each. The self-induction was now nearly 20 times greater, and as the resistance was the same $\dfrac{Lp}{R}$ was to be much greater. It was feared though that the effect of distributed capacity which was largely increased would be detrimental to the rise of potential on the end. This proved to be the case so that it appears again imperative to overcome also in the receiving circuit the distributed capacity. Various ways are now to be experimented upon with this object in view.

Colorado Springs

Sept. 4, 1899

Experimental coil wound on frame made of bicycle hoop 25 1/4″ diam., 16 layers, 28 turns in each, 448 turns total, self-induction about 1/2 henry. This coil was wound very close to study effect of distributed capacity. It was connected for purposes of tuning to water pipe on one end the other being left free. From free end and water pipe short wires were run to a spark gap. The sparking distance was observed from free end to body of experimenter, next from watermain to body, next between the two wires and finally with body of experimenter connected to free end, the spark between the same points. As it was sure that the vibration of coil was too slow for the impressed vibration of approximately 50,000, wire was gradually taken off.

Number of turns on coil	Longest spark from free end to body of experimenter	Longest spark from water-main end to body of experimenter	Longest spark between the two ends of coil without capacity	Longest spark between the two ends of coil with capacity of experimenter on free end	Observation
I.	II.	III.	IV.	V.	VI.
448	1/8″	5/64″	scarcely visible	scarcely visible	
420	1/8″+d	5/64″	,,	,,	
392	1/8″+d	5/64″	,,	,,	
364	3/16″	3/16″	,,	,,	
336	3/16″+d	3/16″	small spark	small brighter spark	
308	1/8″	3/16″	,,	spark larger	
294	5/32″	5/32″	,,	livelier spark	
280	1/8″	1/8″	,,	,,	
266	3/32″	3/32″	,,	,,	

Indications up to present confirm detrimental effect of capacity. A coil was now added in series. This coil was one used often in New York and was wound on a drum 30″ diam. There were about 150 turns total length of wire 1125 feet. To this added the 266 turns of experimental coil giving length of 6.5 feet per turn made length 2100 feet for exp. coil or total 3225 feet. This was very *near quarter* wave length as it ought to be. Now results were

I.	II.	III.	IV.	V.	Observ.
266+150	7/16″	3/32″	3/32″	3/32″ lively	
Exp. additional coil					

Experiments were now continued with experimental coil alone and showed

252	1/8″	1/8″ less d	small spark	small spark livelier	Observ. The distributed
224	3/32″ bright	1/16″ ,,	,,	,,	capacity unmis-
196	1/16″	1/16″ less d	,,	,,	takably prevents
168	1/16″	1/16″ ,,	practically the same.		rise on end.

To observe better the end of coil which was before connected to earth (or water pipe) was now connected to a wire run from one turn of secondary, that is from the turn which was nearest to earth connection of secondary. The connection was in the previous experi-

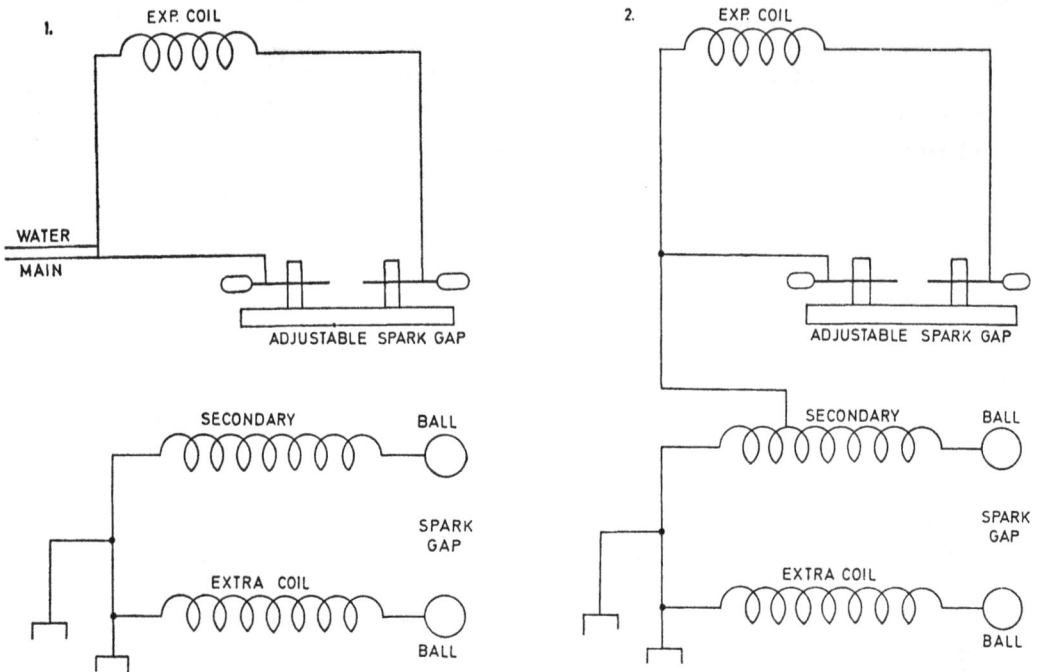

ments as illustrated in one, then it was changed to the connection shown in 2. Nothing was, in principle, changed by this connection, only a higher e.m.f. (initial) was obtained and the tuning was made easier. This I have found to be an excellent way to adopt in tuning coils. The results were as follows:

I.	II.	III.	IV.	V.	Observ.
168	3/4″	3/4″	very small spark	small spark livelier	When the wire was taken off down to
140	3/4″	7/8″	1/16″	,,	19 turns there was no spark between
112	5/8″	1″	stronger	,,	the rods even at a distance of 1/64″
98	5/8″	7/8″	,,	,,	but when the hand was approached to
84	5/8″	1″	,,	,,	free end of coil, sparks would be drawn 3/4″ long
70	5/8″	1″	,,	,,	and then spark
56	3/4″	3/4″	,,	,,	would jump between the rods.
42	3/4″	7/8″	much stronger	much stronger	This obvious and easily explained.
28	3/4″	7/8″	5/16 lively	5/16 still livelier	
19	3/4″	3/4″	almost nothing	small spark	

The general conclusions already arrived at before were still further confirmed by these experiments.

They were: 1) distributed capacity must be done away with at any price; 2) the wire should have one quarter of wave length; 3) the last plan of tuning is the best; 4) harmonics appear prominently even under the conditions of these experiments (in the experimental coil the greatest spark between both ends of the coil was obtained when the wire was 200 feet long, this was just $\frac{1}{16}$ of the lenght of secondary); 5) it is most important to tune secondary and extra coil so that they are of the same period exactly, to avoid beats.

Colorado Springs

Sept. 5, 1899

Experimental coil freshly wound on old drum 4 ft. high. Wire No. 18 and a small part of No. 20 covered with wax.

Turns $\begin{cases} 467 \text{ No. } 18 \\ 49 \text{ No. } 20 \end{cases}$ Total   516

Plan of connections:

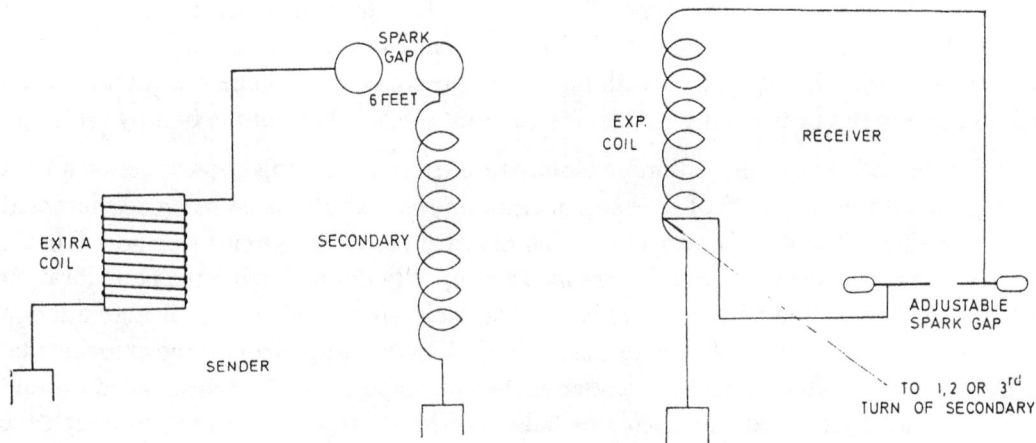

The coil had nearly 1/4 wave length and the response was at once good, a 6″ spark being obtained from the free end, also between both ends of coil. The spark would have been probably longer but this was the limit to which the gap could be adjusted.

Measuring carefully the spark length to body of experimenter it was found that from the oscillator end (connection being made to 3^{d} turn) the spark was 1 1/8″ long, while from the free end of the coil the spark was 5″ long giving more than 4 times the former value.

As it was thought that the body of experimenter was of too large a capacity and affected therefore the vibration of the experimental coil — diminishing the potential, while it did not sensibly affect the powerful oscillator — a ball was fastened to an insulating stand and spark length tried in this way. With a ball of 4″ diam. the spark from the oscillator end was 1/2″ while from the free end of the exp. coil it was 4″. A still smaller capacity was now used in the belief that perhaps the 4″ ball was too large but the experiments showed a contrary result. It was thought that the wave length being estimated from that of the oscillator and extra coil must be longer than that of the wire on experimental coil. This led to consideration of certain advantages of long waves allowing a great length of wire to be wound up on the experimental coil, this in certain instances overbalancing the advantages of the larger magnifying factor which the short waves offer.

The connection was now made to the second and then again to the first turn of oscillator secondary and, as even in this case the effects were inconveniently strong, connection was made to the water pipe to diminish impressed e.m.f. But even now the streamers would go over the spark gap. Several balls were now experimented with. Results were as follows:

Turns exp. coil	Spark from oscill. end	Spark from free end	Spark between terminals	Ball diam.
516	3/8″	1 1/2″	3 3/8″	no ball
516	1/8″	3/8″	5/8″	18″
516	3/16″	9/16″	1 1/2″	8″
516	5/16″	15/16″	1 7/8″	5″

It was now important to get an idea of the magnifying ratio and the spark was tried on the water pipe and on the free end of coil and the lengths compared. On the water pipe it was 1/64″ and on the free end 1 1/2″. This was fair but the coil was not yet quite well tuned. Completing the adjustment with more care the spark on the pipe was found 1/100″ and on the free end of the coil 2″. This was *quite satisfactory* but not the best by far.

Further efforts to tune still more closely resulted in producing a spark between rods 2 1/4″ and with the capacity of the experimenter on *free rod*, the same being disconnected from everything, 3 3/4″. The capacity in the primary oscillating circuit was now 5 tanks on each side and the self-ind. box 7 turns *in*. This capacity did not secure the best vibration of the sender which was a little slower but was suitable for the coil and no further attempt was made to tune still more advantageously by winding up more turns on the experimental coil. An important fact not to be forgotten is that the experimental coil responded without *any spark* passing between the oscillator balls. Obviously it was seen that, although the experimental coil during the tuning was placed so as to avoid induction of the primary system, the same still existed to some extent. To ascertain how much induced e.m.f. was set up spark was first tried between the terminals of the coil without ground connection and the spark obtained was about 1/64″. Now the coil was reversed so that the induced e.m.f. was against the directly communicated e.m.f. through the water pipe and it was found still that a spark of 1″ between the rods was obtained. The same would have been probably longer had it not been for the fact that the end of the coil was influenced by the metal of the sink which was near. As this could not be helped the effect could only be approximately estimated. All this showed that the induced e.m.f. from the primary system was not to any considerable degree responsible for the rise of pressure on free end of experimental coil.

The coil (experimented with) was now taken outside the building and one end connected to a water pipe running across the field. At a distance of 250 feet from shop or rather from the connection of secondary to ground a spark between the rcds 1/4" long was obtained and when the body of the experimenter was connected to the insulated sparkrod the spark was 1". At a distance of 400 feet the spark without capacity was still 1/8" and *with* capacity of experimenter 1/2" although at one place the pipe was buried for 30 feet in the ground. Strong shocks were obtained at that distance before the point of connection.

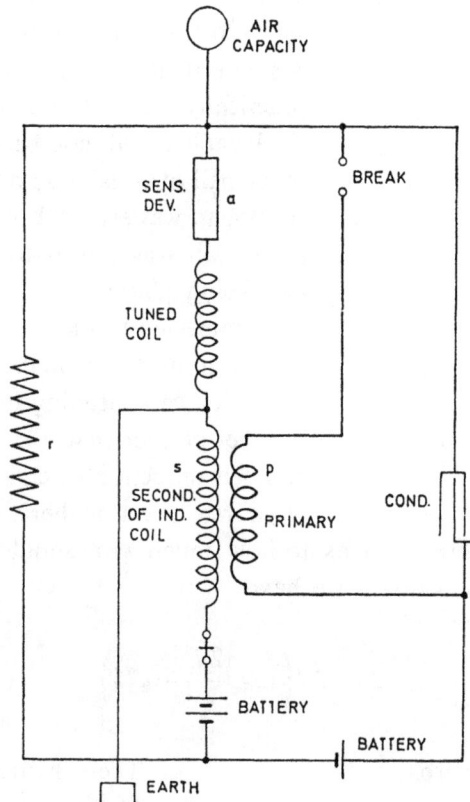

The experiments having shown the effects of distributed capacity to be very hurtful if not fatal to success with tuned coils, for convenience a winding was adopted to give very small capacity and thus the greatest possible length of wire and highest potential on the free end without *any capacity*. Capacity on the end was not needed since the free end is connected to a sensitive device practically without capacity. Since it was desirable to get the greatest possible rise of pressure on this device, it was much better to tune for a condition without capacity on the free end, for any capacity would cause diminution of pressure since the amount of energy was fixed. But wound in this way the tuned coil was not quite suitable to serve at the same time as secondary of the induction coil and, to utilize older apparatus, finally the connection shown in diagram on the left was adopted, which was found to be best.

Experimental coil for receiving apparatus with short waves. These were produced in the following manner: the extra coil, repeatedly described, was connected in series with the secondary of oscillator, both being first tuned to the same period so that there was a nodal point on the place of connection. The tension on extra coil terminal (a ball of 38 cm. capacity) was over 3 million volts, as was evident from streamers from ball. At a distance from the extra coil (8 feet) another ball 38 cm. capacity was supported, and this ball was joined by a heavy cable 400,000 circular mills section to the ground. The cable was 120 feet long and was not straight but made 3 small turns about 4 feet diam. The rest was practically straight. As the ball connected to the thick cable could not be elevated as high as the other ball of equal size on top of the extra coil a spark-gap was established as indicated in sketch. A small ball was joined to the cable leading up this ball being placed at about the height of the large ball connected to the thick cable. In estimating the vibration of the system comprising the large ball and thick cable leading to ground it was assumed for the present that the large cable was straight and self-induction calculated on this basis would, of course, give a smaller value, but this was thought sufficient to give the first idea as to how much wire should be placed on receiving coil. Assuming the cable straight we have

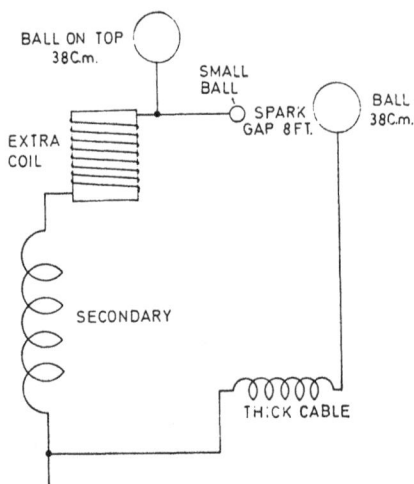

$$L_8 = 2\,l\left(\log_e \frac{2\,l}{r} - 0.75\right)$$

$2\,l = 240$ ft $= 7315$ cm. approx.

$r = 0.64$

$\dfrac{2\,l}{r}$ approx $= 11,600$

$\log_e \dfrac{2\,l}{r} = 9.6$

$n = 3 \times 10^6$

These figures gave approx. $66,000$ cm. $= L_8$

From this

$$T = \frac{2\,\pi}{1000}\sqrt{\frac{38}{9 \times 10^5} \times \frac{66}{10^6}}$$

$$T_{\text{approx}} = \frac{1}{3 \times 10^6}$$

Now λ would be $186,000 : 3 \times 10^6 = 0.062$ miles or 3273 feet. Say $\dfrac{\lambda}{4} = \dfrac{3280}{4} = 820$ feet. This wire is to be wound on a drum $10''$ diam. Therefore, we want 328 turns *at least*.

A new experimental coil wound with 400 turns on same drum 10″ d:am.
66″ long.

The coil when attached to a water pipe gave on free end spark 5/8″. To test whether the wave length is greater 72 turns were added and sparks were decidedly stronger. But adding 50 more turns the effect was weaker. The self-induction was now calculated to get a better idea of the probable wave length and L was 2,000,000 cm. approx. As with this L the capacity would have to be extremely small, far less than the coil evidently had, it was safe to proceed in taking wire off. Gradually shortening the wire increased the spark length until at 405 turns and a capacity of 15 sq. inches tinfoil the longest spark was obtained about 1″. Calculating from wire length $\lambda/4$ was 1010 feet approx. giving $n=245,500$ per sec. *approx.*

As there was a possibility of confounding the true vibration with a harmonic, wire in definite lengths was taken off. With 270 turns and small capacity on end the effect was still good. From that point on the diminution was steady.

The wire *No. 20* was now taken off and wire No. 18 wound in place to study the effect of diminished resistance. New exp. coil wound on drum 10″ diam. used before. It was estimated that for the vibrating system before described, comprising ball 38 cm. capacity and 120 feet cable 400,000 c. mills, about 400—420 turns would be needed. There was wire enough for 495 turns. The spark was taken to the body of the experimenter,

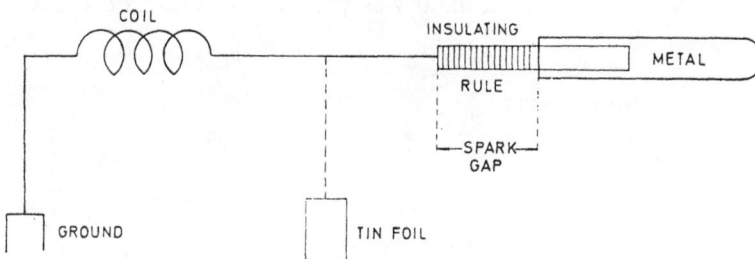

the length being at once read off by a simple arrangement comprising a small rule of insulating material and a metal strip, the position of which was adjustable relative to end of the insulating rule. The metal strip was held in hand and the end of the insulating rule was maintained almost in touch with the wire forming the free terminal of the coil which was carefully placed in the proper position such that there was *no induced* e.m.f. from the primary but only through the ground connection could the coil be excited. The connection of coil and manner of reading off spark-length is indicated in the above diagram.

Results:

Turns	Spark to body of experimenter	Spark to body of exp. with small capacity attached to end of wire
495	3/8"	much less
470	7/16"	,,
460	7/16+1/64	,,
450	7/16+1/32	,,
440	9/16	1/2"
435	9/16	1/2"
430	almost 11/16"	1/2"
425	3/4"	1/2"
420	3/4" full	1/2"
415	7/8"	1/2"
410	1"	9/16"
405	1 3/16"	5/8"

With 405 turns the limit was nearly approached. With 400 turns the spark without capacity was 1 1/4" and with tinfoil on wire 5/8"; with 395 turns the former was 1 1/4" the latter 5/8" and with 390 the same also, with 385 practically the same. The system still needed a small capacity for when a hand was held at a distance of about a foot a spark of 1 3/8" could be obtained.

These data give wave length.

Colorado Springs

Sept. 8, 1899

In some previous experiments coils were used wound on a drum 10" diam. but the inductances were not measured as the changes were made too often. The following data of a new coil built for similar purposes will be useful in connection with the preceding experiments.

The coil was wound on a new drum 10 5/16" diam. and 41 1/4" length. There were 550 turns of No. 18 wax-covered wire.

Data for calculating inductance

diameter of coil $d = 10 \ 5/16'' = 10.3125'' = 26.19$ cm.

$$S = \frac{\pi}{4} d^2 = 0.7854 \times 685.9 = 538.7 \text{ cm.sq.} \qquad\qquad N = 550$$

$l = 41.25'' = 104.77$ cm. $\qquad\qquad\qquad\qquad\qquad N^2 = 302,500$

$$L = \frac{4\pi}{l} N^2 S = \frac{12.5664 \times 302,500 \times 538.7}{104.77} = 12.5664 \times 302,500 \times 5.14$$

$L = 19,538,800$ cm. or **0.019539 henry** approx.

following readings to measure the inductance were taken:

E	I	ω	R	
117	6.2	880	9.586	average of three readings practically the same.

From these data

$$\frac{E}{I} = \frac{117}{6.2} = 19 \text{ approx.} \qquad\qquad \left(\frac{E}{I}\right)^2 = 361 \quad R^2 = 91.89$$

$$\left(\frac{E}{I}\right)^2 - R^2 = 269.11$$

$$L = \frac{\sqrt{269.11}}{880} = \frac{16.41}{880} \text{ or}$$

$L = 0.01865$ henry $= 18,650,000$ cm.

Colorado Springs

Sept. 11, 1899

Experiments were continued with apparatus before described and the effects outside at a distance investigated, the chief object being to establish nodal points on earth's surface. The transmitting apparatus was one giving more rapid vibrations and was improvised as indicated in the left sketch.

The apparatus for investigation comprised the ten'' drum, before referred to, wound with 395 turns wire No. 18 B. & S. and to increase magnifying factor another layer was wound on top, thus doubling the section. It was found that the scheme of double windings is not a good one because the e.m.f. in both wires are apt to be unequal and it is more difficult to make adjustment. The connections of apparatus were as indicated in the right sketch.

The secondary of the induction coil was connected between the two legs of the receiver, this being convenient for eventually reversing. A high self-induction L was provided to give initial excitation but the apparatus worked also without it. The batteries B and B' were connected both in the same way and opposite, the former giving best results. The tests showed that without *any* capacity or wire l the disturbances were recorded about one mile away; only the ground connection was essential as the waves were still fairly long, about 4000 feet (approx.)

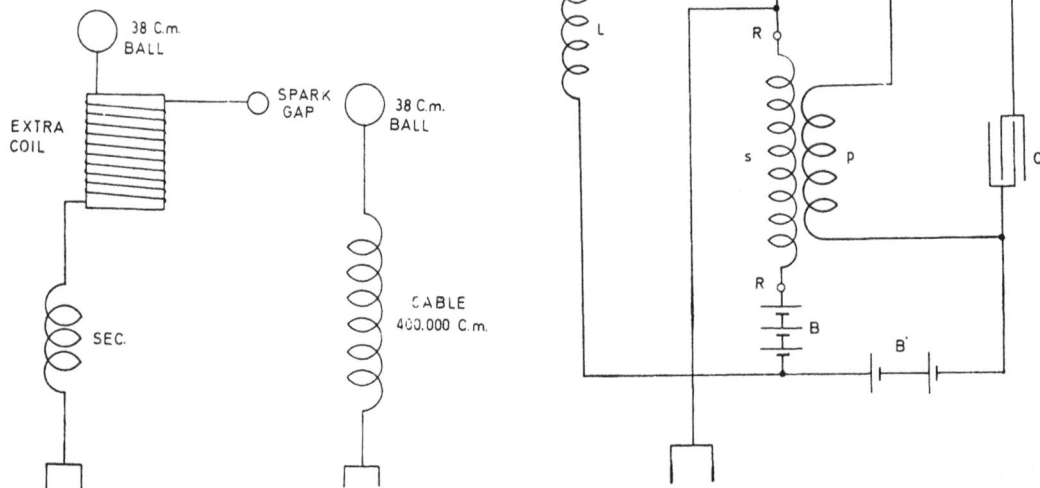

Colorado Springs

Sept. 12, 1899

Experiments were again resumed after some changes for the better had been made. The secondary was reduced to 26 turns and the adjustment was so made that but little self-induction remained in the self-ind. box and all tanks were used. The best condition was obtained with 8 tanks on each side and self-ind. on sixth turn. The extra coil was now adjusted to the same vibration. As with the ball lifted up the vibration was somewhat too slow for the secondary, the ball was lowered to about half the height when resonance was secured with nearly the same capacity and self-induction in primary circuit as corresponded to the vibration of the secondary. Although the agreement was not quite close, the effects were remarkable. The streamers went from ball 38 cm. capacity on top as freely as though it were a small one, this showing that the e.m.f. was far in excess, possibly many times the 3 million volts which theoretically are necessary to produce streamers from a ball of this radius of curvature.

On Westinghouse Transformer the following change was made. The wire was cut in the middle and the two parts connected as shown in diagram: the end of the first half was connected to the tank which as before remained connected to the ground. The second half was left intact. This mode of connecting afforded the advantage of connecting the two parts in multiple arc for 22,500 volts or 30,000 volts — thus providing double current capacity. This was recognized as necessary as the one half previously used did not load the jars quite fast enough as was evident from the measurement and calculation of constants. When the parts are used in multiple arc the connection is as illustrated by - - - - - - lines. When the old connection was desirable the dotted connections were taken off and the connection indicated by — · · — · · — line.

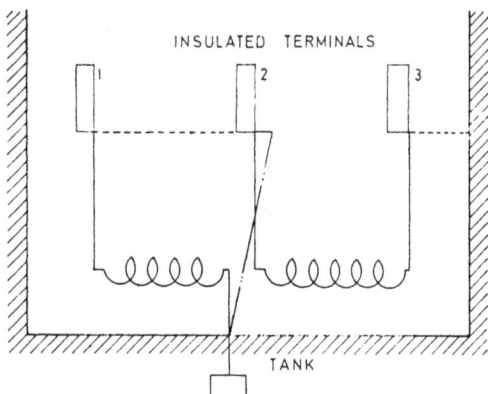

A further change was made today by substituting for the 5″ pulley on the alternating motor another pulley of 6″. This gives now $\dfrac{6}{2} \times \dfrac{2100 \times 40}{60} = 4200$ breaks per sec. The tests showed best results with secondary of oscillator, 8 tanks on each side self-ind., 6 turns in. Streamers were all along on the top wire which was raised today, *very strong*, more so than before.

Experiments continued with the object of completing adjustments of secondary and extra coil with *ball elevated* to the highest position.

It was found that the capacity necessary in primary for resonance with extra coil was increased 25% when ball was lifted about half way. This gave the basis for the calculation showing that about 10 turns from the extra coil had to be taken off to give the same vibration with ball elevated to top position. All in all nearly eleven turns were taken off and it was found that the vibration came out very closely as estimated. The resonance of secondary was obtained with all jars or tanks (8 on each side) and *6 turns* self-induction, while resonance of extra coil with ball elevated took place with all tanks likewise and 4 turns self-induction. The coil was still a little faster than the secondary. When both were connected in series the effects were magnificient, streamers up to 12 feet from the ball. To get best effect a middle value of self-induction had to be inserted, but although with this number of turns (about 4) both the secondary and extra coil were weakened individually, their joint effect was much stronger. This showed importance of very *close tuning*.

As it was impracticable in the form of apparatus used in some of these experiments to insulate the sensitive device from the break, a number of arrangements were adopted to dispense with this necessity. Some are illustrated below:

These diagrams are self-explanatory. In all of them both the secondary coil and synchronized coil have their ends free for the purpose of enabling great size of pressure. This has been found a great advantage as has also the construction of a resonating coil in which distributed capacity is reduced to minimum.

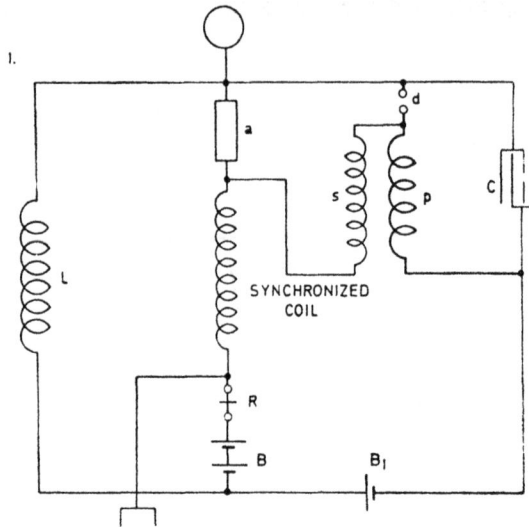

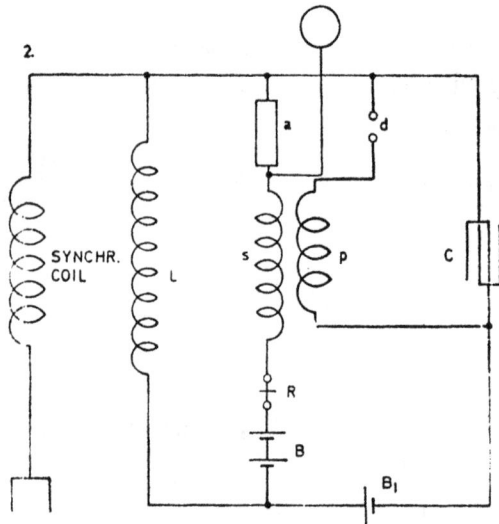

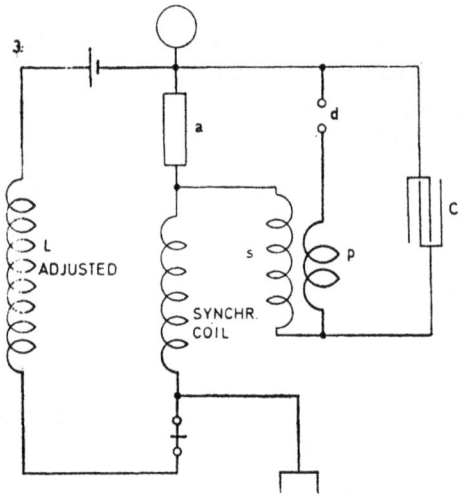

In some of the above arrangements the secondary coil was dispensed with and a part of the synchronized coil utilised to give initial excitation. It was found in these experiments that the primary must be for the best results always on the side near the ground connection as otherwise the influence of the primary is detrimental to a great rise. In one instance results were remarkably good with ratio of transformation 1 : 250, that is two turns of primary and 500 turns in synchr. coil.

Further experimentation led to adopting one of the two arrangements illustrated according to whether an independent induction coil was used or not. The induction coil secures the advantage that the synchronized coil need not be touched and the apparatus is made suitable for any coil. On the other hand to use the synchronized coil itself has the chief advantage of having the coil entirely open. This latter advantage is secured to a large extent also when an independent induction coil is used, as in following diagram (1): The lettering is as in previous diagrams. A small condenser C_1 is connected to secondary to allow easy passage to the high frequency currents from the ground through the synchronized coil to the sensitive device and wire or capacity in the air.

Diagram two shows manner of connecting when the synchronized coil itself is used as the secondary of the induction coil. In this case the primary consisting from 1—5 turns or so is placed near the ground and the tuning is effected with *all apparatus* mounted together except sensitive device.

This seems best so far judging from tests.

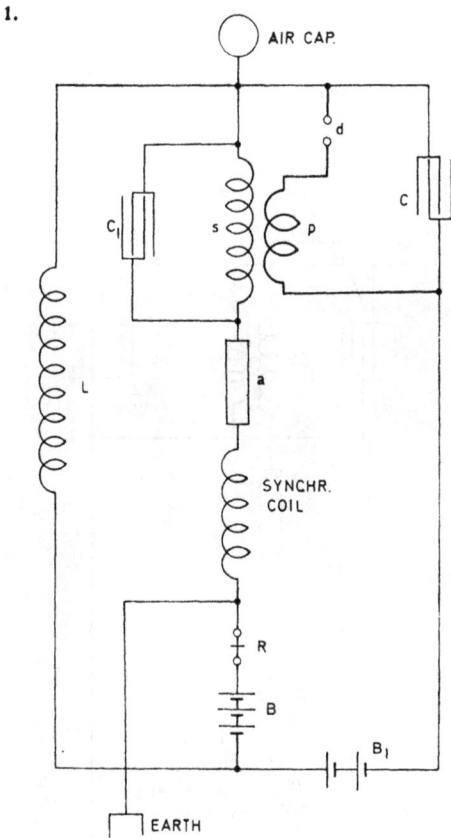

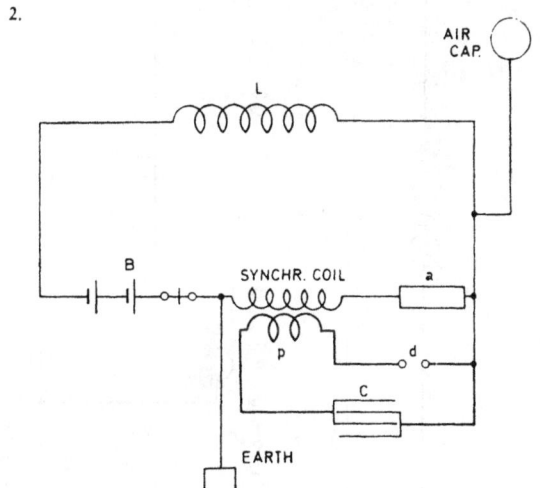

Remark: Another battery is sometimes placed in synchr. coil circuit (this is not shown).

187

Sept. 17, 1899

To suit the two boxes which were made some time ago for the reception of the receiving instruments in easily portable form a number of connections were adopted. These boxes are 9" wide, 14" long and 10" high overall. In the lower part was placed the induction coil, batteries, condenser, resistances and cell. A board was provided to close up this part and on the board was mounted: sensitive relay, clockwork driving break and sensitive device, also a special circuit interrupting device. These boxes were merely made for the

BOX LOOKED FROM TOP
WITH COVER OPEN

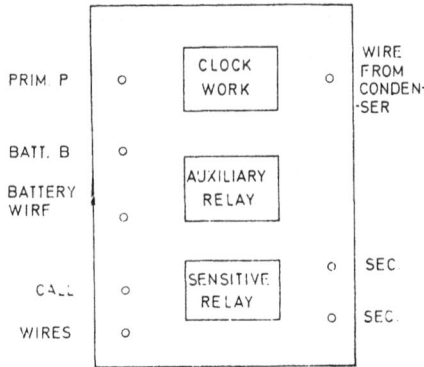

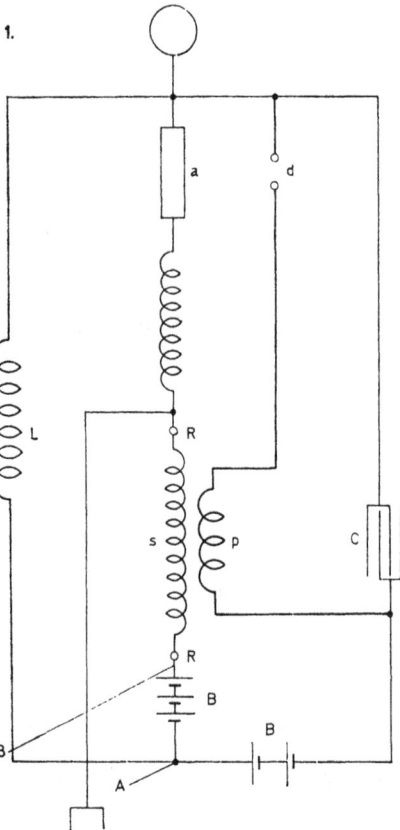

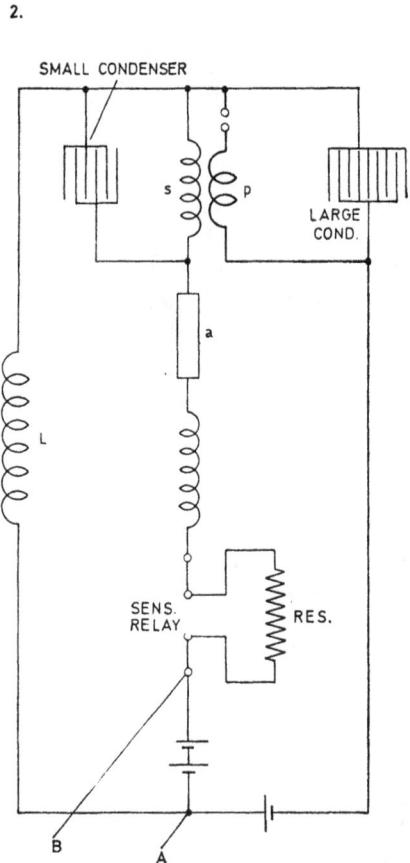

investigation outside and such use. The synchronized coil was wound in one instance around a drum 10″ in diam. and about 4 feet high from the ground and carrying on top a board for placing the box with the instruments and supporting a light rod for air or capacity wire. In another form of apparatus the synchronized coil was wound on drum of 2 foot diam. and 18″ high, which was supported on a tripod of photographic outfit.

These two connections illustrated in Diagrams 1. and 2. were found best suitable. The small condenser around secondary s comprised only a few sheets of mica and tinfoil sufficient to let the currents of a frequency of 50,000 per sec. pass through easily.

Colorado Springs

Sept. 18, 1899

Experiments were resumed with all transformers in place, high speed break and connection in multiple arc of West. Transformer. The object was to further test the intensity of the vibrations produced particularly without spark. The connection was as in diagram. It was though that in this arrangement, which was dwelt upon before, the disturbances were produced more economically than when using a spark discharge. The experi-

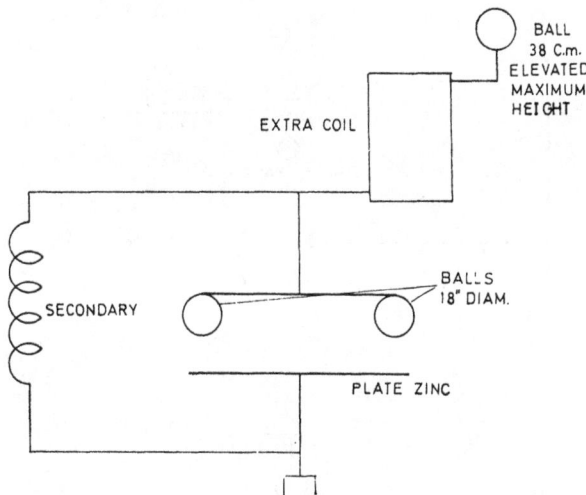

ments fully confirm this. In the tests the capacity of the two balls of 18″ diam. did not very materially derange the adjustment and period of the circuit. This is to be expected; as for the secondary the capacity was far too small and on the other hand the independent vibration of the extra coil could not be materially interfered with since the condenser formed by the two balls and zinc plate allowed free passage of currents to earth. Now the important thing was to decide whether it is better to make length of extra coil one half or one quarter of wave as before. This to be thoroughly investigated. The working was excellent with 1/4 wave length.

Colorado Springs

Various arrangements with oscillator and extra coil for production of most powerful disturbances.

All of these arrangements have been experimented with and described before, and so far the plan illustrated in Diagram 4. seems to be best. In Fig. 1. the extra coil is merely a means of increasing pressure on the end. In Figs. 2. and 3. the vibrations through the

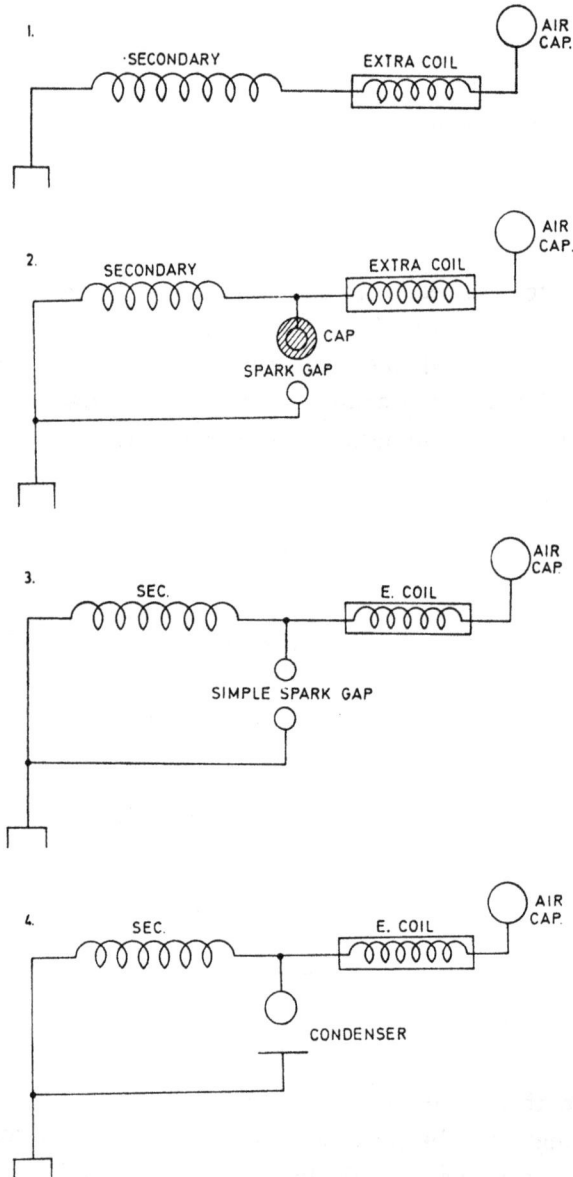

ground are intensified by the extra coil working directly on the ground either through a gap with capacity or without same. In all cases it has been found important to have the two systems vibrate in synchronism. The same considerations apply to Diagrams 4, 5 and 6. The plan in Fig. 4. being found best, the question is what is the best length to give to the wires.

With each secondary and extra coil having one quarter of a wave length the action on the condenser is not most intense. With the extra coil 1/2 wave length and the secondary

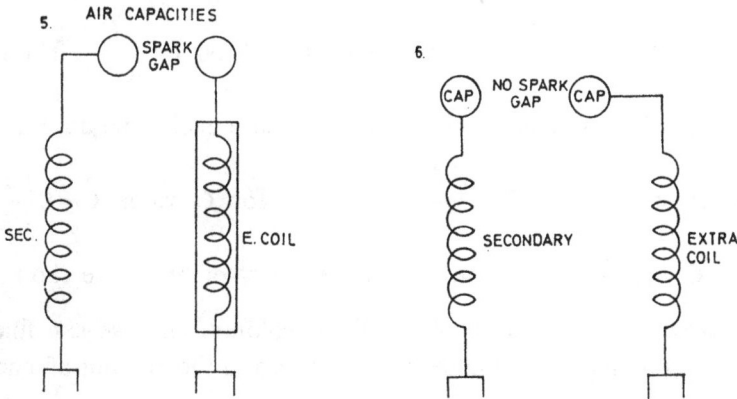

AIR CAPACITIES

1/4, they both cooperate on the condenser producing on the ball a much greater pressure. This appears the best relation in Fig. 4. In Fig. 5. and 6. it is found best to make extra coil 3/4 wave length and the secondary 1/4 for obvious reasons.

Colorado Springs

Sept. 20, 1899

Consider a form of oscillator of great simplicity particularly adapted for telegraphy similar to type exhibited before Am. Ac. of Science. A coil of high self-induction is connected in series with a condenser and across the condenser is placed a break generally in series with the primary of the coil. Very sudden discharges are produced when using a fine stream of electrolyte or mercury to effect shortcircuit. The stream is broken by condenser current. The plan followed for some time was to produce the stream automatically by a magnet worked by a key. The connections are shcematically indicated in the diagram.

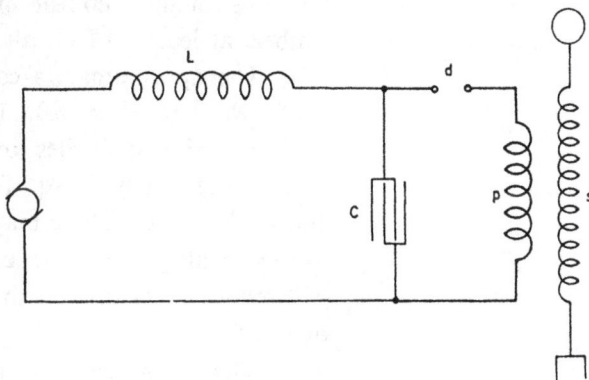

The question is to get the proper capacity of condenser, the amount of self-induction and other particulars. The secondary of the oscillator may be connected as shown to the ground and elevated object of capacity or else a spark gap may be used.

If R be the resistance of charging coil and L the self-induction and E the e.m.f. of the generator the maximum current that could flow through the coil would be $\dfrac{E}{R}$, but as the stream has a resistance r the maximum current will be $I=\dfrac{E}{R+r}$. The moment of the coil will be $\dfrac{1}{2}I^2L$. The condenser will be able to store each time an amount of energy $=\dfrac{E^2C}{2}$ and we must have $\dfrac{1}{2}LI^2=\dfrac{E^2C}{2}$. This gives for C value $C=L\left(\dfrac{I}{E}\right)^2$ but $\dfrac{I}{E}=$ $=\dfrac{1}{R+r}$ hence $C=\dfrac{L}{(R+r)^2}$. Now to obtain frequency of break we should calculate the time sufficient to evaporate a portion of the liquid column. This we can find easily from dimensions of column and its resistance, specific heat and the amount of energy which is passed through it.

(This to be followed out.)

Colorado Springs

Sept. 21, 1899

Proposed structure to elevate terminal to a height of 140 feet from ground.

On a telegraph post very strong and reaching nearly to the roof of the building is to be placed a cap consisting of a pipe 10″ diam., about 2 feet long with a coupling for a 6″ pipe. The cap will widen out at the bottom so as to keep the wood safe against the streamers.

The pipes come in lengths of 20 feet. This will give roughly 120 feet of pipe plus cap and timber, at least 20 feet, all in all 140 feet.

The approximate area of pipe above the roof will be: $\pi\times20\times12\,(6+5+4+3+3+2)=17{,}332$ sq. inches or 120 sq. feet. The ball having nearly 20 sq. feet. We shall have 140 sq. feet+cap. There may be possibly 150 sq. feet with all joints. The electrostatic capacity will reduce turns on coil to probably less than one half.

The wind pressure will be considerable but except for an unusually strong wind it will be fairly safe.

The construction of new secondary for oscillator was begun this morning. The plan of tapering coil before used was abandoned as it was decided to obtain effects by the extra coil, this making it desirable to obtain better energy transfer from primary to secondary and if possible increased impressed e.m.f. on extra coil. The mutual induction will be much better and the oscillator more efficient.

The diameter of the new coil is to be exactly 15 meters inside of wire or about 49.25 feet. Two turns of primary are to be used as before, generally connected in multiple. Provision is made for 48 turns of secondary. Twenty two of the new turns will be equivalent to the 25 turns used last on tapering frame. The frame is being built up as in sketch. The primary cables separated by pieces 1 1/8″ thick. The secondary wire No. 10 used before to be wound in grooves provided in mouldings as shown. Two groovers were provided in each moulding this making the work simplest. Space was provided for two wires in each groove as it might be found later necessary to double copper. Primary and secondary are to have same amount of copper.

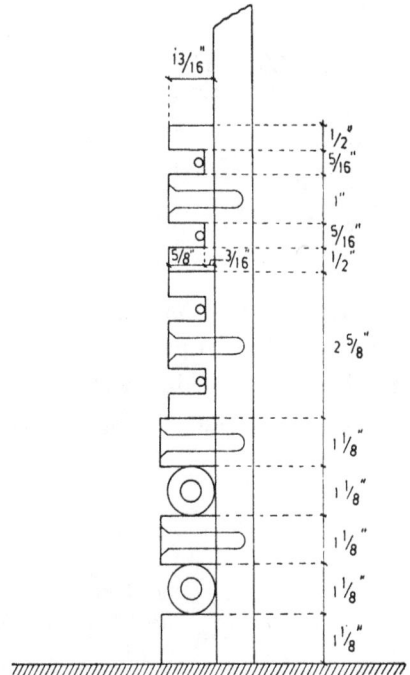

For best results the copper masses in primary and secondary of reconstructed oscillator should be equal.

There are two primary cables generally connected in multiple arc. These cables each have 37 wires No. 9 B. & S. The area in mills from table of wire No. 9 is 13,090, the total section of one cable being therefore

$$\left. \begin{array}{r} \underline{13{,}090 \times 37} \\ 39270 \\ 91630 \\ \hline 484{,}330 \end{array} \right\} = 484{,}330 \text{ c. mills}$$

or

$$\left. \begin{array}{r} \underline{484{,}330 \times 0.0005067} \\ 2421650 \\ 2905980 \\ 3390310 \\ \hline 2454100110 \end{array} \right\} = 245.41 \text{ mm. square}$$

Taking one wire in each groove (No. 10 B. & S.) we have for 48 turns 10,380 mills being the section of wire No. 10 total section reduced to one turn

$$\left.\begin{array}{r}10380\times 48 \\ \hline 41520 \\ 83040 \\ \hline 498240\end{array}\right\}=498,240 \text{ c. mills or}$$

$$\left.\begin{array}{r}498,240\times 0.0005067 \\ \hline 2491200 \\ 2989440 \\ 3487680 \\ \hline 2524582080\end{array}\right\}=252.46\,\text{mm. square. That is very nearly the section of primary cables.}$$

Therefore, if two primary cables are used there should be two wires in each groove of the secondary as provided for. The total section of primary will then be 490.82 sq. mm. and of secondary reduced to one turn: 504.92 sq. mm., but as the wire is slightly stretched the sections or masses respectively will be more equal.

Colorado Springs

Sept. 24, 1899

Note relative to Westinghouse Transformer.

The iron core is of dimensions indicated in sketch. The insulation from core 1″ thick blocks of wood. Insulation between fibre paper about 1/2″ thick.

There are three primary coils and 4 secondary coils. For a transformation to 60,000 volts a part of primary is left out. Best way of working is to use smaller transformation ratio to 45,000 approx.

The secondary coils are connected up alternatively, that is, beginning with the first on the left, then to third, then to second and from there to the fourth coil.

The transformer gives every evidence of having a high efficiency. The leakage current is remarkably small for so large a machine. The alternate connection and disposition of coils is evidently to reduce stray field and also for convenience to enable use of similarly wound coils, that is, coils wound on one form.

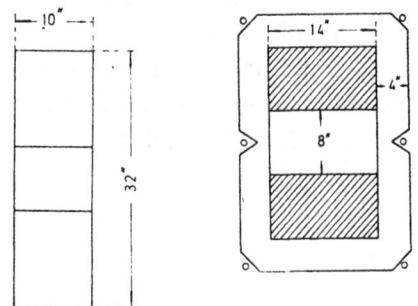

Testing of mineral oil for a ceiling of 300° show it to be oil of excellent quality penetrative, of high flashing point and good insulating properties.

Measurement of self-induction of primary of oscillator and regulating self-in. coil.

Readings:

Conductor measured	Current	Voltage across conductor	p
2 primary turns	33.2	6.4	Computed from revolutions
2 primary turns	58.9	11.7	of synchronous single
1 primary turn	58.9	5.85	phase motor speed 35 per
Self-in. coil	33.2	2.5	sec. 8 pole motor gives
Self-in. coil	58.9	4.45	$n=140$
			$p=880$

Of the above readings the one showing 58.9 amp. was taken repeatedly and is very probably closer than the other reading with smaller current. Taking this as the basis I find, neglecting resistance of both primaries and self-in. coil being very small L_{2p} of two primaries:

$$\frac{E \times 10^9}{I\,p} = \frac{11.7 \times 10^9}{58.9 \times 880} = 225,730 \text{ cm.}$$

L of one primary will be $L_p = 56,432$ *approx.* as the coils are practically one.

$$L_c = \frac{E \times 10^9}{I \times 880} = \frac{445}{1170} \times 225,730 = 85,855 \text{ cm.} \quad \text{approx.}$$

As there are 24 turns we may take as a rough approximation when quickly computing: 3600 cm. per turn when there are a considerable number in.

Following method for determining period of vibration, inductances and capacities is simple and convenient. The vibrating system is formed by a continuously variable and exactly determinable inductance and a capacity standard, or by an inductance standard and continuously adjustable condenser or by a system in which both these elements are continuously adjustable and can be exactly determined in one way or another. This system is then excited by a primary vibrating system in a convenient manner and one or both of the elements of the excited system is varied until resonance is obtained. This gives the period of the primary system and if in this only one more element is known all the others can be easily determined. The excitation is conveniently secured and graduated by connecting the wire leading to the system to be excited to the ground through an adjustable spark gap, which is generally very small. This method was applied to determining the period of the primary system used in these experiments in the following manner: a standard self-induction coil made long ago and used in experiments in N.Y. about 1560 turns wound on a drum 3 1/2″ diam. was shunted by the adjustable condenser, also frequently used and consisting of two brass plates 20″ diam., and this system was connected to one of the terminals of the Westinghouse transformer as illustrated in diagram below. By varying the length of spark at b the degree of excitation was varied to any value desired, the spark at α serving to determine maximum rise of potential on terminals of excited system.

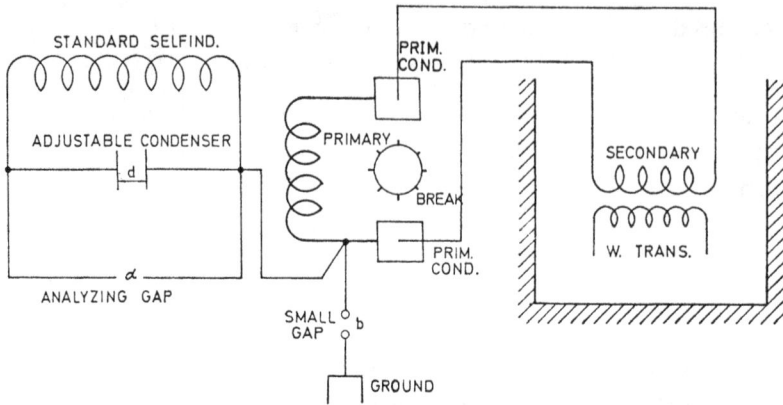

Particulars: $L=0.0176$ H. Res. of coil$=59.457$ ohms, drum 3 1/2″, turns 1560 approx.

Readings in one case:

Capacity in primary total
144 bottles$=0,1526$ mfd.

Inductance primary
0.000025 H

$$T_{\text{approx}} = \frac{2\pi}{10^3} \sqrt{0.000025 \times 0.1526} = \frac{1225}{10^8}$$

Resonance was obtained with the plates nearly 0.8 cm. apart, the period of excited system being slightly slower.

$$C_{\text{sec}} = \frac{A}{4\pi d} = \frac{2027}{4\pi \times 0.8} = \text{approx. } 200 \text{ cm.}$$

$$T_s \text{ approx. } = \frac{1230}{10^8}$$

Colorado Springs

Sept. 27, 1899

Determination of inductance of coil used in series with extra coil when no ball was used on latter, with old secondary.

160 turns No. 10 B. & S. wire rubber-covered Habirshaw, drum 2 feet diam. $=$ $=60.96$ cm. Length$=42.5″=107.95$ cm.

First measurement average of readings:

$$I=5.9 \qquad E=38.25 \qquad R=1.054 \qquad \omega=880$$

$$\frac{E}{I} = 6.483 = \sqrt{(1.054)^2 + (880\,L)^2}$$

$$L = \frac{\sqrt{6.483^2 - 1.054^2}}{880} = \frac{6.4}{880} = 0.00728 \text{ H or } 7{,}280{,}000 \text{ cm.}$$

Second measurement average of readings:

$$I = 6.77 \qquad E = 43.25 \qquad R = 1.054 \qquad \omega = 880$$

$E = 6.39$ from this $L = 0.00716$ H or $7{,}160{,}000$ cm.

This variation probably due to change in ω and it is probably safest to take average

$$\left\{ \begin{matrix} 0.00716 \\ 0.00728 \end{matrix} \right\} \quad \text{or} \quad \left. \begin{matrix} 0.00722 \text{ H} \\ 7{,}220{,}000 \text{ cm.} \end{matrix} \right\} \quad \begin{matrix} \text{as best} \\ \text{values.} \end{matrix}$$

The calculated value from above dimensions is:

$$L = \frac{4\pi N^2 S}{l} = \frac{4\pi \times 25{,}600 \times 2919}{107.95} = 8{,}700{,}000 \text{ cm. or } 0.0087 \text{ henry}$$

$$N^2 = 25{,}600$$

$$S = \frac{\pi}{4} 60.96^2 = 2919 \text{ cm. sq.}$$

The difference must be due to the internal capacity of the coil or possibly inexactness of the dimensions above.

Colorado Springs

Sept. 28, 1899

One of the difficulties in telegraphy and a drawback in practical introduction is the elevated terminal of capacity as proposed by me. It is difficult to elevate a structure to the desired height and keep the same insulated from the ground and with very powerful vibrations, as here produced even an insulated wire leading up offers difficulties because of streamers which reduce the force and the effect at a distance.

I propose to overcome this by the following plan. A structure is to be elevated from the ground up to the desired height and an insulated wire from the oscillator is to be brought

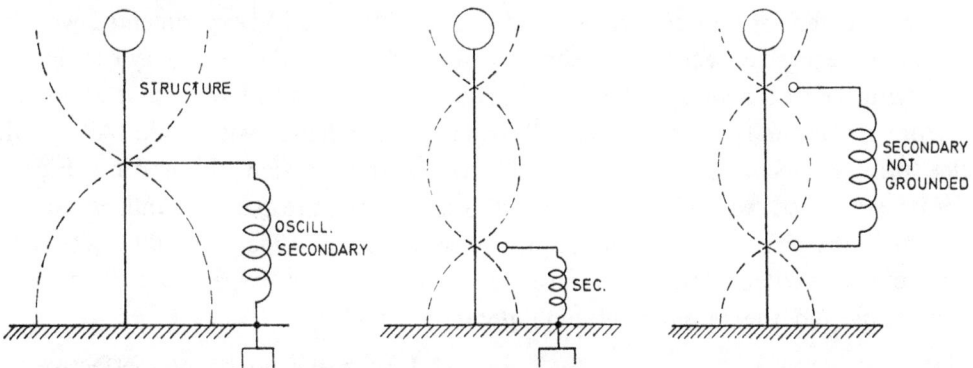

up to a point on the structure and connected to same or else brought into proximity, as for instance when a spark is used. This point should now be so located that there is at the same moment, say, a position maximum on top and a negative on the bottom or ground, that is the top and bottom should be one half of the wave apart or a multiple thereof.

When secondary is grounded as usual it will probably be advantageous to make the wave length in both systems so that they work in unison on the ground.

Colorado Springs

Sept. 29, 1899

Various advantageous arrangements of oscillating circuits for producing disturbances in the natural media.

The object of these arrangements is to produce especially in conjunction with an "extra coil", as before explained, disturbances in the most effective and economical manner. In such a coil the e.m.f. is raised to an extremely high value by the "magnifying ratio". The arrangements furthermore contemplate doing away with the spark which consumes energy, although in many respects it possesses advantages giving, in particular, a very high rate of energy delivery. In the diagrams three such arrangements which have been experimented with are illustrated.

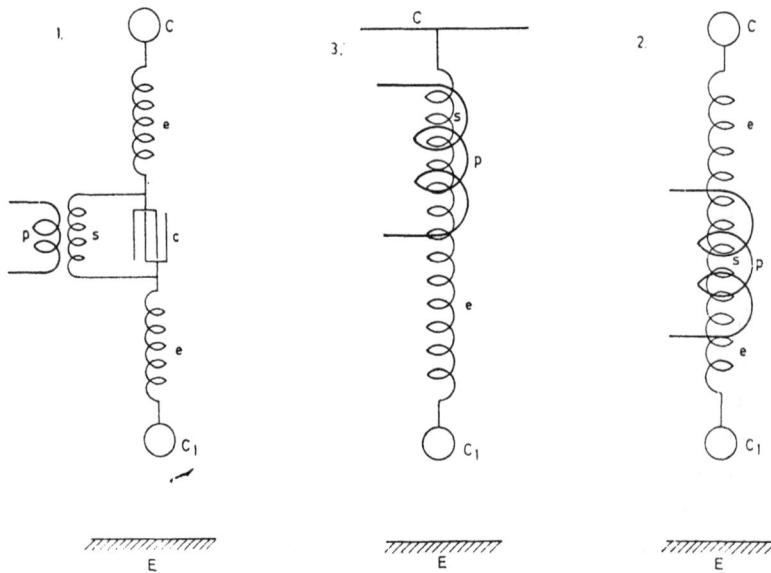

In Fig. 1. the form of connection is shown most frequently experimented with here. The primary p energizes secondary s shunted by condenser c, the secondary exciting extra coils e e with their capacities C C_1 at the free terminals, one of which, C_1, is at some distance from the ground or groundplate E forming a condenser with same. All the three systems, primary, secondary and extra coil have the same period of vibration. Fig. 2. illustrates a simplified way; in this instance the extra coils are partially influenced by induction from the primary p. In Fig. 3. again the extra coil e may be only electrically or also inductively excited. The upper terminal is here a very large capacity as the roof of a building and the terminal of high potential is C_1.

This seems to be very effective.

Colorado Springs Notes

Oct. 1—31, 1899

Following items partly completed omitted for want of time:

Oct. 5. More complete description of photographs taken.

Oct. 14, 22, 23 $\left\{\begin{array}{l}\text{page 5}\\\text{page 6}\end{array}\right\}$, 25 and 29 corrected results deduced from experimental data recorded.

Patent matters nearly completed:

a) Method securing excessive e.m.f. momenta.

b) Various ways of avoiding use of elevated terminal in power transmission etc.

Colorado Springs Sept 19. 1899.

Various arrangements with oscillator
and extra coil for production of more
powerful disturbances

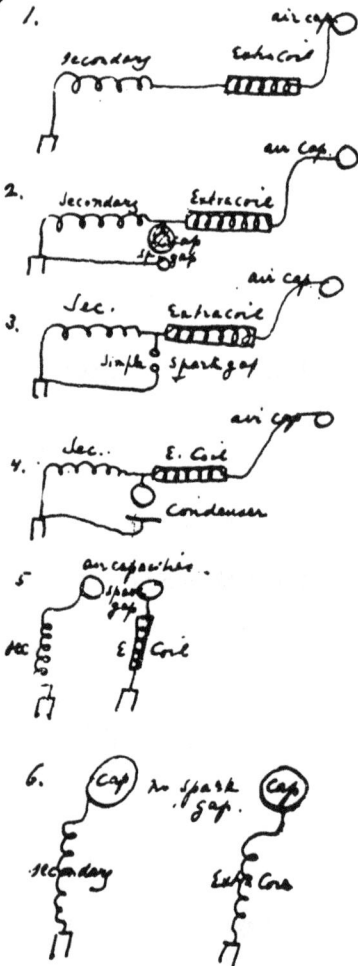

1.
secondary   extra coil   air cap

2.
Secondary   Extra coil   air cap
gap   spark gap

3.
Sec.   Extracoil   air cap
Simple spark gap

4.
Sec.   E. Coil   air cap
Condenser

5.   air capacities.
spark gap
Sec   E. Coil

6.   Cap   no spark gap.   Cap
secondary   Extra Coil

All of these arrangements have
been experimented with and described before,
and so far the plan illustrated in
diagram 4. seems to be best.
In fig 1. extra coil is merely a means of
increasing pressure on end.
In fig. 2 and 3. the vibrations through
the ground are intensified by the
extracoil working directly on the ground
either through a gap with capacity or
without same. In all cases it has been
found important to have the two systems
whole in synchronism.
The same considerations apply to the
diagrams 4, 5, and 6.
The plan in fig 4. being found best
it is the question what is the best
length to give to the wires.
With each secondary and extra coil having one
quarter of wavelength. The action on condenser
is not most intense. With the extreme
1/4 wavelength and secondary 1/4 they work together
on condenser producing a bell, much greater pressure. This applies
best relation in fig 4. In fig 5. and 6. it is found best
to make extra coil 3/4 wavelength and secondary 1/4 for obvious reasons.

The new secondary was wound with 22 turns in all. The last turn was placed on top on porcelain insulators to prevent injury to the wood and breaking through on the last turn where danger greatest. It was evident that the coil could not stand the strain as there was only 1″ between the turns. But the winding was tried for trial. The total length of coil was now — disregarding the last turn which was on top of frame — 27″ approx. Now the length of the old coil was 63″ (last turn excepted). Therefore, if the turns would have been of the same area the self-induction of the new coil would have been increased by a ratio of $\frac{63}{27}$ because of length and diminished by a ratio of $\left(\frac{21}{25}\right)^2$ because of turns. Now the average turn of the old coil was 44 feet dia. and of the new 49 feet approx. This made each of the new turns about 10″ larger than the average of the old. On the whole then the self-induction of the new coil should have been roughly $\frac{63}{27} \times \left(\frac{21}{25}\right)^2 \times \left(\frac{11}{10}\right)^2 L$, where L is the self-induction of the old coil, that is $L_1 =$ (nearly) 2.06 L.

The storing capacity will also be increased by a ratio of $\frac{3.125}{1.25}$ or nearly 2.5 capacity of the former. Consequently the period of the new coil should be nearly $\sqrt{2.5 \times 2.06}$ times longer or nearly 127% longer. With this distribution of turns it will be necessary to use a much smaller number.

As expected the coil wound before was unable to stand the strain and a different distribution of turns of the secondary was made. The ten turns nearest to the ground were left as before and the remaining were placed one turn in each second groove making the distance between the upper turns 2 5/8″ and lower ones 1″. Even with this arrangement, some of the upper and also some of the lower turns would break through.

A change was again made and only 4 of the turns — the lowest — were left 1″ apart and all the others were placed in every second groove. The tests showed that the turns could now withstand the full charge. Perhaps up to six first turns might have been left at a distance of 1″ — but it was thought that the distribution was good enough as it was.

The last change reduced the number of turns to 18 (the uppermost not counted). Compared with the last form of tapering coil of 25 turns the period of the new was now approximately estimated from rough data below:

diameter of one new turn	49 f.
,, average old ,,	44 ,,
turns of new coil	18
,, old ,,	25
average distance of new turns	2 5/8″
,, ,, old ,,	3 1/8″
length of new coil	42″
,, old ,,	63″

Calling L_1 self-ind. of new coil and L_2 that of old tapering coil, we would have approximately:

$$L_1 = \frac{63}{42} \times \left(\frac{18}{25}\right)^2 \times \left(\frac{49}{44}\right)^2 L \text{ or } L_1 = L \textbf{ nearly}$$

Now the capacity of new coil will be greater by a ratio of $\dfrac{3.125}{2.625} = \dfrac{25}{21}$ because of distance of turns and smaller by a ratio of $\dfrac{l}{l_1}$, l being length of new and l_1 that of old coil. Now

$$\frac{l}{l_1} = \frac{18 \times 49 \times \pi}{25 \times 44 \times \pi} = \frac{88}{110} = \frac{44}{55}.$$

Hence the capacity of new coil to that of old will be as

$$\frac{25 \times 44}{21 \times 55} \text{ or as } \frac{110}{116}$$

or very nearly equal.

Therefore the distribution of the turns will secure nearly the same period of vibration as in the old coil.

Useful data in estimating possible errors due to the proximity of the ground or conductors or other causes.

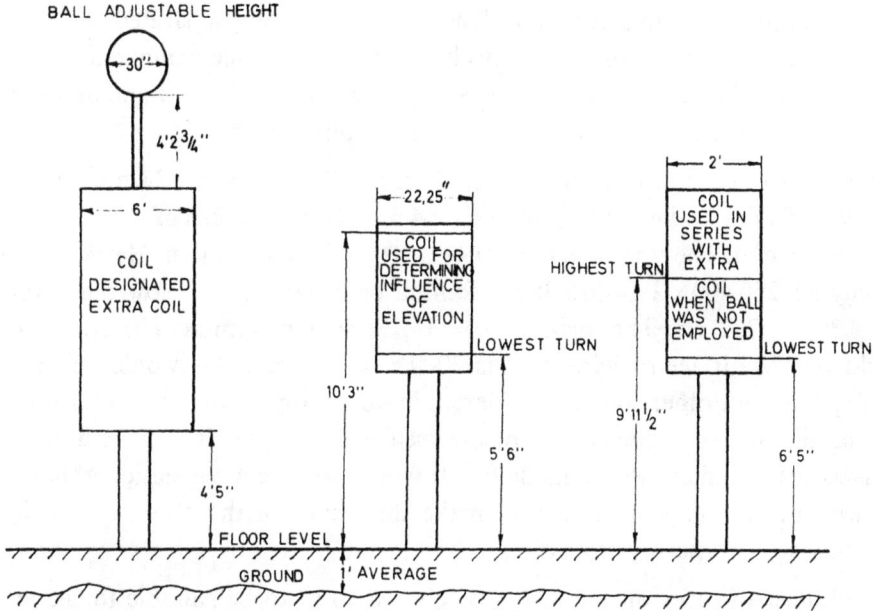

BALL ADJUSTABLE HEIGHT

←30″→

4′2¾″

6′

COIL DESIGNATED EXTRA COIL

4′5″

FLOOR LEVEL

GROUND   1′ AVERAGE

←22,25″→

COIL USED FOR DETERMINING INFLUENCE OF ELEVATION

LOWEST TURN

10′3″

5′6″

2′

COIL USED IN SERIES WITH EXTRA

HIGHEST TURN

COIL WHEN BALL WAS NOT EMPLOYED

LOWEST TURN

9′11½″

6′5″

SECONDARY OF OSCILLATOR

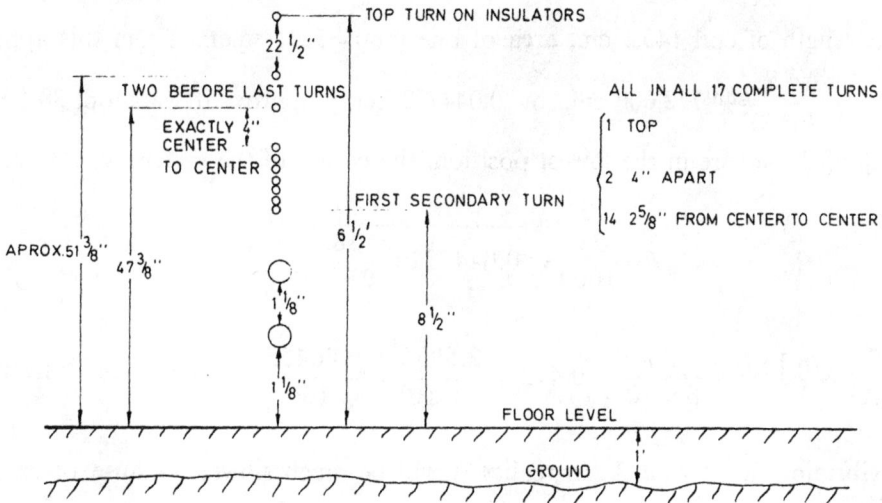

TOP TURN ON INSULATORS

22 ½″

TWO BEFORE LAST TURNS

EXACTLY 4″ CENTER TO CENTER

FIRST SECONDARY TURN

6 ½′

APROX.51 ⅜″

47 ⅜″

1 ⅛″

1 ⅛″

8 ½″

ALL IN ALL 17 COMPLETE TURNS

1 TOP
2 4″ APART
14 2⅝″ FROM CENTER TO CENTER

FLOOR LEVEL

GROUND   1′

Test to determine more exactly influence of elevation on capacity of an insulated body.

A coil was wound to suit best the special conditions of this test. On a drum of 2 feet and 1 1/4″ diam. were wound 400 turns, wire cord No. 20. The insulation was very thick but of small specific inductive capacity. This wire with widely separated turns was used in order to make the capacity of the coil itself as small as possible compared with capacity of insulated body. The latter was in this case the one sphere of 30″ diam. arranged to be elevated at will to a height up to nearly 40 feet from the ground.

The wire had a diam. approx. No. 20 B. & S.=0.032″ or 0.8128 mm. The circumference of wire if solid would have been about 2.4 mm. Now one turn of wire was $\pi \times 25.25″$ or about $\pi \times 641$ mm. The total length $\pi \times 641 \times 400 = 256,400 \times \pi$ mm. Hence total surface of wire only $\pi \times 256,400 \times 2.4 = 615,360 \ \pi$ sq.mm. or 6154π sq.cm. Now the diam. of the sphere was 30″ or 76.2 cm. The surface $\pi \ d^2 = 18,231$ sq. cm. approx. The utmost we could take would be 1/2 surface of wire, that is 3000 π sq.cm. and this would be only 1/2 of surface of sphere. But other things considered, it would appear that the error due to electrostatic and distributed capacity of the coil itself would be small. This is however to be further investigated and allowance made for. It would be, of course, desiderable to entirely do away with the capacity of the coil to make the results of the observations rigorously true, but this will be hardly possible.

The number of turns on the coil was selected so as to be suitable to the apparatus used normally. The total length of the coil so far as wound was 57 1/8″ or 145.1 cm, diam. of drum 25.25″=64.14 cm. To calculate self-induction we have then the following data:

Turns: 400, length of coil 145,1 cm, area of one loop=3231 sq.cm. From this approximately $L = \dfrac{4 \pi N^2 S}{l} = 44,772,000$ cm, or 0.044772 henry approx. If we adopt 38.1 cm as the capacity of the sphere in the lowest position, the period of the system would be:

$$T = \frac{2 \pi}{1000} \sqrt{0.044772 \times \frac{38.1}{9 \times 10^5}} =$$

$$= \frac{2 \pi}{3 \times 10^5} \sqrt{0.1706} = \frac{2 \pi}{3 \times 10^5} \times 0.413 = \frac{2.5936}{3 \times 10^5} = \frac{0.86455}{10^5} \ \text{or} \ n = 115,668 \ \text{nearly.}$$

This is a vibration far too quick, in reality it will be much slower because of capacity in the coil.

Diagram shown illustrates arrangement used in experiments: The coil of 400 t. was excited by secondary of oscilator, and capacity and self-induction in primary were varied until resonance of free system comprising coil of 400 t. and ball of 38.1 cm cap. was obtained. This was evident from spark length and other indications, as streamers.

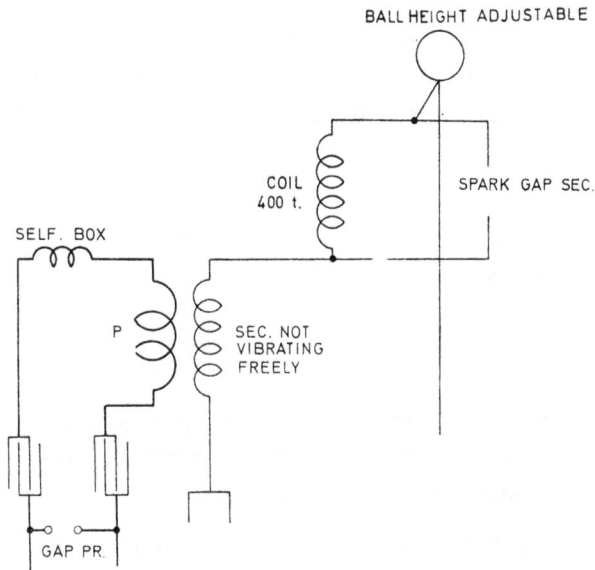

Results:

Ball position	Capacity in primary on each side	Turns in self. box	Spark gap primary	Spark second.	
lowest	1 old+3 2/3 new tanks	22	1/2″+1 turn	53″	The experiments were interrupted here because of darkness
2 feet higher	1 old+4 1/3 ,,	22	,,	,,	

From this computed found that by elevating 2 feet, capacity was increased by a ratio of $\frac{5\ 1/3}{4\ 2/3}=\frac{16}{14}=\frac{8}{7}$ or about 15% nearly, really 14.3%. This is a value found nearly the same before in previous tests.

To be followed up.

Oct. 5, 1899

Test of secondary last pattern, 17 turns in all.

To ascertain the period a spark gap adjustable was used in the secondary from end to earth, as usual, and the capacity and self-ind. of primary was varied until maximum spark length in secondary and other indications showed maximum resonant rise.

Results:

Capacity in primary	Turns in self. box	
7 tanks on each side	13 ,,	⎱ 1 turn
8 ,, ,,	10 ,,	⎰ primary (2 multiple) approximate

Now a ball 38.1 cm. was added on the end and placed near the earth plate at a distance of about 3 1/2 feet but very little affecting the vibration. This shows that distributed capacity in secondary is very large as before.

Capacity of one tank being 0.03816 mfd. Taking approx. L_p=56,400 and 13 turns box 46,800

We have:

$$T = \frac{2\pi}{1000} \sqrt{3.5 \times 0.03816 \times \frac{103,200}{10^9}} =$$

$$= \frac{2\pi}{1000} \sqrt{\frac{13.784}{10^6}} = \frac{2\pi}{10^6} \sqrt{13.784} = \frac{23.36}{10^6} = 0.00002336 \text{ and}$$

$$n = 42,800 \ approx.$$

Some data to be preserved:

Res. of cord used in measuring resistances:	0.596	ohm
Res. of secondary 17 turns	2.804	,,
Res. of 2 primary t. in series	0.004	,,
Res. of Regulating self-ind. coil	0.054	,,

Coil wound for W.T. outfit.: Aug. 24, 1899

Spool about 8 1/2″ long: 399 turns per layer ⎱

                                              22 layers ⎰

Latest secondary in receiving apparatus, No. 30 wire, 90 turns per layer, 35 layers. Res. 424 ohms.

Tests showed best result with 56—68 primary turns. 62 turns, No. 20 cord, are now used. Res. of cord 0.39 ohm.

———————

Test with special coil for determining more accurately the law of variation of capacity with elevation. The coil with 400 turns No. 20 cord was used as previously and the adjustable ball of 38.1 cm capacity. The ball was normally above the extra coil repeatedly referred to, which on account of its great internal capacity was not used. Although unconnected some error was necessarily caused by the presence of the coil near the ball when the latter was near its lowest position.

A further error was anticipated from the influence of the roof at the points when the ball, which was being gradually lifted during the test, was nearest to it. The smallest distance or nearest point was about 11 feet. Nevertheless some action was bound to occur although the wood is very dry here and it is proposed in later tests to eliminate these erors as much as possible. The roof was covered with some sort of tar paper the influence of which can only be conjectured at present. The most reliable data will be those obtained with ball at the highest points when it is above all structures. The connections used are

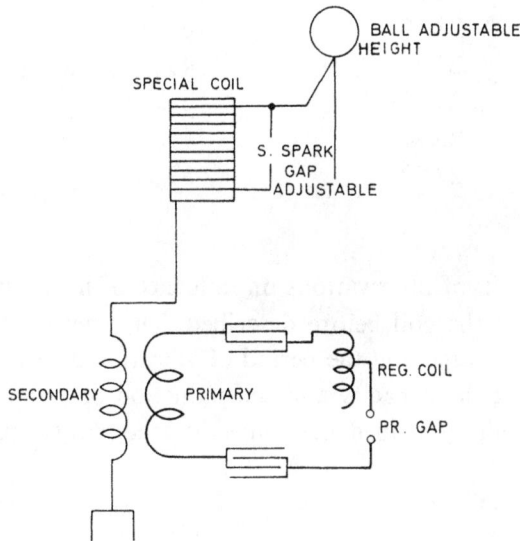

shown in diagram. The ball was lifted by steps of *one* foot each and the primary self-induction and capacity was adjusted until maximum resonant rise was observed on special coil, and adjustable secondary spark gap serving as analizer, besides streamers which served even as a better guide. The results of the test are shown on the following table:

Table showing results. The capacities are all reduced to the same inductance so that as the primary capacity changes, the secondary is changed in the same ratio.

Height from ground feet	Capacity in tanks	Percent of increase	Figures reduced to self-ind. of one primary:
20.66	4.36		
21.66	5.05	16%	
22.66	5.77	14%	
23.66	6.27	9.5%	
24.66	6.00	0	
25.66	6.24	4%	

two primary turns reduced

Height from ground feet	Capacity in tanks	Percent of increase	Figures reduced to self-ind. of one primary:
25.66	6.64	10%	To correct later
26.66	8.87	33%	
27.66	11.48	40%	factor $\dfrac{6.24}{6.64}$ to be used
28.66	14.84	30%	
29.66	18.64	26%	
30.66	21.08	13%	
31.66	22.36	6%	
32.66	23.28	4%	
33.66	24.72	6%	
34.66	25.68	4%	
35.66	27.00	5%	

Table showing results of observations on influence of height in determining capacity of a sphere connected to the coil before described. The sphere of 38.1 cm. electrostatic capacity was gradually elevated and the period cf vibration determined for every position of the sphere, by varying the capacity and self-induction of primary circuit. As the self--inductions of both circuits remained the same the secondary capacity varied exactly as the primary.

Height of sphere from ground in feet	Capacity of primary expressed in tanks on each side	Turns in regulating self-ind. coil	Self-ind. in reg. coil		Total self-ind. in primary circuit	Arc		Capacity in primary on one side reduced to same value of self-induction in primary; minimum 106,800 m or 137,200 c	Increase of capacity of primary or sec. in percents	
			measured data cm.	calculated data cm.		in primary	in secondary		rate	absol.
1	2	3	4	5	6	7	8	9	10	11
20.66	4.66	21.5	77,400	103,200	133,800 m. / 173,200 c.	1/2″ + 1 turn	52 1/4″	5.84 m. / 5.88 c.		
21.66	5.33	22	79,200	105,600	135,600 m. / 175,600 c.	″	″	6.77 m. / 6.82 c.	16 m. / 16	16 / 16
22.66	6	22.5	81,000	108,000	137,400 m. / 178,000 c.	″	″	7.74 m. / 7.78 c.	14.3 / 14.1	32.5 / 33
23.66	7	20	72,000	96,000	128,400 m. / 166,000 c.	″	″	8.41 m. / 8.47 c.	8.6 / 8.9	44 / 44
24.66	8	14	50,400	67,200	106,800 m. / 137,200 c.	″	″	8.00 m. / 8.00 c.	—5 / —5	37 / 37
25.66 }	8 one pr. turn	15.5	53,800	74,400	112,200 m. / 144,400 c.	″	″	8.42 m. / 8.42 c.	5.25 / 5.25	44.5 / 43
	1.66* two pr. turns	24	86,400	115,200	312,000 m. / 395,200 c.	1/8″ + 1 turn	20″	8.04 m. / 8.07 c.	0.5 / 0.9	37.7 / 37.2

1	2	3	4	5	6	7	8	9	10	11
26.66	2.33 / 2.48 r	20	72,000	96,000	297,600 m. / 376,000 c.	"	27.75	10.71 m. / 10.71 c.	33.2 / 32.7	83.4 / 82.2
27.66	3 / 3.2 r	19	68,400	91,200	294,000 m. / 371,200 c.	"	32 1/4	13.63 m. / 13.7 c.	27.26 / 28	133.2 / 133
28.66	4 / 4.26 r	18	64,800	86,400	290,400 m. / 366,400 c.	"	little less / "	17.93 m. / 17.98 c.	31.55 / 31.24	207 / 205.8
29.66	5 / 5.32r	18.5	66,600	88,800	292,200 m. / 368,800 c.	2"+ 1 turn	still less / "	22.56 m. / 22.6 c.	25.8 / 25.7	286.3 / 284.3
30.66	6 / 6.35 r	14	50,400	67,200	276,000 m. / 347,200 c.	2.5"+ 1 turn	still less / "	25.4 m. / 25.4 c.	12.6 / 12.4	335.8 / 332
31.66	6 / 6.35 r	18	64,800	86,400	290,400 m. / 366,400 c.	2.5"+ 3 t	still less / "	26.72 m. / 26.8 c.	5.2 / 5.5	357.5 / 355.8
32.66	6 / 6.35 r	21.5	77,480	103,200	303,000 m. / 383,200 c.	3"+ 3 t	still less / "	27.94 m. / 28.7 c.	4.6 / 7.1	378.4 / 388
33.66	6.33 / 6.86 r	22	79,200	105,600	304,800 m. / 385,600 c.	"	still less	30.29 m. / 30.46 c.	8.41 / 6.13	418.6 / 418
34.66	6.66 / 7.1 r	21	75,600	100,800	301,200 m. / 380,800 c.	"	still less	30.96 m. / 31.24 c.	2.21 / 2.56	430 / 431
35.66	7 / 7.5 r	21	75,600	100,800	301,200 m. / 380,800 c.	"	still less	32.70 m. / 33.00 c.	5.6 / 5.6	460 / 460

Observations:

Tuning was very sharp from line 27.66 to end. The arc in the primary was getting stronger and stronger. This probably because of lower frequency. The flaming of primary arc also because of this. For same reasons secondary arc was getting continually weaker as the ball was elevated.

m read measured data

c read calculated data

r read reduced data.

From the sign* two primary turns were used in series as the vibration got too low and not enough primary capacity was available with one turn. This was thought preferable to tuning to harmonic. From the figures on the side of 25.66 line it will be seen that the capacity with two turns was not exactly one quarter of 8, that is 2 tanks, but less. This was probably due to the fact that the tanks are not exactly equal and more so because when two primaries are used in multiple as one single turn the self-induction is less than 1/4 of that of two turns in series. To reduce to same self-induction as with one turn the capacities obtained with two turns should be multiplied by $\dfrac{6.64}{6.24}$

Colorado Springs

Oct. 6, 1899

Measurement of coefficient of self-induction and mutual induction.

Secondary last form 17 turns of which total 16 turns disposed on frame described before, one turn in every second groove, one groove near the primary *free*.

Average of readings: $p=880$

Voltage across secondary $E=122.5$ V, $I=13.8$ amp.; Res. $R=2.804$ ohms. From this

$$L_1^2 = \frac{\left(\dfrac{E}{I}\right)^2 - R^2}{\omega^2} = \frac{70.94}{880^2} \text{ H}^2 \qquad \frac{E}{I} = 8.877; \quad \left(\frac{E}{I}\right)^2 = 78.8$$

$$R^2 = 7.86$$

$$L_1 = \frac{1}{880}\sqrt{70.94} = \frac{8.4226}{880} \text{ H or}$$

$$L_1 = \frac{8,422,600,000}{880} = 9,571,140 \text{ cm.} = \text{approx. } 0.00957 \text{ H}$$

For determining M the average of a number of readings was:

Current in secondary $I=15$ amp., e.m.f. across two primary turns 11.2 V$=E$, $\omega=880$.

We have from this:

$$M = \frac{E}{I\omega} = \frac{11.2}{15 \times 880} = 0.000848484 \text{ H}$$

M=848,484 cm.

14*

Colorado Springs

Oct. 7, 1899

The secondary was again changed by displacing the 15th and 16th turn, the rest remaining as before: namely, 14 turns, a turn in each second groove and the 15th and 16th turn each in every third groove and the top turn on porcelain insulators as before. This changed very little the constants of the circuit.

The test was now made closed with the following results:

Capacity in primary (one turn)	Turns in regulating coil	Capacity on terminal of secondary
on each side 7 2/3 tanks= total =0.14615 mfd.	6	none
7 2/3 ,,	7	Ball 38.1 cm capacity at a distance 2 feet from a plate connected to earth terminal of secondary
7 2/3 ,,	nearly same, just a little more	Ball 1 foot from earth plate.
7 2/3 ,,	10	Ball 9″ from earth plate.
7 2/3 ,,	13	Ball 6″ from earth plate.

Tuning of extra coil to suit vibration of the secondary latest design, as specified on another sheet today.

An attempt was made to get the vibration of the coil with ball elevated exactly as that of the secondary. To adjust the vibration the ball was elevated to various positions and soon the adjustment was reached.

With an elevation of 2 feet lower than the highest point (35.66 feet from the ground) — that is 33.66 feet from the ground the maximum resonant rise on coil, as evidenced by the spark, was obtained with 7 2/3 tanks capacity on each side and 13 turns in the self--induction coil. Lowering the ball just a trifle and making capacity on each side 8 tanks the resonant maximum rise in coil took place with 10 turns in regulating coil. This was almost exactly the vibration of the secondary as previously ascertained.

The two were now connected in series and discharges on a spark gap of something like 12 feet were obtained, although the W.T. was not strained to the utmost. When the

212

spark wire was taken off and ball with a rubber covered wire No. 10 specially prepared left alone, streamers formed on top of ball, but little as the wind was blowing. The ball being disconnected and rubber wire alone left, the streamers were very fierce reacting sometimes 16—18 feet. Best results were with *6 turns* in coil in the last experiments.

A more careful tuning of extra coil without ball, only rubber covered wire or cord No. 10 (which was refered to before), the tip of cable being brought out about 2 feet and inclined to horizontal about 45°. This wire to prevent streamers was specially made and was covered with rubber. Composition Habirshaw, 40% pure Para. The tuning of the coil alone with the self-induction coil specially wound in series, gave:

Capacity of primary	Turns in self-ind. regulating coil
8 tanks on each side	6 1/2

The coil used in series with the extra coil was one especially adjusted so as to give the same vibration to the system as when the extra coil was used alone and with ball on top at highest point. The coil has 160 turns and is wound on a drum 2 feet in diam. with wire No. 10, same as used in the secondary and extra coil.

Now the secondary with a ball on the end and elevated at a height of 2.66 feet from the ground gave also exactly, with 8 tanks in primary on each side and 6 1/2 turns in regulating coil, maximum effect. When the two were connected in series the display was magnificent, sparks flying to the ground a distance of over 16 feet. Their curved paths stretched out would be certainly 24 feet long. This was not the maximum of the power of the apparatus as the spark in the primary could have been still lengthened without difficulty.

Colorado Springs

Oct. 8, 1899

Experimental data in connection with tuning of extra coil and secondary as recorded yesterday.

Capacity on each side in primary circuit 8 tanks	Turns in self-ind. Regulating coil 6.5—7	With ball 38.1 centim. on free end of secondary, the ball being 2.66 feet from a grounded zinc plate, the both had exactly the same period.

Now the length of wire in the secondary was 803 meters, namely 17 turns each of a diameter of 15 meters.

The total length of wire in the extra coil circuit was: the extra coil itself 889 meters, namely 149 turns each of diameter of 1.9 meters plus special coil inserted in series: 307 meters, namely 160 turns each of diam. of 0.61 meter. This special coil was used when the

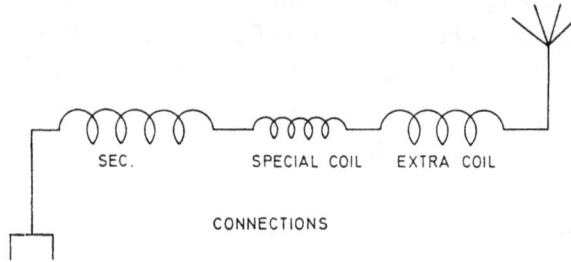

ball on top was not employed as capacity and the coil was so adjusted that the vibration was the same without the ball as with the ball and without the special coil; it being understood that the ball was at its highest position in such case. The total length of wire was therefore:

$$\underbrace{\text{secondary}}_{\text{one system}} + \underbrace{\text{special coil} + \text{extra coil}}_{\text{one system}} = 803 + 307 + 889 = 1999 \text{ meters}$$

with all connections the length was increased to 2030 meters (17 meters rubber wire on top; 13 meters lower connecting wire).

Now capacity in primary was 4 tanks$=4\times0.03816=0.15264$ mfd.

Total self-ind. in primary circuit was approximately	56,400 cm one primary turn
	6600 cm approx. — all connections in primary circuit
89,000 cm	26,000 cm self-induct. coil

As the self-ind. of the regulating coil was rather a little larger we may take total self-induction as 90,000 cm, or $\dfrac{9}{10^5}$ H. This would give

$$T = \frac{2\pi}{10^3}\sqrt{0.15264\times\frac{9}{10^3}} = \frac{2\pi}{10^5}\sqrt{0.137376} = \frac{6.28\times0.37}{10^5} = \frac{2.3236}{10^5} \text{ and } n = 43,000.$$

This would give a wave length approx. $\dfrac{186}{43}=4.33$ miles or $\dfrac{\lambda}{4}=1.08$ miles$=5700$ feet while there is a little more wire. Probably self-induction and capacity or either of them in the primary have been slightly underrated or else the wire is not quite as long as calculated before.

Today an other effort was made to ascertain more closely rate of increase of capacity of a body with the height above the ground. The same adjustable sphere was used and to eliminate some influences which might cause errors the special coil, which was rewound with same wire as before and had now 404 turns, the wire being wound a trifle higher, was connected to the first turn of secondary of oscillator, that is the turn first from the ground. This gave a small initial e.m.f. and reduced the error due to capacity very materially. (Here the distributed capacity is meant). Also, since the pressures developed were much smaller owing to small initial pressure on the special coil the streamers did not appear and did not therefore complicate the observations as in some previous cases. The plan of connections is clear from the diagram below. The object was specially to obtain a number of values which were as closely determined as possible and with the ball in positions entirely above the building. Only three values could be obtained owing to darkness setting in, but these seemed fairly close as the tuning was done over and over with same results. These were as follows:

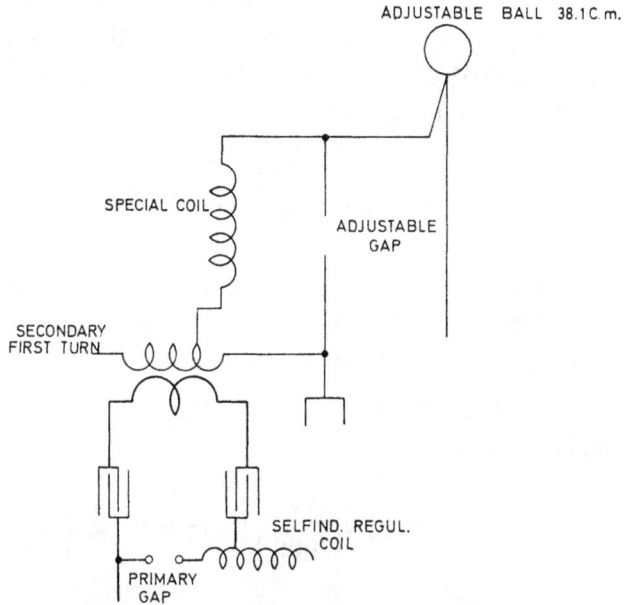

Height of ball above ground	Turns in Reg. coil	Total self-ind. in primary		Total capacity in pr. mfd.
35.66 feet	22	reg. coil 79,200 connections 6600 primary 56,400	142,200 cm.	8 tanks each side 0.15264
34.66 ,,	19	129,400 cm.		0.15264
33.66 ,,	16 (trifle less)	118,600 cm.		0.15264

Now since the capacity in the primary was the same the capacity in the secondary varied directly as the self-induction of the primary. The above figures show that from 33.66 feet to 34.66 feet, that is an elevation of one foot, the increase was 9.1%, while for the next foot higher it was nearly 9.9% on the average, say, 9.5%. At this rate the increase of capacity of the elevated sphere with the rise from the ground would be greater than before found. The absolute rate of increase can be approximately estimated from the period of vibration. As before found the special coil with 400 turns had a self-induction of 44,772,000 cm. Now, however, with 404 turns this would be increased about 1% so that the self-induction would now be $\left.\dfrac{44,772,000}{447,720}\right\}$ 45,219,720 cm. But to this should still be added the self-induction of one turn of the secondary and wire leading to the ball and also the wire leading from the bottom of the special coil to the first turn of the secondary. The total length of these three wires is 240 feet and this is about 12% of the total length of the wire in the special coil which is 2854 feet. But inasmuch as the one secondary turn was very close to the primary and inasmuch as the other two wires were not coiled up, the self-induction of these wires was comparatively small, estimated a little over 200,000 cm so that the total self-induction was with fair approximation: 44,500,000 cm, or about 0,0445 henry (calculated). Now, with a ball of 38.1 cm capacity the period of secondary would have to be:

$$T_s = \frac{2\,\pi}{1000} \sqrt{\frac{38.1}{9 \times 10^5} 0.0445} = \frac{2\,\pi}{3 \times 10^7} \sqrt{1695.45} = \frac{86.19}{10^7} \text{ and}$$

$n = 116,000$. This would be ignoring the internal capacity of the coil itself and its effect in slowing down the vibration. Now this capacity can be approximately estimated as well as the absolute increase of the capacity of the sphere from the primary vibration.

Taking the figures with the ball at a height of 33.66 feet from the ground, the primary vibration was:

$$T_p' = \frac{2\,\pi}{1000} \sqrt{0.15264 \times \frac{118,600}{10^9}} = \frac{26.7}{10^6} \text{ or} = \frac{267}{10^7}$$

This vibration is evidently slower than one of period T_s and it will be easily seen that

$$\frac{T_s}{T_p'} = \frac{86.19}{267} = \sqrt{\frac{38.1}{C+38.1}}$$

where C is the internal capacity of the coil. Assuming for the present this capacity as not being distributed along the coil but in one place, we get

$$C + 38.1 = 38.1 \times \left(\frac{267}{86.19}\right)^2 \quad \text{and} \quad C = \left[\left(\frac{267}{86.19}\right)^2 - 1\right] \times 38.1 \text{ cm}.$$

Following this up we get for C value $C = [(3.098)^2 - 1] \times 38.1 = (11.597 - 1) \times 38.1 = = 10.597 \times 38.1 = 403.75$ cm.

This is not the actual capacity of the coil but that ideal capacity by which the sphere should be increased to give the vibration of the primary.

We can now write

$$T_p' = \frac{2\,\pi}{10^3} \sqrt{0.0445 \times \frac{(403.75 + c)}{9 \times 10^5}} \qquad \ldots\ldots\ldots \text{ 1)}$$

This capacity c is now the actual capacity of the sphere at an elevation above considered, that is 33.66 feet. Namely, the total capacity was that of the sphere at that elevation — that is c — plus the ideal capacity of the coil derived from the computation of the primary vibration. This latter capacity has been called C. The inductance of the special coil being as before found, 0.0445 henry, and there being resonance under the conditions of the experiment at the elevation, named T_p', gives the value for the secondary period.

Now at another elevation, say 34.66 feet, we shall have similarly

$$T_p'' = \frac{2\pi}{10^3} \sqrt{0.0445 \times \frac{407.75 + c'}{9 \times 10^5}} \quad \dots\dots\dots\dots \; 2)$$

where c' gives the value of the capacity of the sphere for the elevation of 34.66 feet. Computing I find, from experimental data above given

$$T_p'' = \frac{2\pi}{10^3} \sqrt{0.15264 \times \frac{129{,}400}{10^9}} = \frac{2\pi}{10^6} \sqrt{19.75} = \frac{27.9}{10^6} \; \text{or} = \frac{279}{10^7}$$

Now, from equation 1 and 2, we get:

$$\frac{T_p'}{T_p''} = \frac{267}{279} = \frac{\sqrt{0.0445 \dfrac{403.75 + c}{9 \times 10^5}}}{\sqrt{0.0445 \dfrac{403.75 + c'}{9 \times 10^5}}} \quad \text{or} \quad (1.045)^2 = \frac{403.75 + c'}{403.75 + c}$$

and from this: $1.092 \times (403.75 + c) - 403.75 = c'$, but c being 38.1 we find for $c' = 78.75$ cm, which result shows that by lifting the ball from an elevation of 33.66 feet to 34. 66 feet or *one foot* higher the capacity has been increased from 38.1 to 78.75 cm, or nearly 106.7%. Similarly we find the increase from 34.66 to 35.66 feet by computing the primary vibrations at these elevations. By analogy to the previous we have

$$T_p''' = \frac{2\pi}{10^3} \sqrt{0.15264 \frac{142{,}200}{10^9}} \quad \text{and, since} \quad T_p'' = \frac{2\pi}{10^3} \sqrt{0.15264 \frac{129{,}400}{10^9}} \, ,$$

we get

$$\frac{T_p'''}{T_p''} = \sqrt{\frac{1422}{1294}} = \sqrt{1.099}$$

Now the vibration which corresponded to the primary vibration T_p'' in the special coil was $\dfrac{2\pi}{10^3} \sqrt{0.0445 \times \dfrac{403.75 + 78.75}{9 \times 10^5}}$ and the one which corresponded to T_p''' was

$$\frac{2\pi}{10^3} \sqrt{0.0445 \times \frac{403.75 + c''}{9 \times 10^5}} \, ,$$

where c'' will stand for the capacity of the sphere at the altitude of 35.66 feet.

And now we have:

$$\frac{T_p'''}{T_p''} = \sqrt{1.099} = \sqrt{\frac{403.75 + c''}{403.75 + 78.75}} \quad \text{and from this}$$

$$1.099 \times (403.75 + 78.75) - 403.75 = c'' = 126.51 \text{ cm.}$$

The value at one foot lower was, as before found 78.75 cm, therefore by lifting the sphere from 34.66 to 35.66 feet, the capacity was further increased by 126.51—78.75 = = 47.76 cm, or about 125%. The value which would correspond to the mean would therefore be about 116% per foot. The method followed contains still some possible errors. One of them lies in the assumption that the capacity of the sphere was 38.1 cm at the starting point. Also there may be an error in the estimation of self-induction of the turns of the regulating coil.

Colorado Springs

Oct. 10, 1899

Resistances measured:

Large extra coil	with cord	3.7 ohms
149 t. wire No. 10	cord	0.596 ohms
drum 75"	Coil alone	3.104 ohms

Coil used in series with extra coil. When ball was not used on top of latter:

160 t. No. 10 wire	with cord	1.65
drum 2 feet	cord	0.596
	Coil alone	1.054 ohms

Resistance of coil used in determining influence of elevation on capacity:

400 turns No. 20 cord	with cord	31.20
drum 25.25"	cord	0.596
	Coil alone	30.604 ohms

Resistance of secondary latest:

	with cord	3.36 ohms
	cord	0.596 ohms
	Secondary alone	2.764 ohms

Photographs of streamers were taken late last night again and at an early hour this morning under the following conditions.

First two plates exposed to ten flashes, 1/2 second duration each. These flashes issued from the tip of rubber-covered cable or wire No. 10 which was on top of the extra coil. The tip was inclined about 45° degrees to the vertical and pointing downward. The full front view was taken. A curious observation was made. One of the large streamers, about 22 feet long, disappeared at that length for a space of about a foot and continued again after that for a distance of about 2 feet, so that the total length of it was about 25 feet with a dark interval of one foot. Evidently, the current passed for a distance of a foot through air or dust particles which were better conducting and the path was of a greater section in all probability. Perhaps the air on that spot might have been electrified in such a way as to produce the phenomenon.

The next experiment was made with an exposure of two plates to about forty flashes, the view being the same as before.

After this two plates were exposed to but a single short flash about one second, the view being still the same as before. Now a round sheet zinc disk was fastened to the tip of the wire and two plates were again exposed, there being about twenty flashes. Next, the coil was turned and a side view taken with about forty flashes, two plates being again exposed as in all previous cases, two cameras being used for the sake of safety. Upon this the zinc disk was taken off and a ball of 4″ diam. fastened to the tip of the No. 10 rubber-covered wire. A long exposure of about 50 flashes was again made. The streamers were as expected a little stronger from the ball then from the point as the breaking out took place at a greater pressure. Two plates were used also in this instance.

The next experiment consisted of taking an impression on two plates of the secondary alone in resonating condition. The phenomenon was beautiful to an extraordinary degree. Not only did the top wire glow but from the under wire (turn next below) a steady sheet of streamers of very fine texture issued of an area which must have been many hundred square feet. The free end of the secondary had the ball of 38.1 cm connected to it at a distance of 32″ from the ground but owing to the large radius of curvature sparks did not leap from the ball to the grounded zinc plate below though the distance was small comparatively. During the experiments a short but thick stream issued from the free terminal of the extra coil which had its other end connected to the ground and was thus excited through the vibration of the secondary, having, as stated on a previous occasion, the same period of vibration. The color of the light issuing from the secondary wires, particularly in the neighbourhood of the condensers, was remarkably blue, and should affect the plate strongly, though the intensity was not great. The switch was thrown in fifty times, the duration being about 1/3 of one second, possibly 1/2.

The last experiment consisted of establishing the resonance of the extra coil and secondary in series connected and with the ball of 38.1 centimeter on the free end of the extra coil, the ball being at its lowest position, 20.66 feet from the ground. The other ball of the same size remained as before, connected to the end of the secondary where it was joined to the lower end of the extra coil. As there was great danger of inflaming the roof the power was somewhat reduced, but the display was wonderful in spite of this. This was

Experimental Laboratory, Colorado Springs in early phase of development

Conical secondary high frequency transformer and "extra coil" in action

the most significant experiment showing streamers from the ball of 38.1 cm capacity from which is evident the enormous tension, as well as the inconceivable rate at which the energy was delivered in the vibrating system. Forty flashes were made and afterward the background was illuminated by the arc of the primary circuit to complete the picture.

Colorado Springs

Oct. 12, 1899

Measurement of inductance of 404 turns coil used in determining influence of elevation.

$l=57.125''=145.1$ cm diam. of drum $25.25''=64.14$ cm.

Area of one loop 3231 square cm.

Calculated before for 400 turns when the wire was wound first a little less tight, the value was:

$$L=\frac{4\pi N^2 S}{l}=44,772,000\,\text{cm}.$$

After making corrections for l which upon more careful second measurement was found to be 46 7/8'' only, and taking $N=404$ the value was $L=45,670,000$ cm. The actual measurements gave the following data:

$$I=4.28 \qquad E=194 \qquad R=30.604 \qquad \omega=880 \qquad \frac{E}{I}=45.327$$

and from this

$$L=\frac{\sqrt{45.327^2-30.604^2}}{880}=\frac{33.435}{880}=0.03788\,\text{H}$$

or $L=37,880,000$ cm.

The measured value should be smaller than the calculated but not so much. The internal capacity may be responsible but very likely the current was not quite exactly measured. Corrections to be taken after calibration.

Colorado Springs

Oct. 13, 1899

Measurement of inductances of extra coil and secondary, latest design:
Readings for extra coil:

$$I=5.9 \qquad E=119 \qquad R=3.104 \qquad \omega=880$$

$$\frac{E}{I} = \frac{119}{5.9} = 20.17, \text{ from this}$$

$$L = \frac{\sqrt{20.17^2 - 3.104^2}}{880} = \frac{19.93}{880} = 0.022648 \text{ H}$$

or

$$L = 22,648,000 \text{ cm.}$$

Readings for secondary single wire; latest:

$$I = 6.77 \qquad E = 60 \qquad R = 2.764 \qquad \omega = 880$$

$$\frac{E}{I} = \frac{60}{6.77} = 8.8626 \text{ and}$$

$$L = \frac{\sqrt{8.8626^2 - 2.764^2}}{880} = \frac{\sqrt{70.9060}}{880} = \frac{8.42}{880} = 0.0095682 \text{ H}$$

or $L = 9,568,200$ cm.

Colorado Springs

Oct. 14, 1899

Determination of the natural period of the secondary with and without capacity on free terminal, also of the extra coil and coil of 404 turns used in investigations before described.

The tuning was effected in a manner later to be more fully dwelt upon, which secured closer readings than when exciting, as in some experiments before made, by the primary current or from a turn of the secondary of oscillator.

The excitation was effected in these tests by connecting directly one of the terminals (the lower) of the coil to be tested to one of the terminals of the primary condensers (the one connected to the tank of W. Transformer).

The results of the test are given below:

For secondary of oscillator:

Capacity in primary circuit Total	Self-ind. in primary Turns of Reg. coil + conn.	Observation:
$\frac{(8 \times 36) - 2}{2}$ bottles = 0.126 mfd approx.	15 + conn.	No capacity on free terminal

$$\frac{(8 \times 36) - 2}{2} \text{ bottles} = 0.126 \text{ mfd} \qquad 15 \ 1/4 + \text{conn.} \qquad \text{Ball of 38.1 cm diam.}$$

approx.

$$\frac{(8 \times 36) - 2}{2} \quad ,, \quad == \quad ,, \qquad 19 \ 1/4 + \ ,, \qquad$$ Structure to be described later of iron pipes with ball of 38.1 cm. diam. at an elevation of 141 feet from ground approx.

For extra coil:

$$\frac{(8 \times 36) - 2}{2} \quad ,, \quad = \quad ,, \qquad 19 \ 1/4 + \ ,, \qquad$$ with above structure connected to free terminal.

For coil 404 turns used in preceding investigations:

$$\frac{(6 \times 36) - 2}{2} \text{ bottles} = 0.11342 \text{ mfd.} \qquad 8 + \text{conn.} \qquad$$ Tuning very sharp, no capacity on end.

Remark: With excitation from secondary (first turn) obtained in a previous test of this coil:

$$\frac{(5 \times 36) - 2}{2} \text{ bottles} = 0.094255 \text{ mfd.} \qquad 17 \ 1/2 + 2 \text{ primary turns} + \text{connect.} \qquad$$ Tuning was not quite sharp.

Colorado Springs

Oct. 15, 1899

Continuing the considerations made Oct. 9 on the influence of elevation upon the capacity and taking, instead of the calculated value of the self-induction of the coil used in the experiments, the value ascertained by experiment which, with corrections for one turn secondary and connecting wires, may be put at 0.04 henry, we have, assuming now the coil to have a capacity entirely negligible:

$$T_s = \frac{2\pi}{1000} \sqrt{0.04 \times \frac{38.1}{9 \times 10^5}},$$

as secondary period of the coil with ball or

$$T_s = \frac{0.8164}{10^5} = \frac{81.64}{10^7} \quad \text{Now } T_p \text{ was} = \frac{267}{10^7}, \text{ from this}$$

$$\frac{T_s}{T_p} = \frac{81.64}{267} = 0.306 = \sqrt{\frac{38.1}{38.1+C}} \quad \text{from previous analogy — and from this}$$

$$0.306^2 = \frac{38.1}{38.1+C} \text{ or } C = \frac{38.1}{0.09364} - 38.1 = \frac{38.1 \times 0.90636}{0.09364} = 38.1 \times 9.68 = 368.8$$

Now

$$T_p' = \frac{2\pi}{10^3}\sqrt{0.04 \times \frac{368.8+c}{9 \times 10^5}} \text{ and } T_p'' = \frac{2\pi}{10^3}\sqrt{0.04 \times \frac{368.8+c'}{9 \times 10^5}}$$

but $T_p'' = \dfrac{279}{10^7}$ and from this

$$\frac{T_p'}{T_p''} = \frac{267}{279} = \sqrt{\frac{0.04 \times \dfrac{368.8+c}{9 \times 10^5}}{0.04 \times \dfrac{368.8+c'}{9 \times 10^5}}} \quad 1.045 = \frac{279}{267} \text{ and from this}$$

$$(1.045)^2 = \frac{368.8+c'}{368.8+c} \quad \text{now } c = 38.1$$

$$1.092 \times (368.8+38.1) = 368.8+c' \text{ and}$$

$$c' = 368.8 \times (1.092-1) + 38.1 \times 1.092 = 75.53 \text{ cm.}$$

This value is but little smaller than that before found. From this result it would then appear that by lifting the sphere from 33.66 to 34.66 feet the capacity was increased from 38.1 to 75.53 cm, or approximately 98.3%, a trifle less than before found.

Colorado Springs

Oct. 16, 1899

Inductance of secondary modified by winding another wire No. 10 in multiple with the first. All other particulars remaining:

Readings: $I = 7.1$ $E = 60.5$ $R = 1.382$ $\omega = 880$

$$\frac{E}{I} = 8.521$$

$$L = \sqrt{8.521^2 - 1.382^2} = \frac{\sqrt{70.6975}}{880}$$

$$L = \frac{8.41}{880} = 0.0095568 \ H$$

or

$$L = 9{,}556{,}800 \ \text{cm.}$$

This is closely agreeing with first measurement of a few days ago, the difference being only 1/10%.

Colorado Springs

Oct. 17, 1899

Structure for capacity of extra coil, for investigation of earth vibrations chiefly.

This structure was erected on a pole $10'' \times 10''$ square, tapering on top. Dry fir was used because of toughness and also resinous quality. Pipes of steel of diameters 7″, 6″, 5″, 4″, 3 1/2″, 3″, and 2 1/2″ were used. They were shoved one into the other and riveted, four rivets were used on each joint, the lap being 2′.

The lengths of pipes were as follows:

7″	diam.	23′ 4″
6″	,,	18′ 2″
5″	,,	18′ 4 1/2″
4″	,,	15′ 7 1/4″
3 1/2″	,,	19′ 3 1/4″
3″	,,	18′ 4 3/4″
2 1/2″	,,	8′ 1/4″
Nipples on top		7 3/4″
Total length of pipes		121′ 9 3/4″
Firwood pole		19′ 3″
Total from floor		141′ 3/4″
to ground		1′
Total height from ground to bottom of ball		142′ 3/4″

On the top was supported a ball of 30″ diam. hollow wood covered with tin foil very smoothly and the joints indented so as to have no conducting points sticking out. The joints of the pipes, heads of rivets, etc. were all covered first with sheet rubber pure and then with tape, the latter being finally fastened with strong cord. The ball was shellaced several times and finally covered with weatherproof rubber paint. The pole all along was also painted with the same paint. On the lowest end of 7″ pipe a cap was screwed clearing the wood so as to make it more difficult for the streamers to get to the ground along the pole.

* To prevent lateral play 8 champagne bottles set in beams were used.

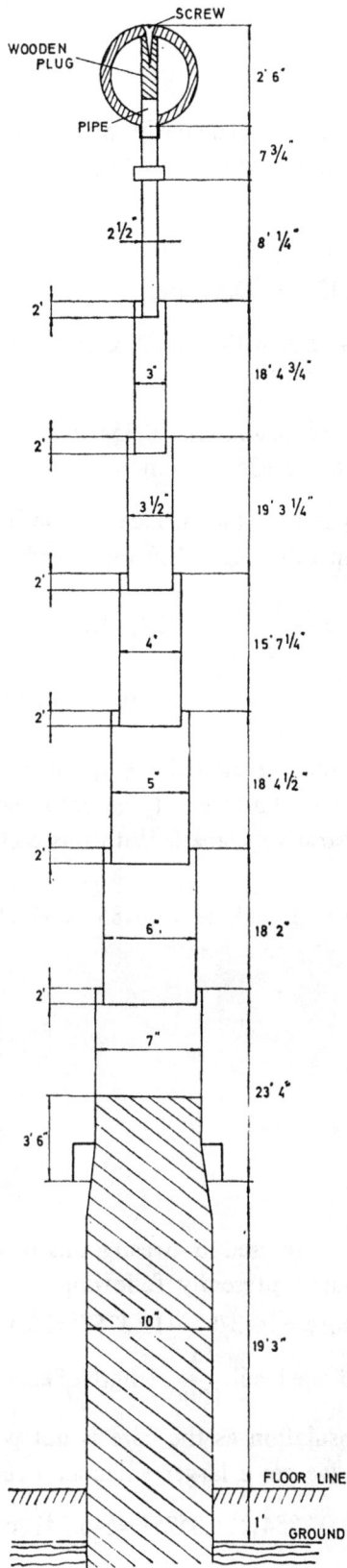

SCREW

WOODEN
PLUG

PIPE

2' 6"

7 3/4"

2 1/2"

8' 1/4"

2'

3"

18' 4 3/4"

2'

3 1/2"

19' 3 1/4"

2'

4"

15' 7 1/4"

2'

5"

18' 4 1/2"

2'

6"

18' 2"

2'

7"

23' 4"

3' 6"

10"

19' 3"

FLOOR LINE

1'

GROUND

Oct. 18, 1899

For special investigations particularly to prosecute further researches on the increase of capacity with elevation, a new coil is to be constructed as nearly as possible to exact dimensions which follow:

diam. of core 14″=35.5 cm
length of core 8×12″=243.84 cm.

The coil is to be wound with cord No. 20, before used in the coil of 404 turns on 25.25″ diam. drum.

Allowance for thickness of insulation: 0.354 cm.
This makes total diameter d=35.854 cm.

For approximate value of the surface of one loop we have then S=1000 cm. sq. There should go on this length 689 turns and this would make

$$L = \frac{4\pi}{l} N^2 S = 24{,}490{,}000 \text{ cm. approx.}$$

or
$$L = 0.02449 \text{ henry.}$$

For the purposes contemplated this coil will be well adapted as the self-induction is large and owing to small diameter of the core and great thickness of insulation the capacity should be comparatively small. With this coil balls of 18″ and 30″ diam, are to be used:

Exact diam. of core should be 14.0485″=35.6825 cm. for S=1000 cm. sq.

Oct. 19, 1899

Experimental coil before used in investigating propagation of waves through ground. This coil was rewound and presently following particulars are good.

Outside diam. of core=10 3/8″=10.375″=26.3525 cm. Turns 550, No. 18 B. & S. wax covered, thickness of one insul $\frac{14″}{1000}$, length of core l=40 7/8″=40.875″=103.8225 cm.

We may neglect insulation as the core is not perfectly round and any irregularity diminishes the area and we get a larger value as a rule. Now, from above data, surface of one loop $S=\frac{\pi}{4}d^2$=0.7854×26.3525²=545.241 cm. sq. This gives

$$L = \frac{4\pi}{l} N^2 S = \frac{12.5664 \times 302{,}500 \times 545.42}{103.8225} = 19{,}970{,}000 \text{ cm, approx. or } 0.01997 \text{ H}$$

This coil is now to be also used in investigations of the variation of capacity with variation of height. With ball of 18″ diam., the capacity of the ball assuming to be: [18″= =18×2.54=45.72 cm]—C=22.86 cm, the period of the system, neglecting for the present the distributed capacity of the coil, would be approximately:

$$T = \frac{2\pi}{10^3} \sqrt{0.01997 \times \frac{22.86}{9 \times 10^5}} = \frac{0.447345}{10^5}$$

or $n = 223,540$ per sec. approx.

In a test resonance was obtained with 42 bottles on each side of the break, that is 21 bottles total or 0.0223 mfd capacity in the primary, which consisted of 6 turns of heavily covered wire No. 10.

As it was inconvenient to measure the primary, L_p was estimated to have a self-induction of:

$$\left[\frac{0.447345}{10^5} = \frac{2\pi}{10^3} \sqrt{0.0223\, L_p} \right], \quad L_p = 22,435 \text{ cm.}$$

This is a convenient and good method but the period must be exactly ascertained. The distributed capacity can never be neglected.

Colorado Springs

Oct. 20, 1899

To ascertain the effect of internal capacity of the coil of 404 turns repeatedly referred to, which was used in the experiments on influence of elevation upon capacity of a conductor, the coil was tuned alone with only a short length of wire attached to it. Resonance with primary was obtained with 118 bottles on each side, one primary turn. For future

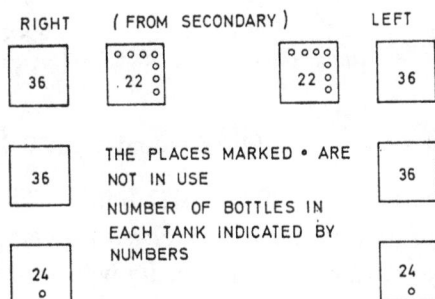

RIGHT   (FROM SECONDARY)   LEFT

| 36 | .22 | 22 | 36 |

36	THE PLACES MARKED • ARE	36
	NOT IN USE	
	NUMBER OF BOTTLES IN	
24	EACH TANK INDICATED BY	24
	NUMBERS	

reference the tanks as used are indicated in diagram. The total capacity in primary was therefore $\frac{118}{2} = 59$ bottles. Since one tank, that is 36 bottles, equals 0.03816 mfd the capacity was $\frac{59}{36} \times 0.03816 = 0.06254$ mfd.

Now the self-induction in the primary was: primary one turn+connections=56,400+
+6600=63,000 and with allowance for one half turn of self-induction coil, we may say
approx. $\dfrac{64}{10^6}$ henry.

The period was

$$T_p = \frac{2\pi}{10^3}\sqrt{\frac{64}{10^6} \times 0.06254} = \frac{8 \times 2\pi}{10^6}\sqrt{0.06254} = \frac{4\pi}{10^6} = \frac{12.5664}{10^6}$$

and from this $n=79,600$ per second. The secondary was vibrating in synchronism hence
$T_s=T_p=\dfrac{2\pi}{10^3}\sqrt{0.04 \times C}$, where C denotes capacity which could be associated with the
coil, if it had no capacity of its own, and this capacity together with the self-induction of
the coil of 0.04 H as before found would give the period $T_s=\dfrac{4\pi}{10^6}=\dfrac{2\pi}{10^3}\sqrt{0.04\,C}$. From
this follows

$$C = \frac{1}{10^4}\ \text{mfd. or}\ \frac{9 \times 10^5}{10^4} = 90\ \text{cm.}$$

This is the value experimentally found and more to be relied upon than that derived
in other ways. Taking this value and adding to the coil a sphere of radius 38.1 cm the
total capacity of the system would be $90+38.1=128.1$ cm and now we should find the
period of the system to be

$$T_s' = \frac{2\pi}{10^3}\sqrt{0.04 \times \frac{128.1}{9 \times 10^5}} = \frac{4\pi}{3 \times 10^6}\sqrt{12.81} = \frac{4\pi}{3 \times 10^6} \times 3.58 = \frac{14.9958}{10^6}$$

or $\dfrac{15}{10^6}$ approx. $=T_s'$ or $n=66,666$ per sec.

We can now find the capacity in primary required for resonance with secondary.
The primary period will be $T_p'=\dfrac{2\pi}{10^3}\sqrt{\dfrac{64}{10^6}C_p}$, where C_p is the capacity required in

primary. Now we have then $T_p'=T_s'=\dfrac{2\pi}{10^3}\sqrt{\dfrac{64}{10^6}C_p}=\dfrac{4\pi}{3 \times 10^6}\times 3.58$ (from above) and

$$\sqrt{\frac{64}{10^6}C_p} = \frac{2 \times 3.58}{3 \times 10^3}\ \text{or}\ \sqrt{C_p} = \frac{3.58}{12}\ \text{or}\ C_p = \left(\frac{3.58}{12}\right)^2 = 0.096\ \text{mfd.}$$

Since one bottle approximates $\dfrac{0.03816}{36}$ mfd we would want $0.096 : \dfrac{0.03816}{36}=90.6$ bottles
total or, on each side of the break $2\times90.6=181.2$ bottles. This would suppose that the
ball could have a capacity of 38.1 cm but this is physically impossible since it would have
to be removed from all objects. In reality its capacity will be always much greater and the
system will vibrate much slower as a rule. In fact, the test showed that resonance was
attained with the ball 10—18 feet from the ground with all bottles in and in addition 5 1/2
turns in the self-induction box. The tuning was naturally not sharp as the capacity was
large and the maximum appeared to be with the ball 10—18 feet from the ground. To
get sharper tuning a smaller ball of 18″ diameter was used and now resonance was obtained
with the ball 10 feet from the ground and all bottles in and no self-induction in the regulating coil.

To ascertain the value of connections in terms of turns of the regulating coil and for other purposes the coil before described, 404 turns with rubber wire attached, was tuned for three different values of the regulating turns. The method was one frequently used in New York and is illustrated in the plan shown. One end of the coil was connected to that end of condensers, or respectively to that end or terminal of transformer which was grounded and the rise of potential of free terminal of tuned coil was observed by an adjustable spark gap. In this manner very close adjustments are easy. The resonating condition indicated by the longest sparks in the adjustable gap was secured with the following three values:

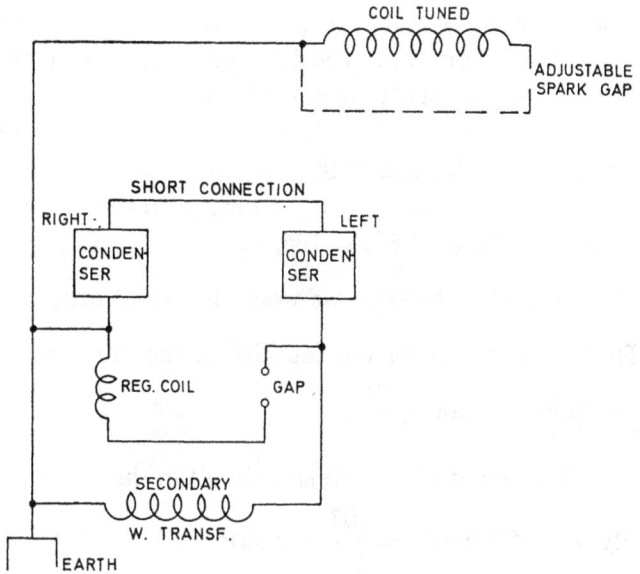

Capacity in the primary in bottles on each side

Turns in regulating coil

total

8 tanks less 2 bottles	=286=0.15157	16.5	The tuning was not
7 less 2 „	=250=0.13249	18=L_1	quite as sharp in the
6 „ 2 „	=214=0.11341	21=L_2	first case,

but it was very sharp in the last two experiments. Taking these and calling x the self-induction of the connections we would have in the case corresponding to L_1 of the regulating coil $T=\dfrac{2\pi}{10^3}\sqrt{0.13249\,(L_1+x)}$ and in the case corresponding to L_2 of the self-induction coil

$$T=\frac{2\pi}{10^3}\sqrt{0.11341\,(L_2+x)}.$$

From this $0.13249\,(L_1+x)=0.11341\,(L_2+x)$ and

$$x=\frac{0.11341\,L_2-0.13249\,L_1}{0.01908}$$

Now L_1 and L_2 should be exactly measured; resonance method will probably suit the purpose best. This is to be more rigorously carried out.

In the experiments before described, for the purpose of ascertaining the influence of elevation on the capacity of a sphere, the latter was connected to the coil of 404 turns by means of a wire No. 10 heavily covered with rubber (3/8" wall). Evidently this wire affected the period of the system and to ascertain to what extent, the wire was placed in the same position as when used in the experiments with the ball at its lowest position — 20.66 feet from the ground. The sphere was omitted but the streamers on the end were prevented by sealing the end with wax, covering with tape and sticking the end of the wire into a glass bottle with very heavy walls. Resonance was attained with (6×36)—2 bottles on each side and 21 turns in the self-induction regulating coil. This was 214 bottles on each side or 107 total. In making the test the primary turn of the oscillator, usually connecting the two coatings of the condensers on the bottom, was left off and the coatings joined by a short wire. The total self-induction in the primary circuit was therefore the 21 turns of the regulating coil plus the connections, or from previous figures $100,800+$ $+6600=107,400$ cm. Now in the test of Oct. 9, with the ball connected to the cable, resonance was obtained with 4.66 tanks on each side or $4.66 \times 36 = 168$ bottles approx. The total self-induction was 21.5 turns of regulating coil $+1$ turn primary $+$ connections $=$ $=103,200+56,400+6600=166,200$ cm, both values from calculated data. Had the self-induction been the same we would have had instead of 168 bottles $\dfrac{166,200}{107,400}=260$ bottles.

Thus with the sphere the capacity in the vibrating secondary system was increased by the ratio $\dfrac{260}{214}$ or $\dfrac{130}{107}$.

The period of the primary circuit in the first case can now be ascertained. The capacity was 107 bottles $=\dfrac{107}{36} \times 0.03816 = 0.11342$ mfd. The self-induction as before stated 107,400 cm. or 0.0001074 H. The period therefore was $\dfrac{2\pi}{1000}\sqrt{0.0001074 \times 0.11342}$. Now the secondary vibrated with the rubber wire attached, the same period which was $\dfrac{2\pi}{1000}\sqrt{0.04 \times C}$, designating by C the ideal capacity associated with the self-induction of the coil. These two were equal and we have $C=\dfrac{0.0001074 \times 0.11342}{0.04}$ mfd, or

$$\frac{10.74 \times 0.11342 \times 9}{0.04} = 30.45 \times 9 = 274 \text{ cm.}$$

This was the capacity with the cable comprised. But before we have found the capacity of the coil alone, with no wire attached, 90 cm or nearly so. The addition of the rubber cable made therefore a considerable difference. It would not have been so high had the streamers been entirely prevented but despite the wax and glass bottle there was a leak which had the same effect as if the capacity had been increased. Since by adding the ball the capacity was increased by the ratio $\dfrac{216}{214}$ we have — calling now c the capacity of the sphere at the initial elevation of 20.66 feet — $\dfrac{260}{214}=\dfrac{274+c}{274}$ and from this

$$c=\frac{274 \times 46}{214} = 58.88 \text{ cm.}$$

This then would be, according to this estimate, the actual value of the capacity of the sphere at that elevation.

Oct. 22, 1899

One of the upper terminals of the condenser (+) usually connected to the ground was joined to the lower end of the coil of 689 turns, the upper end remaining free. The ground was in this case omitted for the purpose of securing a higher initial excitation. The maximum rise was ascertained by an adjustable spark gap as shown. The results of the tests are given in the table below:

Capacity in primary circuit expressed by the number of bottles used	Self-induct. of primary or exciting circuit Turns in regul. coil+connections	Spark length on terminals of the coil	Observation relating to spark
$\dfrac{(8 \times 36) - 2}{2}$	1.125+connections	4 1/8 inch	
$\dfrac{(7 \times 36) - 2}{2}$	2.375+ ,,	4 3/4 ,,	The spark was continually increasing
$\dfrac{(6 \times 36) - 2}{2}$	3.625+ ,,	5 5/8 ,,	and the excitation was reduced as the capacity was getting smaller
$\dfrac{(5 \times 36) - 2}{2}$	4.75 + ,,	5 3/4 ,,	
$\dfrac{(4 \times 36) - 2}{2}$	6.25 + ,,	5 7/8 ,,	reduced
$\dfrac{(3 \times 36) - 2}{2}$	8.25 + ,,	6 1/4 ,,	still more
$\dfrac{(2 \times 36) - 2}{2}$	11.625+ ,,	7 ,,	still more
$\dfrac{(1 \times 36) - 2}{2}$	22 + ,,	9 ,,	still more
appr. $\dfrac{31}{2}$	24 + ,,	about same	

The spark was getting longer because the efficiency of the exciting circuit was increased, as the inductance of this circuit was increased and capacity diminished. There were smaller frictional losses and after each break the system vibrated longer and excited the coil better.

The short stout wire was now substituted by each of the two primary turns separately and joined in multiple with the results indicated below:

$\frac{32}{2}$ 13 1/4 turns+1 primary+connections   The primary used was the upper one.

$\frac{32}{2}$ 14 „ +1 „ + „   „   „   lower one.

$\frac{32}{2}$ 14 1/2 „ +1 „ + „   The tuning was very sharp in the first two cases, not quite so in the last instance.

In order to get useful data as to the self-induction of the connections and also of the various turns of the regulating coil which were most frequently used in the experiments, tests were made as follows. A coil still to be described, built for a special purpose, was used (689 turns, drum 14″ diam., 8 feet long) and was excited from one terminal of the condensers, as indicated in the sketch below. The coil had a definite period which was ascertained with all condensers in and the least possible self-induction; the condensers were taken out and more turns of the self-induction coil inserted until resonance was again attained. Since the period was in all instances the same the self-inductance of the circuit was thus varied inversely as the capacity. When all self-inductance or nearly so was taken out and only the connections remained by a simple ratio between the known capacities and a known inductance, the inductance of the connections was given, or else this quantity was ascertained from the known period which was maintained throughout the experiments (that of the coil before referred to). The lower ends of the condensers, usually joined by the primary, were connected by a short stout conductor of inappreciable resistance and inductance.

The method used in the experiments recorded today for determining experimentally the inductance of the connections is very convenient and secures good results. The coil

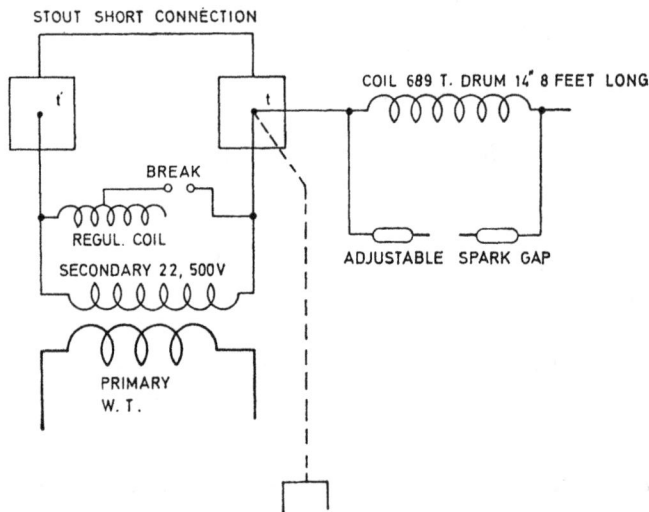

used in the experiments was one of very high self-induction to make the tuning very sharp and it was wound on a drum of relatively small diameter to reduce internal distributed capacity. This likewise improves the sharpness of the adjustment. It was easy to detect variations of one sixteenth of one turn of the regulating primary coil. From the preceding data, calling l the inductance of one turn and l_1 that of 22 turns, and C the capacity in the primary when 1.125 turns were used and C_1 that when 22 turns were employed, we have, since the period was the same:

$$(l \times 1.125 + \text{connections}) \, C = (l_1 + \text{connections}) \, C_1.$$

Now it is not necessary to determine C and C_1 since only the ratio is needed and we may simply take the number of bottles in each case. This gives:

$$(l \times 1.125 + \text{conn.}) \times 143 = (l_1 + \text{conn.}) \times 17.$$

Now in a previous instance l and l_1 were approximately: $l = 4800$ cm, $l_1 = 105,000$ cm. Substituting these values we have:

$$(5400 + \text{conn.}) \times 143 = 17 \times (105,000 + \text{conn.})$$

$$\text{and} \quad \text{connect.} = \frac{(17 \times 1050 - 143 \times 54) \times 100}{126}$$

From this the inductance of the connections would be $= 8040$ cm, or **8000** cm. approx. It would be desirable, however, to eliminate the turns of the coil and so estimate the inductance of the connections directly.

Colorado Springs

Oct. 23, 1899

Experiments to further ascertain the influence of elevation upon capacity.

The coil referred to on a previous occasion was finished with exactly 689 turns on a drum of eight feet in length and 14″ diam. The wire used was cord No. *20* as before stated so that the approximate estimate of self-induction and other particulars holds good. The coil was set up upright outside of the building at some distance to reduce any errors due to the influence of the woodwork. From the building extended a structure of dry pine to a height of about sixty feet from the ground. This framework supported, on a projecting crossbeam, a pulley (wood) with cord for pulling up a ball or other object to any desired height within the limits permitted and this beam also carried on its extreme end and close to the pulley a strong glass bottle within which was fastened a bare wire No. 10, which extented vertically downward to the top of the coil. The bottle was an ordinary Champagne bottle, from which the wine had been poured out! and the bottom broken in. It was forced neck downward into a hole bored into the beam and fastened besides with a cord. A tapering plug of hard wood was wedged into the neck and into this plug was fastened the wire. The bottle was finally filled with melted wax.

The whole arrangement is illustrated in the sketch shown in which b is the bottle with wooden plug P supported on beam B also carrying pulley p, over which passes the cord for pulling up the object, which in this case is shown as the sphere C. The spheres used were of wood and hollow and covered very smoothly with tin foil and any points of the foil were *pressed in* so as to be *below* the surface of the sphere. This is a necessary precaution to avoid possible losses by streamers when the sphere is charged to a high potential. It is desirable to work with strong effects as the greater these are the better the vibration can be determined, but it is necessary to carefully avoid losses and errors owing to the formation of streamers. These have the effect of increasing the apparent capacity so much that a thin wire may often produce results comparable to those obtained with large capacities. The streamers, of course, cause frictional loss and thereby diminish the economy of the system and impair the quality of the results. They also cause a loss of pressure just as leaks in air on water pipes. The ball to be tested or any other object was provided with two metallic bushings, on contacts $c\,c$, consisting of small split brass tubings which were in good connection with the conducting surface and also insured contact

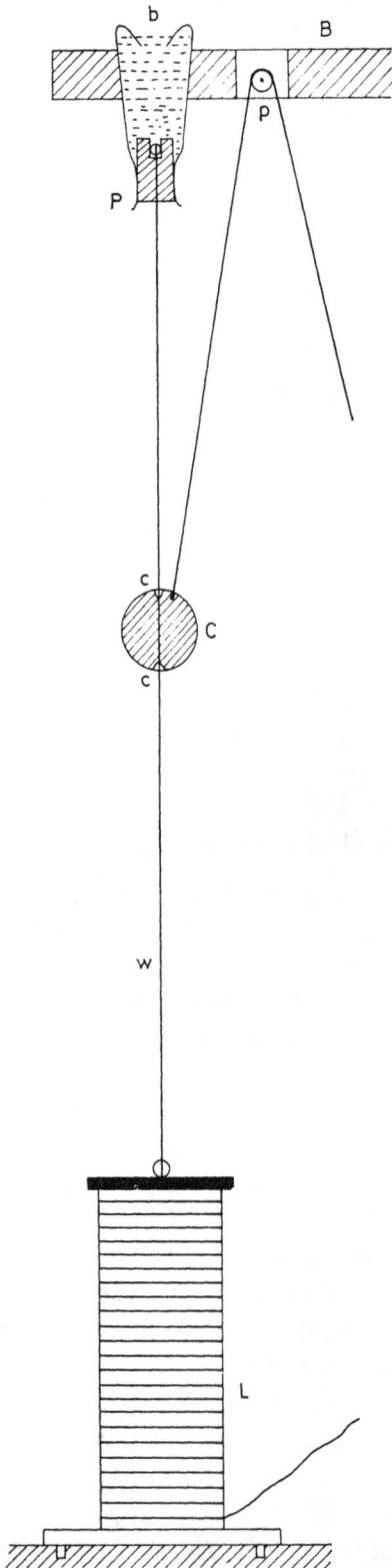

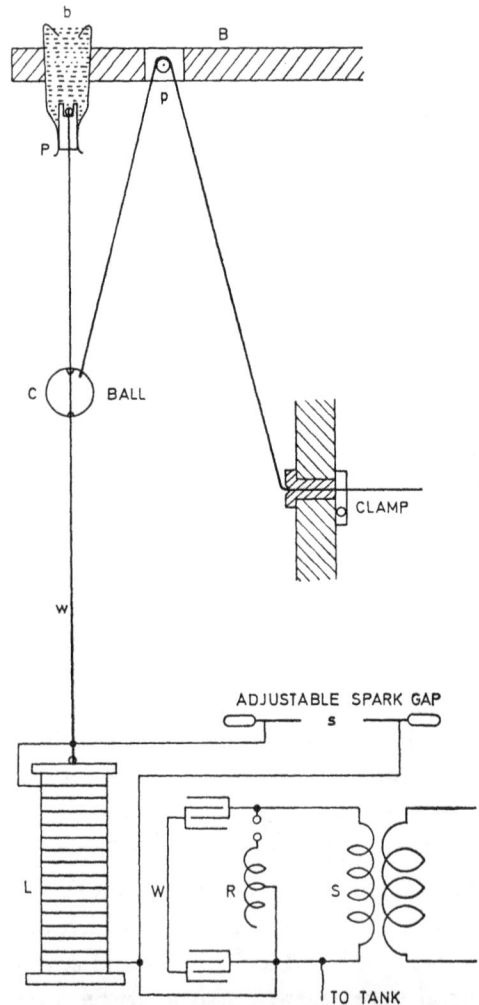

with wire w which at the same time served to guide the ball in its up and down movement. To avoid losses the bushings did not project beyond the surface of the ball and for the same reason the cord was not fastened to a hook but a hole was drilled into the ball, the cord with a knot on the end was slipped in and a wooden plug driven into the hole, so that nothing was sticking out capable of giving off streamers or causing leaks into the air. In the first series of experiments a ball of 18″ diameter was used. The ball was not perfectly round but the error due to a slight irregularity of shape was very small. The plan of connections is shown in the sketch in which the same letters are used to designate the same parts of apparatus as before. The excitation of the coil was effected by connecting the lower end to one of the terminals of the condenser — the one which was connected to that end of the secondary of the 60,000 volt transformer which was in connection with the tank. The tank, as described on a previous occasion, was usually connected to the ground but in these first experiments the ground connection was omitted to secure stronger excitation. From the terminals of the coil two thin and heavily insulated wires were led to an adjustable spark gap s which was manipulated until the maximum rise of potential on the coil was ascertained. The two sets of condensers were joined by a stout short wire W of inappreciable self-induction and resistance and the inductance of the exciting circuit was varied by inserting more or fewer of the turns of the regulating coil R into the circuit through which the condensers were periodically discharged. The wires leading from the coil to the adjustable spark gap s were, as before remarked, very thin, as short as it was practicable to make them and heavily insulated. By observing these precautions the error due to the capacity of these wires themselves was reduced to a minimum, also the loss owing to a possible formation of streamers. To reduce the capacity the wires were led far apart and then brought in line to the spark gap. The lower wire, which was connected to a point of comparatively low potential was of little consequence but on the top wire these precautions were imperative. The procedure was as follows: first the period of the coil L and capacity attached to the free terminal was ascertained by varying the capacity or self-induction, or both, of the primary or exciting circuit until resonance was reached which was evident from the maximum rise of potential. When the period had thus been determined with the capacity, say a sphere, in one position, the position of the body of capacity was varied by shifting it to another place along the wire w and the adjustment of the primary circuit was again effected until resonance was reached, generally by simply varying the length of wire of the regulating coil included in the primary circuit. Now as the self-induction of the coil L remained the same through all experiments, the apparent capacity could be easily determined from the self-induction and the known period of the primary or exciting circuit. By keeping the capacity in the primary circuit the same or, eventually, the self-induction, the procedure was simplified and the capacity in the system including coil L was then at once given by a simple ratio, as in some cases previously described. It was preferable to vary the self-induction as the change of this element could be effected continuously and not step by step, as was the case with the capacity.

The apparatus being arranged as stated, the lowest position of the ball of 18″ diam., which was first used, was 9′ 5″ from center of ball to ground and the highest 58′ 9″.

To ascertain the period of the system L the vertical wire was first disconnected and only the spark wires left on, then the vertical wire was connected and the period again determined by adjusting the primary circuit, then the ball was placed in its lowest position and finally readings were taken with the ball at various heights up to the maximum elevation. The results condensed were as follows:

Coil *L* alone only with spark wires, no vertical wire:

Capacity total in primary circuit	Total primary self- -induction turns of regulating coil+conn.	Height of ball from center to ground	Analyzing spark at *s*
$\dfrac{142}{2}$=71 bottles	9 7/8 turns+connections	*no ball, no wire*	3 3/4"

Coil *with* vertical wire and spark wires:

$\dfrac{142}{2}$=71 bottles	18 3/8 turns+connect.	*no ball*	3 1/4"

Coil with *ball, vertical wire* and spark wires:

$\dfrac{142}{2}$=71 bottles	19 1/2 turns+conn.	9 feet 5"	4"
,, ,,	20 3/8 ,, + ,,	33' 9"	4 3/8"
,, ,,	21 1/4 ,, + ,,	58' 9"	4 1/8"

Owing to the construction of the coil *L* which, as before stated, aimed at sharp tuning it was easy to detect a variation of $\dfrac{1}{16}$ of *one turn* of the regulating primary coil *R*.

It was desirable to take some readings with all the bottles in and the results were nearly the same and could still be read off with fair exactness, although the variation on the regulating coil *R* was reduced to one half, the capacity in the primary being just double that used in the experiments the results of which were just given.

The readings with all the condensers were as follows:

Capacity in primary total	Total self-induction in primary turns of reg. coil+connections	Height of ball from center to ground	Spark on terminals of *L*
286 bottles	10 1/8 + conn.	9' 5"	4 1/4"
,,	10 9/16+ ,,	33' 9"	4 1/2"
,,	11 + ,,	58' 9"	4 3/4"

Note to above experiments:

The vibration of the coil *L* with vertical wire and spark wires was found to be in resonance with the primary in another series of observations with

142 bottles   18 3/4 turns+connections

whereas before it was found to be so with 18 3/8 turns. When the different value 18 3/4 turns was observed the *wind* was *blowing hard* and it would seem as if this would have had the effect of increasing the apparent capacity of the aerial vertical conductor. This is to be followed up.

The experiments seemed to demonstrate clearly that the augmentation of the capacity as the ball was elevated was in a simple proportion to the height, for at the middle position the value found was very nearly the arithmetic mean of the values in the extreme positions.

Colorado Springs

Oct. 24, 1899

Tests continued on the effect of elevation upon the capacity of a body connected to earth.

The same coil was used as in the tests just before recorded, but the ball of 18″ diam. was substituted for by one of 30″ diam.

These readings were in all probability closer as an improvement in the procedure was made.

The results were as follows:

First set of readings.

Capacity in primary circuit	Self-inductance of primary: Turns of reg. coil+ connections	Height of ball above ground from center	Analyzing spark on terminals of excited coil
Bottles total mfd.			
$\dfrac{(8 \times 36)-2}{2}=143=0.1287$ 10 3/8+connections		9′ 11″	4 1/2″
$\dfrac{(8 \times 36)-2}{2}=143=0.1287$ 10 3/4+ ,,		34′ 3″	4 3/4″
$\dfrac{(8 \times 36)-2}{2}=143=0.1287$ 11 1/8+ ,,		58′ 6″	5 1/8″

Second set of readings. Capacity reduced in primary to get greater variation on regul. coil.

$\dfrac{(4 \times 36)-2}{2}=71=0.0639$ 19 3/4+ ,,		9′ 11″	4 3/8″

$$\frac{(4 \times 36) - 2}{2} = 71 = 0.0639 \qquad 20 \ 3/8+ \qquad ,, \qquad 34'\ 3'' \qquad 5\ 1/4''$$

$$\frac{(4 \times 36) - 2}{2} = 71 = 0.0639 \qquad 21 \quad + \qquad ,, \qquad 58'\ 6'' \qquad 5\ 3/8''$$

Note: After these readings had been taken it was found that the ball sliding on the vertical wire did not make a good contact. This might have modified the results slightly.

Colorado Springs

Oct. 25, 1899

Experiments on influence of elevation upon capacity of conductor connected to earth continued:

This time simply a wire No. 10 extending vertically in the continuation of the axis of the coil was used. The coil was the same as before, 689 turns No. 20 cord on drum of 14″ diameter. 36 feet of wire were first taken and after each reading 3 feet were cut off. Each time resonance with the primary was attained and the constants of the primary circuit noted, this giving the necessary data for the determination of the capacity of the vertical wire. The results are indicated below:

Capacity in the primary circuit total Bottles	mfd.	Self-inductance of primary Turns of reg. coil+connect.		Length of vertical wire in feet	Analyzing spark on terminals of excited coil
$\frac{(4 \times 36) - 2}{2} = 71$	0.0639	17	+connect.	36	1/8″
,,	,,	16 3/8	+ ,,	33	,,
,,	,,	15 3/4	+ ,,	30	,,
,,	,,	15 1/4	+ ,,	27	1/16″
,,	,,	14 3/4	+ ,,	24	,,
,,	,,	14 1/4	+ ,,	21	,,
,,	,,	13 3/4	+ ,,	18	,,
,,	,,	13 1/4	+ ,,	15	,,
,,	,,	12 3/4	+ ,,	12	13/16″
,,	,,	12 1/4	+ ,,	9	9/16″

These readings are approximate.

Measurement of inductance of 689 turn coil used in investigations on influence of elevation upon the capacity of a conductor.

Readings were as follows:

Volts av.	Current	ω	Res.
118	3.315	880	28.304

$$\frac{E}{I} = \frac{118}{3.315} = 35.6 \qquad \left(\frac{E}{I}\right)^2 = 1267.36$$

$$R^2 = 800.89$$
$$\overline{\left(\frac{E}{I}\right)^2 - R^2 = 466.47} \quad \textit{Large dynamometer close}$$

$$\omega^2 = 774,400$$

$$L^2 = \frac{\left(\frac{E}{I}\right)^2 - R^2}{\omega^2} = 0.00057654 \quad \text{From this } L = 0.024 \text{ henry}$$
$$\text{or} = \mathbf{24,000,000} \text{ cm.}$$

This is a value slightly smaller than that calculated before. Readings were also taken with small dynamometer. This slightly damaged. The readings are to be revised upon restandardizing.

Volts av.	Current av.	ω
69.25	2.045	880

$$\frac{E}{I} = 33.863 \qquad \left(\frac{E}{I}\right)^2 = 1146.7 \qquad \left(\frac{E}{I}\right)^2 - R^2 = 345.81$$

$$R^2 = 800.89$$

$$\omega^2 = 774,400$$

$$L^2 = \frac{\left(\frac{E}{I}\right)^2 - R^2}{\omega^2} = 0.0004466 \quad \text{From this } L = 0.0211 \text{ henry}$$
$$\text{or } \mathbf{21,100,000} \text{ cm.}$$

Note: This value is decidedly too low owing to dynamometer indicating too large a current. Possibly during the test ω had changed.

Colorado Springs

Test of condensers some of which were recently refilled. The corrections for capacity to be applied to the work of about two weeks ago.

Readings with 7 cells battery showed for *one half*

Microfarad Standard   Defl.   101 ⎫
                                         101 ⎬ average 101°
                                         101 ⎭

For set of condensers all in multiple   51 ⎫
on left side from door   51 ⎬ average 51°
                                         51 ⎭

For set of condensers all in multiple   53 ⎫
on right side from door   53 ⎬ average 53°
                                         53 ⎭

For all condensers in multiple   104 ⎫
                                      104 ⎬ average 104°
                                         104 ⎭

The difference between the two sets is probably due to different heights of solution in the tanks or bottles. In these measurements there were two bottles less on each side in the central tanks.

Taking the above data, the total number of bottles when all were in quantity was $(16 \times 36) - 4 = 576 - 4 = 572$ bottles.

The capacity was $\frac{104}{101} \times 0.5 = $ **0.515 mfd** approx. This would give

for one bottle an average:	0.0009 mfd or	810 cm	⎫
„ two „ „	0.0018 „	1620 „	These are
„ three „ „	0.0027 „	2430 „	frequently
„ 12 „ „	0.0108 „	9720 „	needed
for 36 bottles or one tank	0.0324 „	29,160 „	
„ 18 „ half of one tank	0.0162 „	14,580 „	⎭

Note: Small mica condensers with fibre covers made for resonating circuits have a little less than 1/20 mfd each; two were measured.

Referring to the preceding results the period of the excited system without the vertical wire was as follows:

Capacity in primary 71 bottles$=0.0639$ mfd.

Self-induction of primary connections plus 9 7/8 first turns of the regulating coil measured: 32,700 cm.

$$T_1 = \frac{2\pi}{10^3} \sqrt{0.0639 \times \frac{32,700}{10^9}}$$

With the vertical wire the period corresponding to that of the excited system was T_2, which estimated is:

Capacity in primary as before 0.0639 mfd.
Self-ind. in primary conn.$+18$ 3/8 turns measured: 65,000 cm.

$$T_2 = \frac{2\pi}{10^3} \sqrt{0.0639 \times \frac{65,000}{10^9}}$$

Now calling C the capacity of the excited system with only the spark wires, L the self-induction of the excited coil, before found to be $L=0.024$ H, we have for the period of the excited system:

$$T_{s1} = \frac{2\pi}{10^3} \sqrt{0.024 \times C}, \quad C \text{ being in mfd. Now } T_{s1} = T_1 \text{ or}$$

$$0.024 \times C = 0.0639 \times \frac{327}{10^7} \quad \text{and} \quad C = \frac{327 \times 0.0639}{10^7 \times 0.024} \text{ mfd or}$$

$$C = \frac{9 \times 10^5 \times 327 \times 0.0639}{10^7 \times 0.024} = \frac{0.5771 \times 327}{2.4} \text{ cm.} = \textbf{78.36 cm} = C$$

Calling presently C_1, the capacity of the excited system when the vertical wire was attached, we have by similar reasoning from

$$T_{s2} = \frac{2\pi}{10^3} \sqrt{0.024 \times C_1} = T_2 = \frac{2\pi}{10^3} \sqrt{0.0639 \times \frac{65,000}{10^9}}$$

$$C_1 = \frac{65}{10^6} \times \frac{0.0639}{0.024} \text{ mfd,} \quad \text{or} \quad C_1 = \frac{9 \times 10^5 \times 65}{10^6} \times \frac{0.0639}{0.024} \text{ cm} = \textbf{155.76 cm.}$$

The capacity of the wire alone would from these results be

$$C_1 - C = 155.76 - 78.36 = \textbf{77.40 cm.}$$

Calculated approximately we have the following data for the wire: diam. of wire No. 10$=0.2588$ cm., $r=0.1294$ cm.

$$l=50 \text{ feet approx.} = 50 \times 12 \times 2.54 = 1524 \text{ cm.}$$

From this: $\dfrac{l}{r}=11{,}777.4$ and capacity of wire $=\dfrac{l}{2\log_e\dfrac{l}{r}}=\dfrac{1524}{18.7}=\mathbf{81.5\ cm.}$

The calculated value is a little larger but not much, well within the errors of the adjustment and determination of the quantities by experiment.

Continuing we see that when the ball was at its lowest position the period of the primary was as follows:

Capacity in primary as before 0.0639
Inductance: connections $+19\ 1/2$ turns
measured 68,300 cm or $\dfrac{683}{10^7}$ H. $\left.\right\}\quad T_{pl}=\dfrac{2\pi}{10^3}\sqrt{0.0639\times\dfrac{683}{10^7}}$

Similarly, when the ball was at its highest position, we have:

Capacity in primary circuit as before 0.0639
Inductance: connections $+21\ 1/4$ turns which,
measured, were found: 76,100 cm, or $\dfrac{761}{10^7}$ H $\left.\right\}\quad T_{ph}=\dfrac{2\pi}{10^3}\sqrt{0.0639\times\dfrac{761}{10^7}}$

Now calling C' the total capacity of the excited system when the ball was lowest, we have the period of the excited system:

$$T_{sl}=\frac{2\pi}{10^3}\sqrt{0.024\times C'}=\frac{2\pi}{10^3}\sqrt{0.0639\times\frac{683}{10^7}}\quad\text{from which}\quad C'=\frac{0.0639\times683}{0.024\times10^7}\ \text{mfd,}$$

or C' in $\text{cm}=\dfrac{9\times10^5\times0.0639\times683}{0.024\times10^7}=\mathbf{163.66\,cm=C'.}$

The ball at the lowest position effectively contributed only

$$C'-C_1=163.66-155.76=\mathbf{7{,}90\ cm}$$

but this low value was probably due to the fact that the vertical wire extended above.

Calculating similarly the value of C'', that is the capacity of the excited system when the ball was highest:

$$T_{sh}=\frac{2\pi}{10^3}\sqrt{0.024\times C''}\quad C''=\frac{0.0639\times761}{10^7\times0.024}=\text{in cm.}=182.36\ \text{cm.}$$

Now $C''-C'=182.36-163.66=\mathbf{18.70\ cm}$ gives the actual value for the ball on top. This is an increase of nearly 137% for 49′ 4″ or about 2.76% per foot.

Experiments with coil 689 turns on drum 14″ diam. were again continued to further study the influence of elevation upon capacity.

The same arrangement was used which was described on a previous occasion with vertical wire and ball 30″ diam. The excitation of the coil was effected by connecting it to one of the terminals of the condensers, that which was joined to the terminal of the Westinghouse Transformer which was in connection with the tank — but the connection of the latter to the ground was not made directly but through a spark gap 1/8″ long.

The spark wires leading to the spark gap, serving to observe resonant rise, were No. 26 guttapercha coated and each about 25 feet long: It was really only the upper wire leading from the free terminal which was of importance to consider. The vibration of the coil was first ascertained, nothing except the spark wires being attached. Next a sphere of 30″ diam., the tin foil of which was cut through with a sharp knife to prevent formation of eddy currents, was connected to the top of the coil and the vibration again determined. Then the vertical wire was put on and the sphere taken off. After determining the vibration with the vertical wire the sphere was slipped on the same and readings taken with the sphere at different heights. The results were as follows:

Coil with spark wires alone.

Capacity in primary Bottles total mfd	Self-inductance of primary Turns of reg. coil+conn.	Analyzing spark on terminals of excited coil
$\frac{(4 \times 36) - 2}{2} = 71 = 0.0639$	10 1/2+conn.	3″

Coil with ball of 30″ diam. and spark wires.

Capacity in primary circuit Bottles total mfd	Self-inductance of primary. Turns in reg. coil+connections	Analyzing spark on terminals of excited coil
$\frac{(4 \times 36) - 2}{2} = 71 = 0.0639$	13+conn.	3 1/2″

Coil with vertical wire No. 10, fifty feet and spark wires.

$\frac{(4 \times 36) - 2}{2} = 71 = 0.0639$	19 3/4+conn.	3 7/8″

Readings with ball of 30″ diam., vertical wire and spark wires connected to coil.

Capacity in primary circuit Bottles total mfd	Self-inductance of primary. Turns in reg. coil+conn.	Height of ball from center to ground	Analyzing spark on terminals of excited coil
$\dfrac{(4 \times 36) - 2}{2} = 71 = 0.0639$	20 3/4+conn.	9′ 11″	4 3/8″
$\dfrac{(4 \times 36) - 2}{2} = 71 = 0.0639$	21 3/8+conn.	32′ 8″	4 3/8″
$\dfrac{(4 \times 36) - 2}{2} = 71 = 0.0639$	22 +conn.	55′ 7″	4 3/8″

Readings continued with primary capacity modified.

$\dfrac{(8 \times 36) - 2}{2} = 143 = 0.1287$	11 5/8+conn.	10′ 3″	1/4″
$\dfrac{(8 \times 36) - 2}{2} = 143 = 0.1287$	12 1/4+conn.	33′ 3″	1″
$\dfrac{(8 \times 36) - 2}{2} = 143 = 0.1287$	12 7/8+conn.	55′ 9″	1 1/2″

The capacity in primary was now still further reduced.

The readings were as follows:

Capacity in primary Bottles total mfd	Inductance in primary. Turns in reg. coil+conn.	Height of ball above ground from center	Analyzing spark on terminals of excited coil
$\dfrac{(5 \times 36) - 2}{2} = 89 = 0.0809$	16 1/2+conn.	10′ 3″	4″
$\dfrac{(5 \times 36) - 2}{2} = 89 = 0.0809$	17 3/8+conn.	33′ 3″	4 1/4″
$\dfrac{(5 \times 36) - 2}{2} = 89 = 0.0809$	18 1/4+conn.	55′ 9″	4 3/8″

In these experiments the excitation of the coil was varied by adjusting the small spark gap separating one terminal of the Westinghouse transformer from the ground. The tuning was not very sharp as the ball was large and the magnifying factor of the coil rather small. Taking n approx. 60,000 we have $p = 360,000$ approx., $R = 28.3$, $L = 0.024$ we have for $\dfrac{pL}{R}$ value $\dfrac{36 \times 10^4 \times 0.024}{28.3} = 300$ approx. Not so small after all.

It was desirable to take some readings with the self-induction in the primary remaining the same, the capacity only being varied.

Following results were obtained:

Primary capacity Bottles total mfd	Primary self-induction. Turns in reg. coil.+conn.	Height of ball from center to ground	Analyzing spark on terminals of excited coil
$\dfrac{133}{2}=66.5=0.05985$	22+conn.	10' 3''	4 3/8''
$\dfrac{137}{2}=68.5=0.06165$	22+conn.	33' 3''	4 3/8''
$\dfrac{142}{2}=71\ \ =0.0639$	22+conn.	55' 9''	4 3/8''

These data are to be worked out.

Colorado Springs

Oct. 30, 1899

Measurement of Inductances

The primary of the oscillator, regulating coil and connections to spark gaps and condensers as generally used were joined in series, the gaps and condensers being bridged by short pieces of wire No. 2 of inappreciable resistance. The object was to determine again: the inductance of the primary turns of the oscillator, of the regulating coil for the various turns as employed and also of the connections. The two primary turns were joined in series and readings taken across both to get better values.

The generator, a 1500 light machine smooth armature Westinghouse type, was specially run, the speed being kept constant. The frequency was determined as before by taking the speed of synchronous motor running without any load and with strong excitation of field. There was no slip under these conditions as frequently ascertained.

The average speed was very closely 2070 per minute on a motor which was a 4-pole, this giving 8280 cycles per minute or 138 per second. From which $\omega=867$.

The connections as used are shown in diagram below:

The connections were all left exactly as used generally with the exception that, as before stated, the gaps and condensers were bridged by short pieces of wire No. 2.

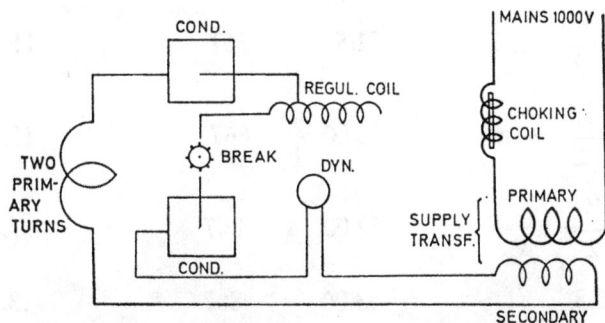

Readings for: *two primary turns, regulating coil 23.5 turns* and *all connections in series:*

E	I	ω	R
14.7	53.00	867	
14.8	53.22	,,	
15.1	54.00	,,	
15.1	54.00	,,	

This gave an average of:

E	I	ω	R
14.925	**53.555**	**867**	

Readings for *two primary turns in series:*

E	I	ω	R
10.45	52.1	867	
10.35	52.1	,,	
10.40	52.1	,,	

This gave average values:

E	I	ω	R
10.4	**52.1**	**867**	

Readings were also taken across each one of the primary turns, the e.m.f. of course being exactly one half but the e.m.f. across the *upper* turn was *slightly greater*, probably because of the greater distance from the ground, possibly because of being nearer to the open secondary.

Readings for *regulating coil 23.5* turns and *connections* of *break* and *condensers* together in series:

E	I	ω	R
4.1	52.1	867	
4.1	52.1	,,	
4.1	52.1	,,	

This gives average values:

E	I	ω	R
4.1	**52.1**	**867**	

Readings for *regulating coil alone* for *various turns:*

E	I	ω	Number of turns across which e.m.f. read	Resistance of the turns included
$\dfrac{7.85}{2} = 3.925$	52.1	867	23.5	
$\dfrac{6.3}{2} = 3.15$	52.1	867	19.5	
$\dfrac{4.8}{2} = 2.4$	52.5	867	15.5	
$\dfrac{3.3}{2} = 1.65$	53.00	867	11.5	
$\dfrac{2}{2} = 1$	53.00	867	7.5	
0.5	53.00	867	3.5	

To get better readings for the separate turns of the regulating coil the voltage was taken again as first across the whole: 2 primary turns, connections and regulating coil all in series and the turns of the latter were then varied, readings being taken for each case. The results are indicated below:

E	I	Number of turns of the regul. coil included in the circuit	ω	Resistance total of circuit included
11.55	55.8	0.5	867	
11.675	55.8	2.5	,,	
11.90	55.8	4.5	,,	
12.15	55.3	6.5	,,	
12.45	55.3	8.5	,,	
12.80	55.3	10.5	,.	
13.20	54.9	12.5	,,	
13.50+a little	54.9	14.5	,,	
13.90	54.9	16.5	,,	
14.25	54.4	18.5	,,	
14.60+a little	54.4	20.5	,,	
14.15	51.1	22.5	,,	
14.30	51.1	23.5	,,	

To facilitate estimation of the inductance of the various turns the readings were reduced to the same value of current and are as follows:

11.55	55.8	0.5	867
11.675	55.8	2.5	,,
11.90	55.8	4.5	,,
12.26	55.8	6.5	,,
12.562	55.8	8.5	,,
12.915	55.8	10.5	,,
13.416	55.8	12.5	,,
13.721	55.8	14.5	,,
14.128	55.8	16.5	,,
14.617	55.8	18.5	,,
14.976	55.8	20.5	,,
15.452	55.8	22.5	,,
15.616	55.8	23.5	,,

Measurement of inductances: New coil for further investigating effect of altitude upon capacity of body connected to earth.

The drum of 14″ diam. and 8 feet long was again used, the cord No. 20 being taken off and wire No. 10 rubber- covered used instead. Most of the data from before remained, only the number of turns was reduced from 689 to 346.

The readings of e.m.f., current and frequency were as follows:

E	I	ω		R	
85	15.635	880		coil with cord	1.752
84.75	15.6	,,		cord	0.596
82.5	15.3	,,		coil alone	1.156 ohm.

Average values:

84.38	15.512	880	

This gives: $\dfrac{E}{I}=5.44$ $\left(\dfrac{E}{I}\right)^2=29.5936$

$$R^2 = 1.3364$$
$$\omega^2 = 774{,}400$$

$$L^2 = \frac{\left(\dfrac{E}{I}\right)^2 - R^2}{\omega^2} = \frac{29.5936 - 1.3364}{774{,}400} = \frac{28.2572}{774{,}400} \ \text{henry} = 0.00003649$$

or $L = \sqrt{0.00003649} = \textbf{0.00604 H}$ or $L = \textbf{6,040,000 cm.}$

Now for calculating with reference to note Oct. 18 we have the following data:

diam. of core 14″=35.5 cm

$$L = \frac{4\pi}{l} N^2 S = \frac{12.5664 \times 119{,}716{,}000}{243.84}$$

length of core 96″=243.84 cm=l
S=1000 sq. cm.
N=346
N^2=119,716

From this L=6,169,600 cm
or 0.00617 henry approx.
This is nearly 2% greater than measured value — fairly close.

Colorado Springs Notes

Nov. 1—30, 1899

Want of time compelled omission of following items partly worked out:

Nov. 27, 28 Corrected and completed results of experiments of

Nov. 3 $\left\{ \begin{array}{l} \text{with wire of different} \\ \text{lengths and ball} \end{array} \right\}$, 19 and 21

Nov. 29, 30 a) Extra coils in series exciting one another, $\left. \begin{array}{l} \\ \\ \end{array} \right\}$ complete
        b) Methods of tuning by telephone, description
        c) Exciting receiving circuit through small sensitive arc

Patent matter worked on Nov. 1—30:

Exclusion of messages in telegraphy:

a) Two or more synchronized receiving circuits $\left. \begin{array}{l} \\ \\ \end{array} \right\}$ text to be
   controlling receiver completed.
b) Key or safety combination

Colorado Springs Nov 1. 1899

Measurements of inductance of new extra coil built chiefly for investigating propagation of waves through the ground and similar objects also to investigate further the behaviour of strong streamers.

The frame was made of light notched woodwork fastened to three strong wooden rings. Provision was made for 106. turns. The rings were 8 feet in diam. and taking further 1½" on each side for the cross pieces the total diameter of coil inside was 8'3". The length was 8 feet less 1½" on top and 1" on bottom making total length of coil 7 feet 10". The data are as follows:

Length of coil 7'10" = 94" = 238.76 C.M. = l ; diam. 8'3" = 99" = 251.46 C.M. = d
Turns wound on 106 = N (really 105 turns wound + one turn loose)
Area $S = \frac{\pi d^2}{4} = 49662.52$ C.M.Sq. From these data we

Have $L = \frac{4\pi}{l} N^2 S = \frac{12.566 \times 11236 \times 49662.52}{238.76} = \frac{208 \times 12.566 \times 11236}{...}$

$= 141196 \times 208 =$

$L = 29,368,764$ C.M. ~ 0.029369 H. approx   Calculated

Now the readings were.

E	$\ddot{U}$	ω	R		
110	6.05	880	coil with cord 3.26	$\frac{E}{J} = 18.1818$	$\left(\frac{E}{J}\right)^2 = 330.58$
110	6.05	880	cord 0.596	$R^2 = 7.097$	
110	6.05	880	coil alone 2.664	$\left(\frac{E}{J}\right)^2 - R^2 = 323.483$	
				$\omega^2 = 774400$	

This gives $L^2 = \frac{323.483}{774400} = 0.000417721$

$L = \sqrt{0.000417721} = 0.020402$ H ~ 20,420,000 C.M.

Measurement show a value much lower but this was to be expected as the turns are large also far apart ½". However this measurement is to be made again to be sure of the errors.

Measurement of inductance of new extra coil built chiefly for investigating propagation of waves through the ground and similar objects. Also to investigate further the behaviour of strong streamers.

The frame was made of light notched woodwork fastened to three strong wooden rings. Provision was made for 106 turns. The rings were 8 feet in diam. and taking further 1 1/2″ on each side for the crosspieces the total diameter of coil inside was 8′3″. The length was 8 feet less 1 1/2″ on top and 1/2″ on bottom making total length of coil 7 feet 10″. The data are as follows:

Length of coil 7′ 10″=94″=238.76 cm=l; diam 8′ 3″=99″=251.46 cm=d

Turns wire No.10 106=N (really 105 turns wound+one turn loose)

Area $S=\dfrac{\pi d^2}{4}=$49,662.52 cm.sq. from these data we have

$$L=\frac{4\pi}{l}\,N^2S=\frac{12.5664\times 11,236\times 49,662.52}{238.76}=208\times 12.5664\times 11,236$$

$$=141,196\times 208$$

$$L=\textbf{29,368,768 cm, or 0.029369 H}\text{ approx.}$$

Now the readings were:

E	*I*	ω	*R*	
110	6.05	880	Coil with cord	3.26
110	6.05	880	cord	0.596
110	6.05	880	Coil alone	2.664

Calculated

$$\left(\frac{E}{I}\right)=18.1818 \quad \left(\frac{E}{I}\right)^2=330.58$$

$$R^2=7.097$$

$$\left(\frac{E}{I}\right)^2-R^2=323.483$$

$$\omega^2=774,400$$

This gives $L^2=\dfrac{323.483}{774,400}=0.000417721$

$$L=\sqrt{0.000417721}=\textbf{0.02042 H or 20,420,000 cm.}$$

Measurement shows a value much lower but this was to be expected as the turns are large also 1/2″ apart. However this measurement is to be made again to be sure of the result.

Readings were again taken today to ascertain the value found yesterday for inductance of new extra coil. The results were as follows:

E	I	ω	R	
194	10.7	887	2.664	average of three readings very closely agreeing.

$$\left(\frac{E}{I}\right) = 18.13 \qquad \left(\frac{E}{I}\right)^2 = 328.6969$$

$$R^2 = 7.0969$$

$$\left(\frac{E}{I}\right)^2 - R^2 = 321.60$$

$$L = \frac{\sqrt{321.6}}{887} = \frac{17.933}{887}$$

$L = 0.0202176$ henry or 20,217,600 cm.

This is a value a little smaller than the one found yesterday but it is within the limits of variation of ω.

Note: When the turns are far apart the ordinary formulas for calculating do not apply, and the measured value is the more inferior to the calculated value the greater the turns and the farther apart they are. When very far apart it is better to calculate inductance of one turn and multiply or, if they are not all alike, to calculate then separately and add up, making some allowance for mutual induction.

Secondary last form, two wires No. 10 in multiple 17 turns on frame described on another occasion. To test the values before found for inductance and mutual induction coefficient readings were again taken today. For secondary inductance:

E	I	ω	R	
138	16.5	887	1.382	The first two readings
138	16.5			were in all probability
136	16.3	875		the best and they are
134.5	16.2			taken.

From above:

$$\frac{E}{I} = 8.364 \qquad \left(\frac{E}{I}\right)^2 = 69.9565 \qquad \left(\frac{E}{I}\right)^2 - R^2 = 68.0465$$

$$R^2 = 1.91 \qquad \omega = 887$$

$$L = \frac{\sqrt{64.0465}}{887} = \frac{8.003}{887} = 0.009023 \text{ H or } 9,023,000 \text{ cm.}$$

This is smaller than before found because the turn before last was wound a little higher up as at first sparks would break through. It is to be measured once more however.

Readings for *Mutual Induction* Coefficient.

Two primary turns in series:

Current through primary two turns in series	Volts on secondary	
I	*E*	ω
45.4	34	880
45.4	34	,,
45.4	34	,,

From this $E = M\omega I$, $M = \dfrac{34 \times 10^9}{880 \times 45.4} = $ **851,021 cm.**

Readings with current through the secondary gave:

Current through secondary	Volts on primary two turns series	ω
17.8	13.4	872 ⎫ over
17.8	13.4	,, ⎭

From later readings $M = \dfrac{13.4 \times 10^9}{17.8 \times 872} = $ **863,313 cm.**

This is a little larger probably due to variation of ω. Readings were now taken for each of the primary turns separate with the following results:

Upper primary turn, nearer to secondary:

Current through secondary	E.m.f. on primary	ω
17.9	6.9	872 average three readings

From this $M_{\text{upper pr.}} = \dfrac{6.9 \times 10^9}{17.9 \times 872} = $ **442,059 cm.**

Current through primary	E.m.f. on secondary	ω
30.1	11.8	880 average three readings

$$M_{\text{upper. pr.}} = \frac{11.8 \times 10^9}{30.1 \times 880} = \textbf{445,484 cm.}$$

The difference again in all probability to be ascribed to variation of ω.

Lower primary turn. Readings were as follows:

Current through primary	E.m.f. secondary	ω
30.1	11.1	880 average three readings

Current through secondary	E.m.f. primary	
17.9	6.4	880 average three readings

From first set readings:

$$M_{\text{lower pr.}} = \frac{11.1 \times 10^9}{30.1 \times 880} = \textbf{419,060 cm.}$$

From second set readings:

$$M_{\text{pr. lower}} = \frac{6.4}{17.9 \times 880} = \textbf{406,300 cm.}$$

For two primary turns multiple

Current through primaries	E.m.f. secondary	ω
29.2	11.25	880 average three readings

Current through secondary	E.m.f. primaries	
17.9	6.7	880 average three readings

From first set of readings:

$M_{\text{2 prim. multiple}} = \textbf{437,811 cm}$

From latter readings:

$M_{\text{2 pr. multiple}} = \textbf{425,343 cm}$

Colorado Springs

Nov. 3, 1899

Investigation for the purpose of ascertaining the influence of elevation upon the capacity of a conductor connected to a vibrating system as before used, continued:

The coil of 346 turns, No. 10 wire wound on drum 14″ diam., 8 feet long was again used. The spark wires were as before No. 26 guttapercha covered, 24 feet long each. The readings were as follows:

Capacity in primary circuit *total*	Inductance in primary turns reg. coil+conn.	Length of vertical wire attached to free terminal of excited coil	Analyzing spark on terminals of excited coil
$\dfrac{(3 \times 36) - 2}{2}$ =53 bottles= =0.0479 mfd	7 1/8+conn.	45 feet	3 3/8″
0.0479 mfd	6 3/8+ ,,	36 ,,	3 3/8″
,,	5 3/4+ ,,	27 ,,	3 3/8″
,,	5 3/16+ ,,	18 ,,	3 1/4″
,,	4 3/8+ ,,	9 ,,	3 1/8″
,,	3 5/8+ ,,	0 ,,	3 1/8″

Note: As repeatedly observed, the first addition of a small length of wire to a coil generally produces a great effect but a certain small length being bypassed the increase per unit of length becomes more gradual.

Experiments with coil 346 turns on drum 14″ diam. with the view of investigating influence of elevation upon the capacity of a body connected to earth continued. This time in connection with the vertical wire a ball 30″ diam. was used in the manner before described.

Coil with spark wires alone

Capacity in primary circuit *total*	Turns in regulating coil+connections	Induc- tance	Analyzing spark on terminals of excited coil
$\dfrac{(3 \times 36) - 2}{2} = 53$ bottles$=$ $= 0.0479$ mfd	3 5/8 +conn.		3 1/8″

Coil with 50 feet wire No. 10 vertical

0.0479 mfd	7 3/4+conn.		4″

Coil with ball 30″ diam. lowest position 10′3″ from center to ground

0.0479 mfd	8 1/2+conn.		4 1/8″

Coil with ball mean position 33′9″ from center to ground

0.0479 mfd	8 7/8+conn.		4 1/4″

Coil with ball highest position 57′3″ from center to ground

0.0479 mfd	9 1/4+conn.		4 3/8″

Note: The excitation in this and previous case was as before described by connecting the lower terminal of coil to that terminal of the condensers (primary) which was connected to small adjustable spark gap, specially made for this purpose, to earth.

It would be desirable in these tests to do away with the spark wires as these are apt to introduce errors in the estimates of capacity. Experiments were made with new extra coil to see how close the maximum rise could be ascertained without any spark wires, merely by observing the streamers. For this purpose the extra coil was excited from the secondary of oscillator as in some previous instances. First the extra coil was tuned with spark wires, then the upper wire which is the only one of importance was taken off and the system again tuned. The results were as follows:

New Extra coil with Spark wires Guttapercha No. 26, 24 feet long

Capacity in primary circuit total		Inductance in primary circuit Turns of regul. coil+connections
bottles	mfd	
$\dfrac{(8 \times 36) - 2}{2} = 143$ bottles	$=0.1287$	7 5/8+connections

New Extra coil without upper Spark wire

$$\frac{(8 \times 36) - 2}{2} = 143 \text{ bottles} = 0.1287 \qquad\qquad 5\ 5/8 + \text{connections}$$

Colorado Springs

Nov. 4, 1899

Measurement of inductance of primaries.

Another series of readings were taken with the object of closely determining the inductance of the primary turns. This time a different dynamo was supplying the current. The speed was kept very constant. The readings were as follows:

Current	Electromotive force across two primaries in series	ω
$345° = 58.8$	12 a trifle less	880
$345° = 58.8$	12 ,,	,,
$345° = 58.8$	12 ,,	,,

Allowing a little for zero displacement on voltmeter the average is very closely:

I	E	ω	R two primary turns in series
58.8	11.95	880	0.004 ohm.

This gives $\dfrac{E}{I} = 0.2032$ $\qquad \left(\dfrac{E}{I}\right)^2 = 0.04129$ $\qquad R^2 = 0.000016$

Since R^2 is entirely negligible against $\left(\dfrac{E}{I}\right)^2$ we have:

$$L = \frac{E}{I\omega} = \frac{11.95}{51,744} = 0.000230945 \text{ henry}$$

or **230,945 cm.**

This would give for one primary turn approximately

0.000057736 henry or **57,736 cm.**

The value previously found was 56,400 cm.
Reading of today would appear more reliable.

Capacity of structure of iron pipes, before described, computed:

7″ pipe: Outside diam. 7.625″=19,3673 cm=d

length of pipe with cap=l=23′ 4″=280″=811.2 cm=l

$$C_1 = \frac{l}{2 \log_e \frac{l}{r}} = \frac{811.2}{2 \times 4.42313} \qquad r=9.6837 \qquad \frac{l}{r}=83.77$$

$$C_1 = \textbf{91.7 cm.} \qquad\qquad \log_e \frac{l}{r}=1.9231 \times 2.3 = 4.42313$$

6″ pipe: Outside diam.=6.625″=16.8275 cm=d

length of pipe 18′ 2″=218″=553.72 cm=l

$$C_2 = \frac{l}{2 \log_e \frac{l}{r}} = \frac{553.72}{2 \times 4.182} \qquad r=8.4138 \qquad \frac{l}{r}=65.81$$

$$C_2 = \textbf{66.2 cm.} \qquad\qquad \log_e \frac{l}{r}=1.818292 \times 2.3 = 4.182$$

5″ pipe: Outside diam. ‵5.563″=14.13 cm=d

length of pipe 18′ 4 1/2″=220.5″=560.07 cm=l

$$C_3 = \frac{l}{2 \log_e \frac{l}{r}} = \frac{560.07}{2 \times 4.368} \qquad r=7.065 \qquad \frac{l}{r}=79.27$$

$$C_3 = \textbf{64.11 cm.} \qquad\qquad \log_e \frac{l}{r}=1.89911 \times 2.3 = 4.368$$

4″ pipe: Outside diam. 4.5″=11.43 cm=d

length of pipe 15′ 7 1/4″=187.25″=475.615 cm.=l

$$C_4 = \frac{l}{2 \log_e \frac{l}{r}} = \frac{475.615}{8.832} \qquad r=5.715 \qquad \frac{l}{r}=83.22$$

$$C_4 = \textbf{53.85 cm.} \qquad\qquad \log_e \frac{l}{r} = 1.92 \times 2.3 = 4.416$$

17*

3 1/2″ pipe: Outside diam. 4″=10.16 cm=d

length of pipe 19′ 3 1/4″=231.25″=587.375 cm=l

$$C_5 = \frac{l}{2 \log_e \frac{l}{r}} = \frac{587.375}{9.49} \qquad r = 5.08 \qquad \frac{l}{r} = 115.6$$

$$C_5 = 61.9 \text{ cm.} \qquad\qquad \log_e \frac{l}{r} = 2.062958 \times 2.3 = 4.745$$

3″ pipe: Outside diam. 3.5″=8.89 cm=d

length of pipe 18′ 4 3/4″=220.75″=560.7 cm=l

$$C_6 = \frac{l}{2 \log_e \frac{l}{r}} = \frac{560.7}{2 \times 4.83} \qquad r = 4.445 \qquad \frac{l}{r} = 126.1$$

$$C_6 = 58.05 \text{ cm.} \qquad\qquad \log_e \frac{l}{r} = 2.1 \times 2.3 = 4.83$$

2 1/2″ pipe: Outside diam. 2.875″=7.3 cm=d

length of pipe 8′ 1/4″ $\Big\}$ 8′ 8″=104″=264.16 cm=l
   „   nipples 7 3/4″

$$C_7 = \frac{l}{2 \log_e \frac{l}{r}} = \frac{264.16}{2 \times 4.276} \qquad r = 3.65 \qquad \frac{l}{r} = 72.37$$

$$C_7 = 30.89 \text{ cm.} \qquad\qquad \log_e \frac{l}{r} = 1.859 \times 2.3 = 4.276$$

from above we have *total capacity* of structure

$C:$		
	7″ pipe with cap	91.7 cm.
	6″ pipe	66.2 „
	5″ pipe	64.11 „
	4″ pipe	53.85 „
	3 1/2″ pipe	61.9 „
	3″ pipe	58.05 „
	2 1/2″ pipe	30.89 „
	Ball 30″ diam.	38.1 „
	Total Capacity	$C = 464.8$ cm.

Note: This supposes of course that all these capacities are connected in multiple. To be true it must be assumed that only long waves are used. With short waves the calculated value would not be experimentally borne out.

Colorado Springs

Nov. 6, 1899

Determination of el. st. *capacity* of the structure of iron pipes *by measurement.*

The method previously employed for such a purpose was again used. The coil described on a former occasion, wound on a drum of 10 5/16" and having 550 turns was excited from a vibrating primary system and the maximum resonant rise obtained with only the spark wires attached to the terminals of the coil. Then the upper terminal of the latter was connected to the structure and the maximum rise again obtained. From the two known periods of the primary system and the self-induction of the coil the capacity of the structure was then computed. In order to avoid errors due to the capacity of the wires connected to the coil the precaution was taken to make no change which would in any considerable way affect the result. Thus, when the vibration of the coil with only the spark wires was determined the wire which was to be later connected to the structure was likewise fastened to the upper terminal of the coil in a position such that in the second experiment or series of experiments it was only necessary to tilt the wire to bring it in contact with the structure. If this precaution would not be observed the error introduced by the connecting wire might be — and in fact it would generally be — considerable. The adjustments were first made with the spark wires and the wire which was to be connected to the structure, this wire being placed vertically and at a distance of about four feet from the latter. The maximum rise was observed on the terminals of the excited coil with:

Capacity in primary circuit total	Turns in regulating coil	Spark analyzing on terminals
$\frac{(8 \times 36) - 2}{2} = 143$ Bottles or 0.1287 mfd	2 1/8	2 5/8"

Now the wire was tilted and brought in contact with the structure and resonant maximum rise was observed with:

$\frac{(8 \times 36) - 2}{2} = 143$ Bottles = 0.1287 mfd	17 1/2	2 5/8"

To determine more satisfactorily the self-induction in both the primary vibrations readings were taken in the following manner. The two primary turns, connections to breaks and condensers and the regulating coil were all connected in series — the breaks and condensers being of course bridged by stout and short wires (No. 2 being used) and readings of e.m.f. across the two primary turns were first taken and then across the two primaries plus connections and turns in the regulating coil as were used in the two instances. Since the resistances were entirely negligible with respect to the inductances it was only necessary to make the ratio of the e.m.f. in two instances to determine the inductance of the connections and turns from the known inductance of the two primary turns which was carefully determined before. As the readings were taken practically at the same moments across the primaries alone and across the primaries connections and turns ω could not vary perceptibly, and to make sure of that the readings were taken repeatedly. The current passed through the inductances also remaining the same during the two consecutive readings, the results ought to be therefore more reliable then those obtained otherwise.

The results in the first case were as follows:

$$E_1 = 10.4 \text{ volts} \atop E = 10 \text{ volts} \Bigg\} \quad \begin{array}{l} I \text{ and } \omega \\ \text{the same.} \end{array}$$

In the second case:

$$E_1 = 13.15 \text{ volts} \atop E = 10.3 \quad ,, \Bigg\} \quad \begin{array}{l} I \text{ and } \omega \\ \text{the same.} \end{array}$$

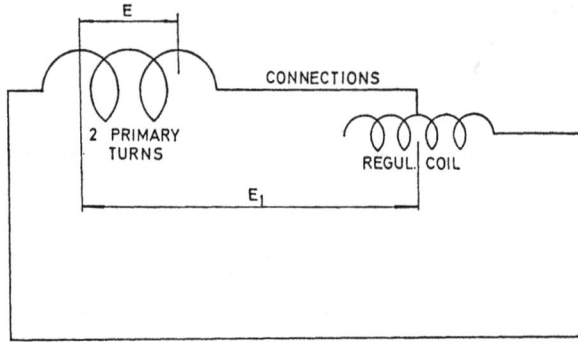

Calling now L the inductance of the two primary turns in series and L_1 that of the two primaries+connections+2 1/8 turns of the regulating coil we have in first case:

$$\frac{E_1}{E} = \frac{L_1}{L} \text{ and } L_1 = \frac{E_1}{E} L$$

Now L was previously determined to be 230,945 cm, hence

$$L_1 = \frac{104}{100} \times 230{,}945 = 240{,}183 \text{ cm.}$$

From this inductance of the connections+2 1/8 turns of the regulating coil is $L_1 - L = $ =9238 cm.

In the second case we have similarly

$$L_2 = \frac{13.15}{10.3} \times 230{,}945 = 294{,}847 \text{ cm.}$$

L_2 being inductance of two primaries+connections+17 1/2 turns of regulating coil.

Hence the inductance of the connections and the turns (17 1/2) included is $L_2 - L = $ =63,902 cm.

From these data the capacity of the structure can now be estimated as follows: In the first case the primary vibration was

$$T_1 = \frac{2\pi}{10^3} \sqrt{\frac{9238}{10^9} \times 0.1287}$$

In the second case the primary vibration was

$$T_2 = \frac{2\pi}{10^3} \sqrt{\frac{63{,}902}{10^9} \times 0.1287}$$

Calling now C_s the capacity of the excited system when the structure was not attached to it, and C_s' that when this was the case we have:

$$T_1 = \frac{2\pi}{10^3}\sqrt{C_s L'} = \frac{2\pi}{10^3}\sqrt{\frac{9238}{10^9} \times 0.1287}$$

and

$$T_2 = \frac{2\pi}{10^3}\sqrt{C_s' L'} = \frac{2\pi}{10^3}\sqrt{\frac{63,902}{10^9} \times 0.1287}$$

L' being the inductance of the excited coil. L' was previously measured and found to be $L' = 18,650,000$ cm. Now from above:

$$C_s = \frac{9238 \times 0.1287}{10^9 \times \dfrac{18,650,000}{10^9}} = \frac{9238 \times 0.1287}{18,650,000};$$

and

$$C_s' = \frac{63,902 \times 0.1287}{18,650,000}; \quad \frac{C_s'}{C_s} = \frac{63,902}{9238}; \quad C_s' = C_s\frac{63,902}{9238}$$

C_s expressed in centimeters is:

$$C_s = \frac{9238 \times 0.1287 \times 9 \times 10^5}{1865 \times 10^4} = \frac{11.583 \times 9238}{1865} = 57.375$$

from this $\quad C_s' = \dfrac{63902}{9238} \times 57.375 = 6.92 \times 57.375 = \textbf{397.03 cm.}$

$$C = C_s' - C_s = \textbf{339.655 cm.}$$

This is a result inferior to the calculated value but it was to be expected as before stated since it can not be correct to assume that all pipes are connected in multiple unless the vibration is very slow. Another coil is to be used with an inductance much higher so as to examine the truth of this opinion. There is, however, a possibility that the reading, when the structure was attached to the excited coil, was too low. In this case namely, the tuning is *not sharp* owing to the large capacity of the system, but when the structure is *not* attached it is quite sharp, hence if there is any error in the adjustment of the circuits it can be only then when the structure was connected. This is to be investigated also. A slight error might have been also caused by the wire which connected the coil to the structure, for although this wire was placed at a distance of 4 feet with its nearest point there might have been enough influence exerted by the structure to make the reading with the spark wires alone larger. This will be ascertained. From previous tests on the increase of capacity with elevation I should expect to find the capacity of the structure much larger than the calculated value.

On the present occasion readings were also taken with the view of determinig the inductance of the connections alone+flexible cable on regulating coil+1/2 turn of regulating coil. This namely is the lowest value which it is possible to give with the regulating coil in circuit. In this case the readings were $E_1 = 10.5$, $E = 10.2$ across the 2 primaries+ +all these connections and across the two primaries alone, respectively. With reference to foregoing and calling L_3 the inductance of the two primaries+connections+flexible cable+1/2 turn of regulating coil:

$$L_3 = \frac{105}{102} \times L = \frac{105}{103} \times 230{,}945 = 237{,}738 \text{ cm.}$$

Hence the inductance of all these mentioned connections is $L_3 - L = 237{,}738 - 230{,}945 =$

$= \textbf{6793 cm,}$ for $\begin{cases} \text{connections proper} \\ \text{1/2 turn regulating coil,} \\ \text{flexible cable in reg. coil.} \end{cases}$

Note: A small error is often caused by the changing position of the flexible cable which makes the readings for a small number of turns *larger* (slightly).

Colorado Springs

Nov. 7. 1899

Further experiments for the purpose of ascertaining the capacity of the structure of the iron pipe by resonance analysis. Two sets of readings were taken: one set with new extra coil the other with coil 346 turns wire No.10 on drum 14″ diam. The readings were as follows:

With new extra coil

Capacity in primary exciting circuit *total:*	Inductance in primary circuit. Turns of regulating coil+conn.	Analyzing spark on terminals of excited coil.
$\frac{(8 \times 36) - 2}{2} = 143$ bottles $= 0.1287$ mfd	22+conn.	2 5/16″
$\frac{(8 \times 36) - 2}{2} =$ „ $= 0.1287$ mfd	8+ „	3 3/4″

With experimental coil described

$\frac{(3 \times 36) - 2}{2} = 53$ bottles $= 0.0477$ mfd	15.75+conn.	4″
$\frac{(3 \times 36) - 2}{2} =$ „ $= 0.0477$ mfd	3.5+ „	3 1/8″

*Note: The first readings are of course *with,* the second *without* structure in each case.

*Note: Readings were also taken on this occasion with coil wound on 10 5/16″ drum before described (550 turns) but merely for the purpose of comparing the inductances of the primary and secondary circuits. With the same capacity as in the last case resonance was obtained with spark wires alone:

$$\frac{(3 \times 36) - 2}{2} = 35 \text{ bottles} = 0.0477 \text{ mfd} \qquad\qquad 8\ 7/8 \qquad\qquad 3\ 1/4''$$

This for future reference.

Returning to the two sets of observations it is to be noted that *one* turn of the new extra coil had been taken off and allowance should be made for this.

Let the primary vibration in the first case be T_{p1} and the corresponding secondary vibration T_{s1} then $T_{p1} = T_{s1}$. Similarly for the second reading with the extra coil when the structure was not attached to the coil. Calling the respective vibration T_{p2} and T_{s2} we have $T_{p2} = T_{s2}$. Now

$$T_{p1} = \frac{2\pi}{10^3}\sqrt{L_{p1} \times C} \text{ and } T_{p2} = \frac{2\pi}{10^3}\sqrt{L_{p2} \times C}$$

where L_{p1} and L_{p2} designate the inductances of the primary circuit in the two cases. From this

$$\frac{T_{p1}}{T_{p2}} = \sqrt{\frac{L_{p1}}{L_{p2}}}$$

as useful relation to remember.

While the self-induction was varied in the primary circuit, and the capacity remained the same, in the secondary it was just the opposite, the self-induction remaining the same and the capacity being varied. Calling now the capacity of the excited system with the structure C_{s1} and without the structure C_{s2} (that of coil with spark wires alone) we have by analogous reasoning:

$$\frac{T_{s1}}{T_{s2}} = \sqrt{\frac{C_{s1}}{C_{s2}}} = \frac{T_{p1}}{T_{p2}} = \sqrt{\frac{L_{p1}}{L_{p2}}} \text{ or } \frac{C_{s1}}{C_{s2}} = \frac{L_{p1}}{L_{p2}}.$$

This is also a convenient equation and useful to bear in mind. In cases when the capacity in the excited system is very often varied it is only necessary to determine the first capacity with which the series of experiments was begun to know all the other values from the known inductances of the primary circuit in two consecutive experiments. But when there are only two values to be determined, as in the present instance, they can be at once calculated from the known primary vibrations.

In the present instance, adopting this procedure we have:

$$T_{p1} = \frac{2\pi}{10^3}\sqrt{0.1287 \times 0.000079} \qquad \text{and } L_{p1}\begin{cases} \text{Connections } +22 \text{ turns:} \\ = 79{,}000 \text{ cm.} = 0.000079 \text{ H} \end{cases}$$

$$T_{p2} = \frac{2\pi}{10^3}\sqrt{0.1287 \times 0.00002526} \qquad L_{p2}\begin{cases} \text{Connections } +8 \text{ turns reg. coil} \\ = 25{,}260 \text{ cm.} = 0.00002526 \text{ H} \end{cases}$$

$$T_{p1} = T_{s1} = \frac{2\pi}{10^3}\sqrt{C_{s1} \times 0.02}$$

$$T_{p2} = T_{s2} = \frac{2\pi}{10^3}\sqrt{C_{s2} \times 0.02}$$

Inductance of extra coil before measured was 0.02042 henry. This owing to one turn less without *change of length* should be reduced to ratio $\left(\dfrac{105}{106}\right)^2$ or about 2% making the inductance very approx. 20,000,000 cm. or 0.02 henry.

From the above:

$$\frac{2\pi}{10^3}\sqrt{C_{s1}\times 0.02}=\frac{2\pi}{10^3}\sqrt{0.1287\times 0.000079}$$

and

$$C_{s1}=\frac{0.1287\times 0.000079}{0.02}\ \text{mfd,}$$

or in centimeters:

$$C_{s1}=\frac{9\times 10^5\times 0.1287\times 0.000079}{0.02}=\mathbf{457.5}\ \text{cm.}$$

Similarly we have

$$C_{s2}=\frac{0.1287\times 0.00002526}{0.02}\ \text{mfd, or}$$

$$C_{s2}=\frac{9\times 10^5\times 0.1287\times 0.00002526}{0.02}=\mathbf{146.29}\ \text{cm.}$$

This would give for the capacity of the structure according to this method only $C_{s1}-C_{s2}=$ $=457.5-146.29=\mathbf{311.21}$ **cm.**

This inferior result I attribute to the fact that the capacity is partially to be taken as *distributed*, owing to the length of the structure. But determining by the same method with a vibration which would be much slower this error should be very small.

Taking now the values for the set of readings with the experimental coil of 346 turns we have:

$$T'_{p1}=\frac{2\pi}{10^3}\sqrt{0.0377\times 0.00005486}\quad\text{and}\quad L_{p1}=\text{connections}+15.75\ \text{turns}$$
$$=54,860\ \text{cm.}=0.00005486\ \text{H}$$

$$T'_{p2}=\frac{2\pi}{10^3}\sqrt{0.0377\times 0.00001116}\qquad L_{p2}=\text{conn.}+3\ 1/2\ \text{turns}=$$
$$=11,160\ \text{cm.}=0.00001116\ \text{H}$$

The inductance of the experimental coil measured being 6,040,000 cm, or 0.00604 henry we have:

$$T'_{s1}=\frac{2\pi}{10^3}\sqrt{0.00604\times C'_{s1}}\quad\text{and}\quad T'_{s2}=\frac{2\pi}{10^3}\sqrt{0.00604\times C'_{s2}}$$

from these relations follows:

$$C'_{s1}=\frac{0.0477\times 0.00005486}{0.00604}\ \text{mfd. and}$$

$$C'_{s2}=\frac{0.0477\times 0.00001116}{0.00604}\ \text{mfd.}$$

Hence

$$C'_{s1} - C'_{s2} = \frac{0.0477}{0.00604} \times (0.00005486 - 0.00001116)$$

$$= \frac{0.0477}{0.00604} \times 0.0000437 \text{ mfd, or}$$

$$C'_{s1} - C'_{s2} = \frac{9 \times 0.0477 \times 4.37}{0.00604} \text{ centimeters} = \textbf{310.6 cm.}$$

very nearly the same value.

It is to be expected that the value found with a quicker vibrating system should be smaller since then the structure begins to act not as one condenser but as a series of condensers or distributed capacity, all parts not being charged at the same time.

This method of determining the capacity implies therefore to be quite correct a *slow* vibration and, furthermore, negligible capacity in the vibrating system itself and also that the body the capacity of which is determined should not be of too great length since this must cause errors.

From reading with coil of 550 turns it follows, since the capacities both in the primary and secondary circuits remained the same, that the inductances in the primary and inductances in the secondary or excited circuit bore the same ratio, that is:

$$\frac{\text{Ind. 8 7/8 turns} + \text{conn.}}{\text{Ind. 3.5 turns} + \text{conn.}} = \frac{\text{Ind. coil 550 turns}}{\text{Ind. coil 344 turns}}$$

Ind. of coil 550 turns was: 18,650,000

Ind. of coil 344 turns was: 6,040,000

Hence $\dfrac{\text{Ind. 8 7/8 turns}}{\text{Ind. 3 1/2 turns}} = \dfrac{1865}{604}$ **This for later comparison.**

Table of *inductances* prepared from preceding readings.

					cm.	H
Two primary turns in series					230,945 cm.	0.000230945 H
One of the primary turns					57,736 ,,	0.000057736 ,,
All connections to condensers and breaks *as used*					5004 ,,	0.000005004 ,,
All connections plus one half turn of reg. coil (first turn)					5774 ,,	0.000005774 ,,
All connections plus the whole first turn of reg. coil					6544 ,,	0.000006544 ,,
,,	,,	1 1/2 ,,	,,		7314 ,,	0.000007314 ,,
,,	,,	2 ,,	,,		8084 ,,	0.000008084 ,,
,,	,,	2 1/2 ,,	,,		8854 ,,	0.000008854 ,,
,,	,,	3 ,,	,,		10,009 ,,	0.000010009 ,,
,,	,,	3 1/2 ,,	,,		11,164 ,,	0.000011164 ,,
,,	,,	4 ,,	,,		12,319 ,,	0.000012319 ,,
,,	,,	4 1/2 ,,	,,		13,474 ,,	0.000013474 ,,
,,	,,	5 ,,	,,		15,158 ,,	0.000015158 ,,
,,	,,	5 1/2 ,,	,,		16,842 ,,	0.000016842 ,,
,,	,,	6 ,,	,,		18,526 ,,	0.000018526 ,,
,,	,,	6 1/2 ,,	,,		20,210 ,,	0.000020210 ,,
,,	,,	7 ,,	,,		21,894 ,,	0.000021894 ,,
,,	,,	7 1/2 ,,	,,		23,578 ,,	0.000023578 ,,
,,	,,	8 ,,	,,		25,262 ,,	0.000025262 ,,
,,	,,	8 1/2 ,,	,,		26,946 ,,	0.000026946 ,,
,,	,,	9 ,,	,,		28,870 ,,	0.000028870 ,,
,,	,,	9 1/2 ,,	,,		30,794 ,,	0.000030794 ,,
,,	,, 10	,,	,,		32,718 ,,	0.000032718 ,,
,,	,, 10 1/2 ,,	,,		34,642 ,,	0.000034642 ,,

* This table is close enough for all general estimates.

After 10 1/2 turns the increase is 3850 cm. per turn so that the inductance of 10 1/2+n turns+conn. will be 34,642+$n \times 3850$ cm. With the entire coil in, there are 23 1/2 turns having 84,692 cm. or 0.000084692 H.

In order to test the accuracy of the preceding measurements, readings of the e.m.f. across the two primaries, connections and the regulating coil — all joined in series — were taken repeatedly and in as rapid a succession as was found practicable, the number of the turns of the regulating coil being varied after each set of readings. The diagram below shows the connections of the various inductances while the readings, reduced to the same e.m.f. across: [the two primary turns+connections +1/2 of one turn of the regulating coil+flexible cable] are given in table below:

e

2 PRIMARY TURNS   CONNECTIONS

REGULATING COIL

E

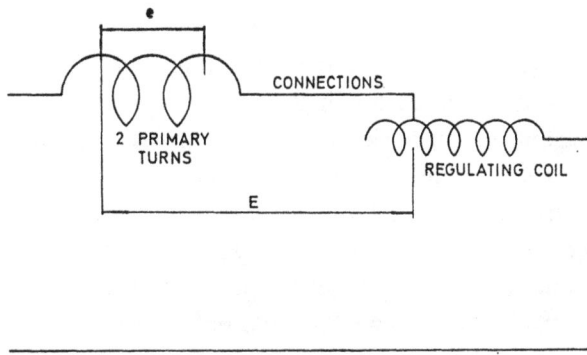

Number of turns of the regulating coil included in circuit.	E.m.f. across two primary turns plus the connections, plus flexible cable+one half of one turn reg. coil					Difference of e.m.f. between successive readings:
	first series of readings	second series	third series	fourth series	average	
23 1/2	15.873	15.928	15.928	15.928	15.914	
22 1/2	15.706	15.706	15.706	15.706	15.706	0.208
20 1/2	15.318	15.263	15.34	15.318	15.310	0.396
18 1/2	14.929	14.8185	14.929	14.929	14.901	0.399
16 1/2	14.4855	14.4855	14.4855	14.4855	14.4855	0.4155
14 1/2	14.119	14.0415	14.1525	14.097	14.102	0.3835
12 1/2	13.764	13.7085	13.764	13.764	13.75	0.352
10 1/2	13.3755	13.3755	13.3755	13.3755	13.3755	0.3745
8 1/2	13.0425	12.9537	13.009	13.0425	13.012	0.3635
6 1/2	12.7095	12.654	12.7095	12.7095	12.6956	0.3164
4 1/2	12.3765	12.3765	12.3765	12.3765	12.3765	0.3191
2 1/2	12.1545	12.1212	12.1875	12.1545	12.1544	0.2221
1/2	12.00	12.00	12.00	12.00	12.00	0.1544

ω was in these readings smaller than 880.

Note: When reduced to the same e.m.f. across the two primaries and connections, 1/2 turn and flexible cable the average values agree fairly well with the readings before recorded. The table prepared on the bases of the values before found will be accurate enough for all ordinary estimates. Both sets of readings show that there is about 0.2 volt variation per turn, the few first turns excepted.

Following readings were taken today for the purpose of putting together a table of the inductances of the various turns of the regulating coil. The machine was specially run and all care was taken to get the readings as close as practicable. The method used in a previous case was again adopted which consisted in reading the e.m.f. across the two primary turns in series and simultaneously the e.m.f. across the two primary turns+ +connections+the turns in the regulating coil. The resistances as before stated being entirely negligible, the inductance in each case was given by the ratio of the e.m. forces and the known inductance of the two primary turns. By this method the error which might have been caused by a variation of ω which could only be determined by taking the speed of the generator, the apparatus for the more exact determination of this quantity being unfortunately left in New York. The results are indicated in the following table:

E.m.f. across two primary turns in series	E.m.f. across two primary turns+ connections+turns of the regulating coil	Number of turns reg. coil	I	ω	Increase of e.m.f. from step to step
12.00	12.3	1/2	58.8	880	
12.00	12.45	2 1/2	58.8	,,	0.15
12.00	12.70	4 1/2	58.8	,,	0.25
12.00	13.05	6 1/2	58.8	,,	0.35
12.00	13.40	8 1/2	58.8	,,	0.35
12.00	13.80	10 1/2	58.8	,,	0.40
12.00	14.20	12 1/2	58.8	,,	0.40
12.00	14.60	14 1/2	58.8	,,	0.40
12.00	15.00	16 1/2	58.8	,,	0.40
12.00	15.40	18 1/2	58.8	,,	0.40
12.00	15.80	20 1/2	58.8	,,	0.40
12.00	16.20	22 1/2	58.8	,,	0.40
12.00	16.40	23 1/2	58.8	,,	0.20

This shows an increase per turn of about 0.20 V, except the few first turns.

Colorado Springs

Nov. 9, 1899

In some experiments it was necessary to use vibrations of lower frequencies and this made it necessary to insert additional inductances in the condenser discharge circuit. In such cases it was convenient to use the two primary turns only; in order to prevent a strong sparking on the secondary and changing reaction on the primaries it was necessary to join the ends of the secondary. Readings were taken to determine more closely the inductance of the primaries with the secondary closed.

The results were as follows:

Current	E.m.f. across two primary turns in series	ω
58.80	8.75	880
58.40	8.5	,,
58.00	8.33	,,
Average 58.40	Average 8.45 (with allowance for zero displacement)	880

With the secondary open the readings were exactly as before:

58.8	11.95	880

Reduced to same current for both cases the readings with secondary closed become:

58.8	8.5	880

The inductance of two primary turns as before found 230,945 cm.$=L$. We have for their inductance with *secondary closed* $\dfrac{8.5}{11.95}L=\textbf{164,270 cm.}$

With both primaries in multiple it ought to be **41,068 cm.** approx.

According to previous estimates the mutual induction coefficient with two primaries in series was: approximately 850,000 cm. The inductance of the secondary was found: 9,568,000 cm. last time, say average of two last determinations 9,560,000 cm. From this data we have for inductance with secondary closed:

$$L-\frac{M^2}{N}=230,945-\frac{85^2\times 10^8}{956\times 10^4}=230,945-\frac{85^2\times 10^4}{956}=230,945-75,575=\textbf{155,370 cm.}$$

These readings above do not quite agree with the result calculated, but I think this only indicates some action of secondary on the primary when the former *is open*, or else the mutual induction coeff. measured a little *too high*. This **very likely.**

As it was not always possible to get along with the primaries alone when using them as inductances two self-induction coils were provided, one wound with wire No. 6, the other with wire No. 2 both on a drum of 5″ diam. The particulars relating to both of these coils will be given below. To ascertain approximately their inductances readings were taken by joining them successively in circuit with the two primary cables in series and taking the e.m.f. across, this giving the inductance of each of them approximately from the ratio of the e.m.f. and the known inductance of the primaries, neglecting, of course, the resistance. The readings were as follows:

For coil wound with No. 6 wire:

E.m.f. across two primary turns+coil all in series	E.m.f. across two primaries in series alone	Current	ω
14.5	6.4	30.9	880
14.5	6.4	30.9	880
14.5	6.4	30.9	880

For coil with No. 2 wire:

13.5	8.2	40.1	880
13.5	8.2	40.1	880
13.5	8.2	40.1	880

This would give approximately inductances of coil No. 6 wire:

$$l = \frac{14.5}{6.4} \times 230{,}945 - 230{,}945 = \frac{8.1}{6.4} \times 230{,}945 = \textbf{292,290 cm,}$$

Coil No. 6 wire

and for coil No. 2 wire

$$l_1 = \frac{13.5}{8.2} \times 230{,}945 - 230{,}945 = \frac{5.3}{8.2} \times 230{,}945 = \textbf{149,340 cm,}$$

Coil No. 2 wire

These figures were first utilised then separate readings were taken. All the particulars of these coils and the measured and calculated values are as follows:

Coil wound with No. 6 wire:

length of wound part $38.75'' = 98.425$ cm, drum $5'' = 12.7$ cm. 129 turns

Thickness of wire with insulation $\dfrac{98.425}{129}$ cm. Thickness of bare wire $= 0.162'' = \textbf{0.41148 cm.}$

Thickness of two insulations $\dfrac{98.425}{129} - 0.41148 = 0.763 - 0.4115 = 0.3515$ cm. This is to be added to the core 12.7 cm diam. making total diam. 13.0515 cm.

To calculate inductance we have therefore the following data:

$$d = 13.0515 \text{ cm}, \quad l_1 = 98.425 \text{ cm}, \quad N = 129, \quad N^2 = 16641, \quad S = \frac{\pi}{4} d^2 = 133.786 \text{ cm.sq.}$$

This gives $l = \dfrac{12.5664}{98.425} \times 16{,}641 \times 133.786 = \textbf{284,247 cm.}$

Now the readings to estimate from were:

e.m.f.	Current	ω	R calculated approx. 180 feet wire 2535ft. per ohm	$\dfrac{E}{I} = 0.271$
				$\left(\dfrac{E}{I}\right)^2 = 0.073441$
13.3	49.1	880		
13.3	49.1	880	0.071 ohm	$R^2 = 0.00504$
13.3	49.1	880		$\left(\dfrac{E}{I}\right)^2 - R^2 = 0.0684$

from this:

$$l = \frac{\sqrt{0.0684}}{880} = \frac{0.2615}{880} \text{ H} \quad \text{or} \quad \frac{261,500,000}{880} = \textbf{297,160 cm.}$$

Small correction should have been made for the e.m.f. making it smaller, this would have made the agreement with the calculated value close.

Coil wound with No. 2 wire

Readings were:

e.m.f.	Current	ω	Resistance will be negligible
6.6	49.1	880	
6.6	49.1	880	
6.6	49.1	880	

$$l_1 = \frac{E}{I\omega} = \frac{6.6}{49.1 \times 880} = \frac{6.6}{491 \times 88} \text{ H}$$

$$\text{or} = \frac{66 \times 10^8}{491 \times 88} \text{ cm.}$$

$$= \textbf{152,750 cm.}$$

The dimensions are as follows:

diam. core $5'' = 12.7$ cm, length of core $38.25'' = 97.185$ cm. Turns 91. The diam. of wire insulation is $\frac{97.185}{91} = 1.068$ cm. Diam. of bare wire $0.2576'' = 0.6543$ cm. This gives for 2 thicknesses $1.068 - 0.6543 = 0.4137$ cm.
From this:

$$d = 13.1137 \text{ cm.}; \quad l' = 97.185 \text{ cm}; \quad N = 91; \quad N^2 = 8281;$$

$$S = \frac{\pi}{4} d^2 = 145.0644 \text{ sq.cm.} \qquad l_1 = \frac{4\pi}{l'} N^2 S = \textbf{155,330 cm.}$$

Probably resistance is not quite negligible, but results are close enough for ordinary use of coil.

Colorado Springs

Nov. 10, 1899

Measurements of the effective capacity of a vertical wire as modified by elevation, by resonance analysis and improved method of locating the maximum rise of e.m.f. on the excited system.

In the previous experiments on the same subject the maximum was located by observing a spark, but it was found that this mode of reading has a number of defects. One of these is the necessity of using spark wires, another the impossibility of locating the maximum very closely — except in cases when the tuning is very sharp. But when considerable capacity is in the system, as it must necessarily be when investigating the modification of capacity, the tuning can never be quite sharp. When the pressures on the excited coil are large spark wires also entail considerable loss, which modifies and vitiates the results of the observations. By the spark method it is also impossible to determine the period and capacity of the excited system itself without any attachments.

In the succeeding observations a method practised in New York was resorted to. This consists of employing a small secondary circuit in feeble inductive connection with the excited system and observing in a convenient manner by a suitable instrument the changes of current or e.m.f. in the secondary. A practical and quite convenient means is to insert a minute lamp consuming but a very small fraction of the normal current and observe the degree of incandescence of the minute carbon filament or thin platinum wire. As the small secondary circuit excercises no appreciable reaction on the excited oscillating system owing to the feeble mutual induction and minute amount of energy consumed in the se-

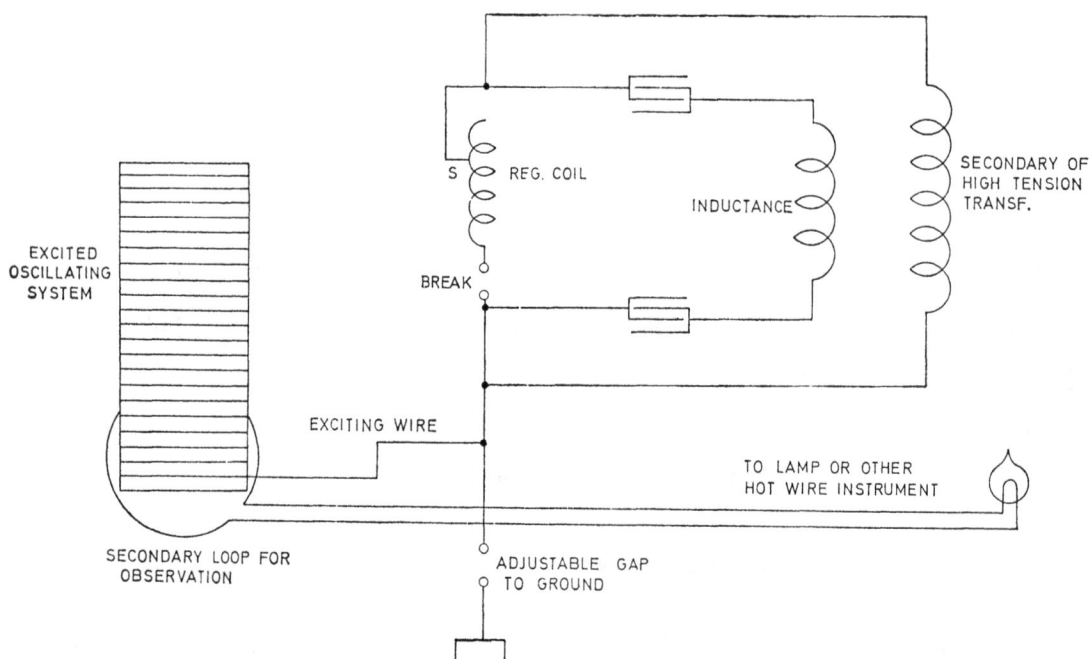

condary, the method is excellent and allows close and reliable readings much more so than the spark wire method. By taking a minute lamp with an exceptionally thin and short filament the energy consumed for the readings is quite insignificant and may be less than one millionth part of the activity of the oscillating system. In the diagram below the arrangement of apparatus as used is illustrated. The excitation was again conveniently varied by an adjustable ground gap. In the secondary circuit feeding the minute lamp it was also of advantage to provide a *continuously* regulable resistance by means of which the brightness of the filament could be reduced to any degree desired. The current from the supply transformer was also regulable as this was necessary in the course of the experiments. Usually I find it advantageous to proceed as follows: first the maximum is located on the proper place of the regulating coil by altering the capacity of the primary circuit until the maximum rise takes place with the contact slide S at the desired point of the regulating coil. A few turns to either side will generally extinguish the lamp. The maximum being thus roughly located, the brightness of the lamp is reduced by inserting resistance or otherwise — as by placing the secondary circuit feeding it farther from the excited system — until the filament is barely visible when the slide S is at the point giving the maximum rise on the excited system. By a little experience it becomes easy to thus locate the maximum within 1/4 of one per cent. By resorting however to ordinary experimental resources it is practicable to reach greater precision still. Of course, the greater the momentum of the excited system the better it is. There are hot wire instruments or detectors of all kinds which allow the method to be refined to any degree desired. A very simple improvement, effective and readily on hand is however to provide a source of energy for bringing the filament or wire just to a point when its luminosity can be detected by the observer. I connect the lamp to a battery of constant e.m.f. through two chocking coils graduating the turns of the latter so that the filament is brought preliminarily to the required temperature. A small amount of surplus energy supplied from the secondary loop is then sufficient to make the filament bright. Thus, less energy is taken from the excited system and the location of the maximum is rendered much more easily. The diagram below illustrates this arrangement in its simplest form. The high frequency currents can not of course pass through the chocking coils. This method is also very suitable for tuning circuits for many purposes as in telegraphy.

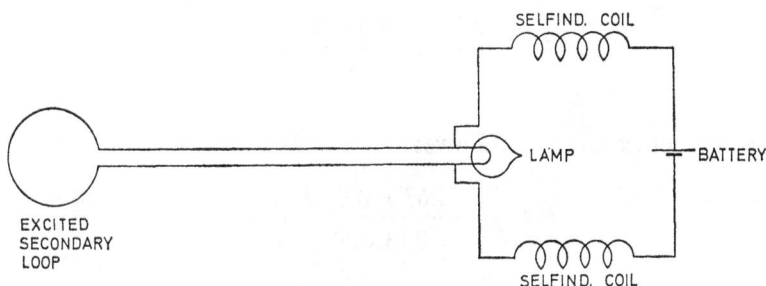

In the present experiments the coil before described: 1314 turns wire No. 18 on drum 14″ diam., 8 feet long was used, the object of the tests being to determine the effective capacity of the vertical wire No. 10, 50 feet long which was used in a number of cases before dwelt upon. The readings were as follows:

Coil 1314 turns with spark wires as before used

Capacity in primary circuit

$$\frac{46}{2} = 23 \text{ bottles} = 0.0207 \text{ mfd.}$$

Inductance in primary circuit

16 Turns of regul. coil+connections+coil No. 6 wire

Note: This reading was taken to test the spark wire method. The agreement was fairly close *15 1/2* turns being found in the previous measurements by spark analysis instead of *16* turns as now. But this was to be expected as with the *spark wires alone*, the capacity being small the tuning is very sharp. The agreement would probably not be quite so close when a large capacity is connected to the excited system.

Coil 1314 turns alone without spark wires

$$\frac{20}{2} = 10 \text{ bottles} = 0.009 \text{ mfd.}$$

12 1/2 turns+conn.+coil No. 6 wire

18*

Coil 1314 turns with vertical wire No. 10 approx. 50 feet long

$\dfrac{70}{2}=35$ bottles$=0.0315$ mfd.   18 1/2 turns+conn.+coil No. 6 wire

The inductance of primary circuit in the first case was:

Coil No. 6 wire   284,000 cm
12 1/2 turns+conn.   42,300 ,, $\Big\}$ total 326,300 cm.

In the second case it was:

Coil No. 6 wire   284,000 cm
18 1/2 turns+conn.   65,400 ,, $\Big\}$ total 349,400 cm.

Calling as before C_{s1} and C_{s2} respectively the capacities of the excited system with and without the vertical wire we have:

$$C_{s1}=\frac{873,000,000}{10^9}=\frac{349,400}{10^9}\times 0.0315$$

* the inductance of excited coil from data obtained before being 87,300,000 cm.

$$C_{s1}=\frac{3494\times 0.0315}{873,000}\ \text{mfd.}$$

Similarly from the preceeding it follows:

$$C_{s2}=\frac{3263\times 0.009}{873,000}\ \text{mfd.}$$

This gives the capacity of the vertical wire:

$$C_{s1}-C_{s2}=\frac{3494\times 0.0315-3263\times 0.009}{873,000}=$$

$$=\frac{110.061-29.367}{873,000}=\frac{80.694}{873,000}\ \text{mfd,}$$

or in centimeters:

$$C=C_{s1}-C_{s2}=\frac{9\times 10^5\times 80.694}{873,000}=\frac{72,624.6}{873}=\textbf{83.2}\ \text{cm.}$$

The calculated value before found was **81.5 cm.** It is assumed in the calculation that the length of the wire was 50 feet exactly, but this might not be so. It will be measured exactly when taken down. The inductance of the coil with wire No. 6 has been taken as 284,000 cm. but the measured values are higher. Taking the average of two measurements we have about 295,000 cm. This would give a higher value for the effective capacity of the vertical wire. It is also possible that the inductance of the excited coil might be a few percent different from that serving as the basis of this estimation.

It is of interest to determine from above data the capacity of the excited coil alone. The same is:

$$C_{s2}=\frac{3263\times 0.009\times 9\times 10^5}{873,000}=\textbf{30.3 cm, approx.}$$

A cylinder of the dimensions of the coil excited would have a capacity $C = \dfrac{l}{2 \log_e \frac{l}{r}}$

Here $l = 8' = 243.84$ cm.　　　　　$r = 7'' = 17.78$ cm.

$$\frac{l}{r} = 13.71 \qquad \log_e \frac{l}{r} = 1.137037 \times 2.3 = 2.6152$$

$$C = \frac{243.84}{2 \times 2.6152} = 46.6 \text{ cm.}$$

Consider now as much of the cylindrical surface as could be covered with the bare wire on the coil:

No. 18 wire diam. $= 0.0403'' = 0.1024$ cm. As there are 1314 turns the wire would cover 1314×0.1024 cm $= 134.55$ cm.

Compared with the cylinder of the length of 243.84 cm the capacity C_1 of the shortened would be in the proportion of 134.55 : 243.84 reduced, that is

$$C_1 \text{ would be } \frac{134.55}{243.84} \qquad C = \frac{134.55}{243.84} \times 46.6 = \text{approx. } \textbf{26 cm.}$$

From this it would seem that a rough estimate of the capacity of such a coil might be obtained by comparison with a cylindrical surface which the bare wire would cover.

Further experiments to ascertain the dependence of capacity upon elevation.

In these experiments the new coil, wound with a much greater number of turns for the purpose of getting a vibration of lower frequency, was used. This coil was wound on the same drum of 14'' diam. and 8 feet length repeatedly used. It had 1314 turns of No. 18 wax covered wire. As the length of the coil and area of the coils remained exactly the same the self-induction was approximately estimated from the inductance of another coil experimented with before. The latter had 689 turns and its measured inductance was 24,000,000 cm. On this basis the inductance of the new coil was $L \left(\dfrac{1314}{689} \right)^2 = ?$, L being the self-nduction of the coil referred to. This would give for $L_1 = \left(\dfrac{1314}{689} \right)^2 \times 24,000,000 = 3.637 \times \times 24,000,000 = 87,288,000$ cm approximately. Comparing it with another coil before described which was wound on the same drum and had 346 turns, and taking the before measured value of the inductance of the latter 6,040,000 cm we get

$$L_1 = \left(\frac{1314}{346} \right)^2 \times 6,040,000 = 14.4225 \times 6,040,000 = 87,111,900 \text{ cm.}$$

which is very nearly the same value.

Rough readings gave:

$$E = 200 \qquad I = 2.5 \qquad \omega = 870$$

from this: $\dfrac{E}{I} = 80$ $\qquad$ $\left(\dfrac{E}{I}\right)^2 = 6,400$ $\qquad$ R calculated: 4816 feet wire No.18
156.9 feet per ohm: $R=30.7$ ohm
(31.68 meas.)

$$R^2 = 942.5$$

$$\left(\dfrac{E}{I}\right)^2 - R^2 = 5457.5$$

$$\sqrt{\left(\dfrac{E}{I}\right)^2 - R^2} = 73.88 \text{ approx.}$$

Inductance nearly **85,000,000 cm.**

For the present investigation the most probably value 87,300,000 cm will be adopted, which is still to be verified.

With the coil before described experiments were made for the purpose of once more determining the capacity of the structure of iron pipes. The adjustments were as follows:

For coil with structure connected to free terminal:

Capacity in primary circuit

Inductance of primary circuit
21 turns regulating coil+conn.+coil wound with wire No. 6 before described

$$\dfrac{(6 \times 36) - 2 + 12}{2} = 113 \text{ bottles} = 0.1017 \text{ mfd}$$

For coil with the spark wires alone:

$\dfrac{46}{2} = 23$ bottles $= 0.0207$ mfd $\qquad\qquad$ 15 1/2+conn+coil No. 6 wire.

In the first case inductance of the primary was $=359,000$ cm. $\qquad \left\{ \begin{array}{ll} \text{Coil No. 6 wire} & 284,000 \text{ cm.} \\ \text{21 turns+conn.} & 75,000 \text{ ,,} \end{array} \right\}$

In the second case $\left\{ \begin{array}{ll} \text{Coil No. 6 wire} & 284,000 \text{ cm} \\ \text{15 1/2 turns+conn.} & 54,000 \text{ ,,} \end{array} \right\} = 338,000$ cm.

Calling C_{s1} capacity of the excited system in first and C_{s2} in the second case we have:

$$\dfrac{2\pi}{10^3} \sqrt{\dfrac{87,300,000}{10^9} C_{s1}} = \dfrac{2\pi}{10^3} \sqrt{\dfrac{359,000}{10^9}} \times 0.1017 \text{ and}$$

$$C_{s1} = \dfrac{\dfrac{359}{10^6} \times 0.1017}{\dfrac{873}{10^4}} = \dfrac{359 \times 0.1017}{87,300} \text{ mfd}$$

or in cm $C_{s1} = \dfrac{9 \times 10^5 \times 359 \times 0.1017}{87,300} = \textbf{376.4 cm.}$

Similarly we have: $\dfrac{2\pi}{10^3}\sqrt{\dfrac{87,300,000}{10^9}} \, C_{s2} = \dfrac{2\pi}{10^3}\sqrt{\dfrac{338,000}{10^9}} \times 0.0207$ and

$C_{s2} = \dfrac{338,000 \times 0.0207}{87,300,000}$ mfd or $C_{s2} = \dfrac{338 \times 0.0207 \times 9 \times 10^5}{87,300}$ cm.

$= \dfrac{3042 \times 20.7}{873} = \textbf{71.67 cm.}$

From this we get *effective capacity* of structure:

$$C_{s1} - C_{s2} = 376.4 - 71.67 = \textbf{304.73 cm,}$$

which is a value very closely before found with *extra coil*.

Note: The readings with spark gap as before practiced are not quite satisfactory and a new method will be tried in the next experiments.

Colorado Springs

Nov. 11, 1899

Experiments for the purpose of ascertaining rate of increase of capacity with elevation continued.

Again the coil with 1314 turns described before was used and the method of locating the maximum rise of potential on the excited system by means of a small circuit inductively connected to the system was resorted to. A few improvements carried out in the mode of using the induced circuit allowed closer readings than it was possible to obtain before with spark observation.

The coil was first tuned alone, without anything being attached to the free terminal. Next the vertical wire No. 10, 50 feet long (approximately) was attached to the free terminal and the tuning again effected, both the primary vibrations being carefully noted. Then a ball 30″ diam. was slipped on to the vertical wire and readings were taken in three different positions of the ball along the wire as before. The results of the observations were as follows:

I. Coil alone

Capacity in primary or exciting circuit

$\dfrac{20}{2} = 10$ bottles $= 0.009$ mfd

Inductance in primary circuit

Turns of reg. 13 coil+conn+coil

No. 6 wire

II. Coil with vertical wire No. 10, 50 feet approx.

$\dfrac{72}{2}=36$ bottles$=0.0324$ mfd   17+conn.+coil No. 6 wire

III. Coil with ball 30″ diam. vertical wire, the ball being at a height of 10′3″ from center to ground.

$\dfrac{86}{2}=43$ bottles$=0.0387$ mfd   13 1/2+ ,, + ,,

IV. Coil with ball 30″ diam. and vertical wire, the ball being at a height of 34 feet from center to ground.

$\dfrac{86}{2}=43$ bottles$=0.0387$ mfd   14 1/2+ ,, + ,,

Note: (it seemed slightly more than 14 1/2 turns)

V. Coil with ball 30″ diameter and vertical wire, the ball being at a height of 57′9″ from center to ground.

$\dfrac{86}{2}=43$ bottles$=0.0387$ mfd   16 1/2+ ,, + ,,

In the first case the inductance of primary circuit was

$\begin{cases} \text{Coil No. 6 wire} & =295{,}000 \text{ cm} \\ \text{13 turns+connections} & = 43{,}300 \text{ ,,} \end{cases}$

total $=338{,}300$ cm.

The primary vibration was therefore:

$$T_{p1}=\frac{2\,\pi}{10^3}\sqrt{0.009\times\frac{3383}{10^7}}$$

Now calling C_{s1} capacity of excited system in the first case we have:

period of excited system

$$T_{s1}=\frac{2\,\pi}{10^3}\sqrt{C_{s1}\times\frac{85}{10^3}}$$

Note: In some estimates before the inductance of this coil was calculated to be a little over 284,000 cm. and this value was taken. But two measurements made before show average of about 295,000 cm. and this value will be assumed in present estimates as being more probable until again careful measurements will be made. The results are then to be corrected.

Note: The inductance for excited coil is taken 85,000,000 cm., this being the value obtained by measurement.

Still to be verified.

From this:

$$C_{s1} = \frac{0.009 \times \dfrac{3383}{10^7}}{\dfrac{85}{10^3}} = \frac{0.009 \times 3383}{85 \times 10^4} \text{ mfd, or in centimeters:}$$

$$C_{s1} = \frac{9 \times 10^5 \times 0.009 \times 3383}{85 \times 10^4} = \frac{0.81 \times 3383}{85} = \textbf{31.84 cm.}$$

This is slightly larger than before found owing to adoption of smaller inductance for excited coil.

In case II. the inductance of the primary circuit was:

$$\begin{cases} \text{Coil No. 6 wire as before:} & 295,000 \text{ cm} \\ 17 \text{ turns+connections} & 59,700 \text{ ,,} \end{cases} \text{ total}=354,700 \text{ cm.}$$

The primary period was:

$$T_{p2} = \frac{2\pi}{10^3} \sqrt{0.0324 \times \frac{3547}{10^7}} \quad \text{and the secondary} \atop \text{corresponding} \quad T_{s2} = \frac{2\pi}{10^3} \sqrt{C_{s2} \times \frac{85}{10^3}}$$

Hence $C_{s2} = \dfrac{0.0324 \times 3547}{85 \times 10^4}$ mfd, or $C_{s2} = \dfrac{9 \times 10^5 \times 0.0324 \times 3547}{85 \times 10^4}$ cm.

$C_{s2}=\textbf{121.68 cm.}$ Hence capacity of the vertical wire will be approximately

$C_{s2}-C_{s1}=\textbf{89.84 cm.}$ This is again larger than before found but probably closer than the former value.

In case III. the primary inductance was:

$$\begin{cases} \text{Coil No. 6 wire} & 295,000 \\ 13\ 1/2 \text{ turns+conn.} & 46,200 \end{cases} \text{ total } 341,200 \text{ cm.}$$

The primary vibration was therefore:

$$T_{p3} = \frac{2\pi}{10^3} \sqrt{0.0387 \times \frac{3412}{10^7}}$$

and the corresponding vibration of the excited system

$$T_{s3} = \frac{2\pi}{10^3} \sqrt{C_{s3} \times \frac{85}{10^3}}$$

from this we have:

$$C_{s3} = \frac{0.0387 \times \dfrac{3412}{10^7}}{\dfrac{85}{10^3}} = \frac{0.0387 \times 3412}{85 \times 10^4} \text{ mfd,}$$

or

$$C_{s3} = \frac{9 \times 10^5 \times 0.0387 \times 3412}{85 \times 10^4} = 139.81 \text{ cm.}$$

The effective capacity of the ball at its lowest position (10′ 3″) from ground was therefore only $C_{s3} - C_{s2} = 139.81 - 121.68 = \textbf{18.13 cm.}$

Now taking cases IV. and V. the primary inductance in the

first of these cases was $\left\{ \begin{array}{ll} \text{Coil No. 6 wire} & \text{295,000 cm} \\ \text{14 1/2 turns} & \text{50,100 ,,} \end{array} \right\}$

the total would be **345,100 cm.** But there is still a doubt whether there have not been 15 turns instead of 14 1/2. This is to be borne in mind. Taking for the present for the inductance 14 3/4 turns as most probable and nearer to the average value of both extreme readings in IV. and V. we have for inductance of the primary **346,000 cm.**

Now in case V. the inductance was: $\left\{ \begin{array}{ll} \text{Coil No. 6} & \text{295,000 cm} \\ \text{16 1/2 turns} & \text{57,800 ,,} \end{array} \right\}$

total **352,800 cm.**

Now since in cases III, IV. and V. the capacity in the primary circuit was not varied we have:

$$C_{s3} : C_{s4} = 341,200 : 346,000 \text{ and } C_{s4} = C_{s3} \times \frac{346}{341} = \textbf{141.78 cm.}$$

and similarly we have:

$$C_{s3} : C_{s5} = 341,200 : 352,800 \text{ and } C_{s5} = C_{s3} \times \frac{3528}{3412} = \textbf{144.56 cm.}$$

The effective capacity of ball at its highest position was:

$$C_{s5} - C_{s2} = 144.56 - 121.68 = 22.88 \text{ cm.}$$

In the mean position the value was: $C_{s4} - C_{s2} = \textbf{20.1 cm.}$ whereas the mean value between 22.88 and 18.13 would be **20.5 cm.** The rise is therefore *linear*. The rise in the effective capacity for 47 feet and 6″ was $\dfrac{18.13}{22.88 - 18.13}$ about 26.2%. Per one *hundred feet* it would be from this: 55.16% or a little over 1/2% per foot.

Colorado Springs

Nov. 12, 1899

Measurements of the effective capacity of the elevated structure of iron pipes were again made today in the manner described before, by means of resonance analysis, the maximum rise of potential on the excited system being determined by a minute lamp included in a secondary circuit without appreciable reaction upon the excited system. The coil with 1314 turns before described was again used, the readings being as follows:

Coil with structure attached:

Capacity in primary circuit

$$\frac{(6 \times 36) + 12}{2} = \frac{228}{2} = 114$$

bottles or 0.1026 mfd.

Inductance in primary circuit
turns reg. coil

15+conn.+coil No. 6 wire

Coil alone, without structure, only connecting wire:

$$\frac{(36 - 6) + 12}{2} = \frac{42}{2} = 21$$

bottles=0.0189 mfd

7 1/2+ ,, + ,,

The inductance in primary in
first case was:

Coil No. 6 wire	295,000
15 turns+conn.	52,000
total	347,000 cm.

The inductance in primary in
second case was:

Coil No. 6 wire	295,000 cm.
7 1/2 turns+conn.	23,600 ,,
total	318,600 cm.

If C_{s1} and C_{s2} be the capacities of the excited system *with* and *without* structure, respectively, then:

$$C_{s1} = \frac{\frac{347,000}{10^9} \times 0.1026}{\frac{87,300,000}{10^9}} = \frac{3470 \times 0.1026}{873,000} \text{ mfd.,}$$

and similarly

$$C_{s2} = \frac{3186 \times 0.0189}{873,000}$$

and

$$C_{s1} - C_{s2} = \frac{3470 \times 0.1026 - 3186 \times 0.0189}{873,000} \text{ mfd} = \frac{356.022 - 60.2154}{873,000} \text{ mfd,}$$

or

$$\frac{9 \times 10^5 \times 295.8066}{873,000} = \textbf{304.95 cm.}$$

This is again a value close to that found with new extra coil. The agreement would be closer still if some connections would be taken in the present instance. I conclude effective capacity is not far from this.

An improvement in the method of locating the maximum rises in the excited system has been effected by taking a lamp with an exceptionally thin filament, consuming only a minute fraction of an ampere, for being heated to redness enough to be perceptible, and furthermore by placing the lamp in a dark box. A "fluoroscope" was used, two holes being drilled in the sides of the box for leading the wires in. By these provisions the readings were made more exact. The new extra coil was again used for trial and the capacity of the structure of iron pipes was again determined. The readings were:

With structure attached

Capacity in primary circuit $\qquad$ Inductance in primary circuit

$\dfrac{(8 \times 36)}{2} = 144$ bottles $= 0.1296$ mfd $\qquad$ Turns+connections

$\qquad\qquad\qquad\qquad\qquad\qquad\qquad$ 18 1/2+ $\qquad$,,

Without structure (only connecting wire)

$\dfrac{(8 \times 36)}{2} = 144$ $\quad$,, $\quad = 0.1296$ mfd $\qquad$ 6 5/16+conn.

Note: In the second case the tuning was, of course, very sharp and it was easy to locate the maximum within 1/16 of a turn of the regulating coil; in the first case, although it was naturally less sharp, it was still easy to locate within 1/4 of a turn; with great care within 1/8 of a turn. This may be said to be within 1/2% which is satisfactory, all the more as the reading is very positive.

The above results give an inductance in the primary circuit, in the first case 65,442 cm, in the second 19,578 cm, computed from the table before prepared. As the capacity in the primary remained the same in both readings we have, calling C_{s1} and C_{s2} capacities of the excited system *with* and without structure and L inductance of the extra coil: $L = 0.02$ henry

$$C_{s1} - C_{s2} = \frac{0.1296\,(65,442 - 19,578) \times 9 \times 10^5}{20,000,000} \text{ cm} = \textbf{267.48 cm.}$$

These readings seem most reliable so far.

In some experiments with coil having 1314 turns wound on drum 14″ diam., 8 feet long the coil was cut in the middle and the two parts, 657 turns each connected in multiple. The self-induction was then practically $\dfrac{1}{4}$ of the self-induction which it had used ordinarily. Readings were taken to determine the inductance when the two parts were connected as stated.

These readings were:

$$\text{e.m.f.} \quad \begin{Bmatrix} 214 \\ 212 \\ 210 \end{Bmatrix} \quad I \begin{Bmatrix} 10.7 \\ 10.6 \\ 10.5 \end{Bmatrix} \qquad \omega = 880$$

Average values:

E	I	ω
212	10.6	880

from this $\left(\dfrac{E}{I}\right) = 20$, $\left(\dfrac{E}{I}\right)^2 = 400$

$R = 7.9$ ohm.
$R^2 = 62.41$

$$\left(\frac{E}{I}\right)^2 - R^2 = 337.59 \qquad \sqrt{\left(\frac{E}{I}\right)^2 - R^2} = 18.375$$

$$L = \frac{18.375 \times 10^9}{880} \text{ cm}$$

$$= 20{,}880{,}682 \text{ cm, approx.} = 20{,}881{,}000 \text{ cm.}$$

The inductance of the coil as ordinarily used would then be approx.

$$= \textbf{83,524,000 cm.}$$

Colorado Springs

Nov. 15, 1899

Experiments with secondary of oscillator to determine capacity of structure, also capacity of secondary.

The readings were as follows:

Secondary alone.

Capacity in primary

$\dfrac{8 \times 36}{2} = 144$ bottles $= 0.1296$ mfd

Inductance in primary

14 3/4 turns + connections.

Secondary with connecting wire leading to structure.

$\dfrac{8 \times 36}{2} = 144$ bottles $= 0.1296$ mfd

15 1/4 ,, + conn.

Secondary with structure connected to free terminal.

$\dfrac{8 \times 36}{2} = 144$ bottles $= 0.1296$ mfd

19 ,, + conn.

* These readings approximate.

In first case inductance of primary was   51,000 cm ⎫
  ,, second ,,   ,,   52,900 ,, ⎬
  ,, third   ,,   ,,   67,400 ,, ⎭

    * All these readings and maybe previous ones to be revised.

Taking the inductance of secondary from measurements before made 9,557,000 cm. we have for C_{s1}, that is, capacity of secondary alone:

$$T_{p1} = \frac{2\pi}{10^3} \sqrt{\frac{51,000}{10^9} \times 0.1296} \;\Bigg| \; C_{s1} = \frac{0.1296 \times 51,000}{9,557,000} = \frac{0.1296 \times 51}{9557} \; \text{mfd.}$$

$$T_{s1} = \frac{2\pi}{10^3} \sqrt{\frac{9,557,000}{10^9} \times C_{s1}} \;\Bigg| \; C_{s1} = \frac{9 \times 10^5 \times 0.1296 \times 51}{9557} = \textbf{622.23 cm.}$$

Now calling C_{s2} and C_{s3} respectively, the capacities of the secondary system with the connecting wire and with wire and structure respectively, since the capacity in the primary was in all cases the same, we have:

$$C_{s1} : C_{s2} = 51,000 : 52,900 \text{ and } C_{s1} : C_{s3} = 51,000 : 67,400$$

and

$$C_{s2} = \frac{52,900}{51,000} \times 622.23 = \frac{529}{510} \times 622.23 = \textbf{645,41 cm.}$$

This gives capacity of connecting wire alone:

$$C_{s2} - C_{s1} = 645.41 - 622.23 = \textbf{23.18 cm.}$$

Similarly we have:

$$C_{s3} = \frac{67,400}{51,000} C_{s1} = \frac{674}{510} \times 622.23 = \textbf{822.32 cm,}$$

and from this the capacity of the structure (the effective capacity) would be:

$$C_{s3} - C_{s2} = 822.32 - 645.41 = \textbf{176.91 cm.}$$

But since

$$C_{s3} : C_{s2} = 67,400 : 52,900 = 674 : 529$$

we have $C_{s3} =$

$$C_{s3} = \frac{674}{529} C_{s2} = \frac{674}{529} \times 645.41 = \textbf{818.54 cm.}$$

This value checks those formerly found and shows that the readings were fairly close. The test shows however that this method of determining capacity will only give a correct value when the distributed capacity is quite negligible. This observation has already been made.

Nov. 16, 1899

Experiments continued on the influence of elevation upon capacity of system connected to earth.

A new coil wound on drum 14″ diam., 8 feet long was used. It had 344 turns No. 10 wire. From the fact that another coil with 346 turns had an inductance of a little over 6,000,000 cm. it is not far away to take the inductance of this coil at that figure.

In the experiments presently described a length of wire No. 12 was used (15 meters long.) The object was to ascertain the capacity of the wire used in connection with the coil. The results of the readings were as follows:

Coil alone without vertical wire.

Capacity in primary circuit

$\frac{36}{2} = 18$ bottles $= 0.0162$ mfd

Inductance in primary circuit

$4\frac{13}{16}$ turns+connections.

Coil with vertical wire No. 12, 15 meters long.

$\frac{36}{2} = 18$ bottles $= 0.0162$ mfd

14 3/4 turns+conn.

The inductance in primary in the first case was 14,530 cm. In second case 51,000 cm, approx.

If C_{s1} and C_{s2} be again the capacity of the excited system in the first and second case respectively, we have by analogy from previous experiments:

$$C_{s1} = \frac{\frac{14,530}{10^9} \times 0.0162}{\frac{6,000,000}{10^9}} = \frac{14,530 \times 0.0162}{6 \times 10^6} \text{ mfd}$$

or

$$C_{s1} = \frac{9 \times 14,530 \times 0.0162 \times 10^5}{6 \times 10^6} = \textbf{35.3 cm.}$$

Since the capacity in the primary circuit remained the same in both experiments, we have:

$$C_{s1} : C_{s2} = 14,530 : 51,000 \text{ and } C_{s2} = \frac{5100}{1453} \quad C_{s1} = \frac{5100}{1453} \times 35.3 = \textbf{121.15 cm.}$$

Hence the capacity of wire alone

$$C_{s2} - C_{s1} = 121.15 - 35.3 = \textbf{85.85 cm.}$$

This is the actual or effective capacity of the wire as used with the coil. But the calculated capacity would be

$$C = \frac{l}{2 \log_e \dfrac{l}{r}}$$

Here $l = 15$ meters $= 1500$ cm.

$r = 0.08081'' = 0.20526$ cm.

$$\frac{l}{r} = 7308$$

$$\log_e \frac{l}{r} = 3.863799 \times 2.3 = 8.887$$

$$C = \frac{1500}{2 \times 8.887} = \frac{1500}{17.774} = \mathbf{84.4 \ cm.}$$

According to this estimate the effective capacity would be only about 1.7% larger than the calculated capacity.

Colorado Springs

Nov. 17, 1899

Experiments to ascertain capacity of various lengths of vertical wire.

Coil with 344 turns and No. 10 on drum 14'' diam., 8 feet long was used. The wire to be tested was No. 12 of a length of 15 meters. The full length was first connected to the free terminal of the coil excited as usual and then 3 meters were cut off each time and the adjustment of the primary circuit made. The results are indicated below:

Capacity in primary circuit	Length of vertical wire	Inductance in pr. cir.
$\frac{22}{2} = 11$ bottles$=0.0099$ mfd	15 meters	21 1/2 turns+conn.
,, ,,	12 ,,	19 ,, + ,,
,, ,,	9 ,,	16 1/4 ,, + ,, 17
,, ,,	6 ,,	13 3/4 ,, + ,, 13 5/8 ,, + ,,
,, ,,	3 ,,	10 1/2 ,, + ,,
,, ,,	0 ,,	7 3/8 ,, + ,,

Approximate estimates from the above readings:

The inductance of coil 344 turns is assumed to be 6×10^6 cm. which is still to be confirmed by close measurement. The inductance of primary when no wire was attached was 7 3/8 turns + conn. = 23,157 cm. With 3 meters wire attached it was 10 1/2 + ccnn. = = 34,642 cm. Hence calling C_{s1} and C_{s2} the capacities of the excited system, in the two cases respectively we have:

$$T_{p1} = \frac{2\pi}{10^3} \sqrt{0.0099 \times \frac{23157}{10^9}} \qquad C_{s1} = \frac{0.0099 \times 23,157}{6 \times 10^6} \text{ mfd, or in centimeters:}$$

$$T_{s1} = \frac{2\pi}{10^3} \sqrt{\frac{6 \times 10^6}{10^9} C_{s1}} \qquad C_{s1} = \frac{9 \times 0.0099 \times 23,157}{60} = \mathbf{34.386} \text{ cm.}$$

$$T_{p2} = \frac{2\pi}{10^3} \sqrt{0.0099 \times \frac{34,642}{10^9}} \qquad \text{and since the capacity in the primary circuit}$$
was the same in both cases:

$$T_{s2} = \frac{2\pi}{10^3} \sqrt{\frac{6 \times 10^6}{10^9} C_{s2}} \qquad C_{s2} = \frac{34,642}{23,157} \quad C_{s1} = \frac{34,642}{23,157} \times 34.386 = \mathbf{51.44} \text{ cm.}$$

The value of effective capacity of the first 3 meters of wire was therefore $C_{s2} - C_{s1} =$ = 51,44 — 34,386 = **17.054 cm.**

Calling now C_{s3} the capacity of the excited system when 6 meters of wire connected to it we have, since in this case the inductance of the primary was 13 3/4 turns + conn. = = 47,154 cm.

$$C_{s3} = \frac{47,154}{23,157} \quad C_{s1} = \frac{47,154}{23,157} \times 34.386 = \mathbf{70.02 \text{ cm.}}$$

Hence the value of effective capacity of the second piece of wire 3 meters long was

$$C_{s3} - C_{s2} = 70.02 - 51.44 = \mathbf{18.58 \text{ cm.}}$$

Now in the case when 9 meters of wire were attached the inductance of the primary was 16 1/4 turns + conn. = 56,779 cm. Calling C_{s4} the corresponding capacity of the excited system we have:

$$C_{s4} = \frac{56,779}{23,157} \times 34.386 = 84.307 \text{ cm.}$$

Hence effective value of the 3rd piece of 3 meters length was

$$C_{s4} - C_{s3} = 84.307 - 70.02 = 14.287 \text{ cm.}$$

Note* In another series of readings for 9 meters the inductance of primary was found to be 17 turns + conn. = 59,665 cm, and on this basis I find:

$C_{s4} = \dfrac{59,665}{23,157} \times 34.386 = \textbf{88.597 cm.}$ According to this the effective value of the 3rd piece 3 meters long would then be

$$C_{s4} - C_{s3} = 88.597 - 70.02 = \textbf{18.577 cm.}$$

When 12 meters wire were attached the inductance of primary was found to be 19 turns+conn.=67,367 cm. Hence similarly $C_{s5} = \dfrac{67,367}{23,157} \times 34.386 = 106.034$ cm, and from this the value of 4th piece of wire $C_{s5} - C_{s4} = 100.034 - 84.307 = \textbf{15.727 cm.}$ But according to second reading it would be:

$$100.034 - 88.397 = \textbf{11.437} \; cm, \text{ only.}$$

Finally, when 15 meters were attached the inductance in primary was: 21 1/2 turns+ +conn.=73,142 cm, and therefore: $C_{s6} = \dfrac{73,142}{23,157} \times 34.386 = 108.609$ cm. and this would give as value of the last piece of three meters

$$C_{s6} - C_{s5} = 108.609 - 100.034 = \textbf{8.575 cm} \text{ only.}$$

Here possibly inductance of wire begins to assert itself. These values as found are still to be considered.

Colorado Springs

Nov. 18, 1899

Experiments were continued to ascertain influence of elevation upon capacity of a system connected to earth as in previous instances. Coil 344 turns referred to before was again used. Also wire vertical No. 10, 50 feet length and ball 30″ diam. The procedure was as in a similar case before. The results were as follows:

Coil without vertical wire.

Capacity in primary $\qquad\qquad\qquad\qquad$ Inductance in primary

$\dfrac{36}{2} = 18$ bottles=0.0162 mfd $\qquad\qquad$ $4 \dfrac{13}{16}$ turns+conn.

Coil with vertical wire No. 10, 50 feet.

$\dfrac{36}{2} = 18$ bottles=0.0162 mfd $\qquad\qquad$ 14 3/4 turns+conn.

Coil with ball 30″ diam. slid on vertical wire.

Capacity primary	Height of ball from center to ground	Inductance in primary
$\dfrac{36}{2}=18$ bottles$=0.0162$ mfd.	10′ 1″	16 1/4 turns+conn.
,, ,,	33′ 8″	16 7/16 ,, +conn.
,, ,,	57′ 3″	16 5/8 ,, +conn.

On the basis of these readings the following results are obtained: In the first experiment when no wire was attached, the inductance in primary was: $4\dfrac{13}{16}$ turns+conn.$= =14,526$ cm. Calling again C_{s1} capacity of excited system we have

$$C_{s1}=\frac{0.0162\times\dfrac{14,526}{10^9}}{\dfrac{6\times10^6}{10^9}}\ \text{mfd}$$

$$C_{s1}=\frac{0.0162\times14,526\times9\times10^5}{6\times10^6}=\textbf{35,298 cm.}$$

In the second case with wire 50 long attached to the excited system the capacity of primary being the same as before and the inductance of primary being 14 3/4 turns+ +conn=51,004 cm, we have:

$$C_{s2}=\frac{51,004}{14,526}\ C_{s1}=\frac{51,004}{14,526}\times35.298=3.511\times35.298=123.931\ \text{cm.}$$

Hence capacity of wire

$$C_{s2}-C_{s1}=123.93-35.298=\textbf{88.632 cm.}$$

Now with ball at its lowest position capacity in primary was as before and inductance 16 1/4 turns+conn.$=56,779$ cm. Hence

$$C_{s3}=\frac{56,779}{14,526}\ C_{s1}=3.9088\times35.298=\textbf{137.973 cm.}$$

From this effective value of ball at the height of 10′1″ was

$$C_{s3}-C_{s2}=137.973-123.931=\textbf{14.042 cm.}$$

With ball at a height of 33′8″ the inductance in primary was 57,502 cm. and at the height of 57′3″ it was 58,223 cm. As the capacity in primary was the same the values for C_{s4} and C_{s5}, respectively, are at once found since

$$C_{s4}=\frac{57,502}{56,779}\ C_{s3}\quad\text{and}\quad C_{s5}=\frac{58,223}{56,779}\ C_{s3}$$

From this we find

$$C_{s4}=1.0127\times137.973=139.725\ \text{cm, and}\ C_{s5}=1.02543\times137.973=141.482\ \text{cm.}$$

19*

The effective value of capacity of ball at the height

of 33′ 8″ was $C_{s4}-C_{s2}=139.725-123.931=15.794$ cm, and

at the height of 57′ 3″ $C_{s5}-C_{s2}=141.482-123.931=17.551$ cm.

From these results it would appear that from the lowest to the highest position there was an increase of about 25% total or per foot of elevation 0.53%. These readings were made under conditions not the best.

Colorado Springs

Nov. 19, 1899

In order to further investigate effect of elevation upon the capacity of a system as before a cylinder of thin sheet iron 4″ in diam. was prepared in sections 2 meter long each, there being 7 sections in all. The separate tubes were slipped one into the other so that when one was taken off each time the total length was shortened by exactly two meters. The cylinder was supported vertically above the coil used in the experiments by means of a cord extending from the wooden structure in previous instances described and the experiments were usually begun with the full length of tube and after each adjustment one length was taken off. The results were as follows:

Capacity in primary circuit			Inductance in primary		Length of cylinder	
$\dfrac{2\times 36}{2}$ =36 bottles=0.0324 mfd			10 3/4	turns+conn.	14	meters
,,	,,	,,	9 7/8	,, + ,,	12	,,
,,	,,	,,	8 3/4+1/16	,, + ,,	10	,,
,,	,,	,,	7 13/16	,, + ,,	8	,,
,,	,,	,,	6 11/16	,, + ,,	6	,,
,,	,,	,,	5 5/16	,, + ,,	4	,,
,,	,,	,,	3 3/4 less 1/32	+ ,,	2	,,
,,	,,	,,	1 3/8+1/16	,, + ,,	0	,,

These results are to be calculated.

Note: coil used 344 turns drum 14″ diam. 8 feet length.

Colorado Springs

Nov. 20, 1899

Experiments with coil 344 turns to determine influence of elevation upon capacity of system as before used were repeated.

The results were as follows:

Coil alone

Capacity in primary

$$\frac{2 \times 36}{2} = 36 \text{ bottles} = 0.0324 \text{ mfd}$$

Inductance in primary

1 3/8 turns+conn.

Coil with vertical wire 50 feet (No. 10)

$$\frac{2 \times 36}{2} = 36 \text{ bottles} = 0.0324 \text{ mfd}$$

8 1/16 turns+conn.

Experiments with ball 30″ diam.

Capacity primary circuit	Height of ball from center to ground	Inductance primary
$\frac{2 \times 36}{2}$ =36 bottles=0.0324 mfd	10′ 1″	8 7/8+conn.
,, ,, ,,	33′ 8″	9 +conn.
,, ,, ,,	57′ 3″	9 1/8+conn.

from this follows:

When coil was alone the inductance of primary was 1 3/8 turns +conn.=7121 cm. Hence, taking inductance of coil=6×10^6 cm, we have similarly to preceding

$$C_{s1} = \frac{0.0324 \times 7121}{6 \times 10^6} \text{ mfd} \quad \text{or} \quad C_{s1} = \frac{9 \times 10^5 \times 0.0324 \times 7121}{6 \times 10^6} =$$

=**34.61 cm,** slightly less than found before.

Now in second experiment with wire connected to vibrating system the inductance was 8 1/16 turns+conn.=25,472 cm. Since the capacity was the same we have as before

$$C_{s2} = \frac{25,472}{7121} C_{s1} = 3.577 \times 34.61 = \mathbf{123.7999 \text{ cm.}}$$

From this follows for capacity of wire alone

$$C_{s2} - C_{s1} = 123.7999 - 34.61 = \mathbf{89.19 \text{ cm.}}$$

When ball was at a height of 10′1″ the inductance was 8 7/8 turns+conn.=28,389 cm; in the middle position it was 9 turns+conn.=28,870 cm, and in the highest position it was 9+1/8 turns+conn.=29,351 cm. From this following values are obtained:

$$C_{s3} = \frac{28,389}{7121} C_{s1} = \frac{28,389}{7121} \times 34.61 = 3.98666 \times 34.61 = \mathbf{137.9783 \text{ cm}}$$

$$C_{s4} = \frac{28,870}{7121} \, C_{s1} = 4.054 \times 34.61 = \mathbf{140.301 \ cm. \ and}$$

$$C_{s5} = \frac{29,351}{7121} \, C_{s1} = 4.1218 \times 34.61 = \mathbf{142.6555 \ cm.}$$

The effective capacity of ball at lowest position was

$$C_{s3} - C_{s2} = 137.9783 - 123.7999 = 14.1784 \ \text{cm}.$$

At the middle position it was

$$C_{s4} - C_{s2} = 140.301 - 123.7999 = 16.5011 \ \text{cm}.$$

and at the highest

$$C_{s5} - C_{s2} = 142.6555 - 123.7999 = 18.8556 \ \text{cm}.$$

Hence from lowest to highest there was an increase of about 33% or very nearly an increase of 0.7% per foot or 70% per 100 feet.

Colorado Springs

Nov. 21, 1899

Investigation on influence of elevation upon the capacity continued: The same coil 344 turns on drum of 14″ diam., 8 feet long was used. The object was to ascertain the relative capacities of a wire in vertical and horizontal position. A wire No. 14, 10 meters long was experimented with. Results were as follows:

**Coil with wire vertical, lowest point
being 8′ 8″ from ground.**

Capacity in primary circuit	Inductance in primary
$\frac{22}{2} = 11$ bottles $= 0.0099$ mfd	17 1/2 turns+conn.

**Coil with same wire horizontal at
distance of 8′ 8″ from ground.**

$\frac{22}{2} = 11$ bottles $= 0.0099$ mfd	18 turns+conn.

The capacity in exciting circuit was now changed and readings again taken, the results being as follows:

Capacity primary	Inductance primary

Coil with above wire vertical as before

$2 \times \dfrac{36}{2} = 36$ bottles $= 0.0324$ mfd   6 5/8 turns+conn.

Coil with same wire horizontal as before

$2 \times \dfrac{36}{2} = 36$ bottles $= 0.0324$ mfd   6 3/4 turns+conn.

Determination of the values of the capacities from preceeding readings:

First set of readings: With wire vertical the inductance in primary circuit was 17 1/2 turns+ +conn.$=61,592$ cm, and with wire horizontal it was 18 turns+conn.$=63,517$ cm. Since in all cases before the capacity of the coil alone was found to be approximately 35 cm$=C_{s_1}$ the capacity of the wire in the vertical and horizontal positions was as follows:

Wire in horizontal position:

$$C_{s2} = \frac{0.0099 \times 63,517}{6 \times 10^6} \text{ mfd or } C_{s2} = \frac{0.0891 \times 63,517}{60} \text{ cm} = \mathbf{94.32 \text{ cm.}}$$

and this gives for capacity of wire in horizontal position:

$$C_{s2} - C_{s1} = 94.32 - 35.00 = \mathbf{59.32 \text{ cm.}}$$

Wire in vertical position $C'_{s2} = \dfrac{0.0099 \times 61,592}{6 \times 10^6}$ mfd or $\dfrac{0.0891 \times 61,592}{60}$ cm$=91.464$ cm, and this gives capacity of wire in vertical position:

$$C'_{s2} - C_{s1} = 91.464 - 35.00 = \mathbf{56.464 \text{ cm}} \text{ or } \mathbf{a \text{ little less.}}$$

From second set of readings we get:

Wire vertical: Inductance in primary was 6 5/8 turns+conn.$=20,631$ cm. Hence

$$C''_{s2} = \frac{0.0324 \times 20,631}{6 \times 10^6} \text{ mfd or } \frac{0.2916 \times 20,631}{60} \text{ cm} = 100.26 \text{ cm}$$

and hence value of capacity in this case was for wire alone

$$C''_{s2} - C_{s1} = 100.26 - 35 = \mathbf{65.26 \text{ cm.}}$$

Wire horizontal: Inductance in primary was 6 3/4 turns+conn.$=21,052$ cm. Hence

$$C''_{s2} = \frac{0.0324 \times 21,052}{6 \times 10^6} \text{ mfd or } \frac{0.2916 \times 21,052}{60} \text{ cm} = 102.31 \text{ cm}$$

and the capacity of the wire in horizontal position was then:

$$C_{s2}'' - C_{s1} = 102.31 - 35.000 = \mathbf{67.31\ cm.}$$

These readings do not agree as well as they ought to.

To be gone over.

Colorado Springs

Nov. 22, 1899

Measurement of small capacities by resonance method.

This method is suitable to determine capacities too small to be measured in other ways conveniently. Coil with 344 turns before described was again used.
Results:

**Coil alone with short piece of stout wire
connected to the free terminal.**

Capacity in primary circuit   Inductance in primary circuit

$$\frac{2 \times 36}{2} = 36 \text{ bottles} = 0.0324 \text{ mfd} \qquad 1\ 1/2 \text{ turns} + \text{connections}$$

**Coil with incandescent lamp 16 c.p. 100 V with
two filaments attached to short thick wire**

$$\frac{2 \times 36}{2} = 36 \text{ bottles} = 0.0324 \text{ mfd} \qquad 1\frac{19}{32} \text{ turns} + \text{conn.}$$

This test gave an idea of the capacity (effective) of the lamp. The primary inductance in first case was 7314 cm. and in the second 7458 cm. from table prepared. From this follows:

$$C_{s1} = \frac{0.0324 \times 7314}{6 \times 10^6} \text{ mfd} \quad \text{or} \quad C_{s1} = \frac{0.2916 \times 7314}{60} \text{ cm} = \mathbf{35.546\ cm.}$$

$$C_{s2} = \frac{0.0324 \times 7458}{6 \times 10^6} \text{ mfd} \quad \text{or} \quad C_{s2} = \frac{0.2916 \times 7458}{6 \times 10^6} \text{ cm} = \mathbf{36.246\ cm.}$$

Here C_{s1} and C_{s2} were respectively the capacities of the system without lamp and with lamp attached. Hence the actual or effective capacity of the lamp in this system was

$$C_{s2} - C_{s1} = 36.246 - 35.546 = \mathbf{0.7\ cm.}$$

An approximate idea is also obtained of the capacity of the short piece of stout wire used to attach the small bodies the capacity of which was to be determined. Namely the capacity of the excited system alone being before determined about 35 cm, the capacity of the wire would be

$$35.546 - 35.000 = 0.546 \text{ cm.}$$

Colorado Springs

Nov. 23, 1899

Measurement of small capacities by resonance method and mode of determining maximum rise before described by means of diminutive circuit was continued. The coil with 344 turns was again used and in order to get better readings on the self-induction regulating coil in the primary, the primary capacity was reduced. The results were:

1.

Coil with short stout wire alone as before.

Capacity in primary circuit

Inductance in primary circuit

$\dfrac{36}{2} = 18$ bottles $= 0.0162$ mfd

$4 \dfrac{13}{16}$ turns + conn.

2.

Coil with lamp same as before.

$\dfrac{36}{2} = 18$ bottles $= 0.0162$ mfd

$4 \dfrac{15}{16}$ turns + conn.

3.

Coil with same lamp seal broken

$\dfrac{36}{2} = 18$ bottles $= 0.0162$ mfd

$4 \dfrac{31}{32}$ turns + conn.

Note: Curious, the increased capacity probably due to absorption.

4.

Coil with one of my Roentgen tubes as described in articles E.R.

$\dfrac{36}{2}$=18 bottles=0.0162 mfd   5 1/8 turns+conn.

5.

Coil with "double focus tube" target connected.

$\dfrac{36}{2}$=18 bottles=0.0162 mfd   5 1/16 turns+conn.

6.

Coil with same tube one of the electrodes connected.

$\dfrac{36}{2}$=18 bottles=0.0162 mfd   5 3/32 turns+conn.

* **Note:** all of these tubes developed rays fairly strong while tested.

7.

Coil with Lennard tube single terminal as described by me E.R., poorly exhausted, streamers passing through it.

$\dfrac{36}{2}$=18 bottles=0.0162 mfd   5 turns+conn.

From these measurements the capacities can now be found.

Calling the inductances in the primary circuit in each succeeding experiment respectively L_1, L_2, L_3, L_4, L_5, L_6 and L_7 we have with reference to prepared table:

L_1=4 13/16 turns+conn.=14,526 cm.

L_2=4 15/16 ,, +conn.=14,947 ,,

L_3=4 31/32 ,, +conn.=15,053 ,,

L_4=5 1/8 ,, +conn.=15,579 ,,

L_5=5 1/16 ,, +conn.=15,368 ,,

L_6=5 3/32 ,, +conn.=15,473 ,,

L_7=5 ,, +conn.=15,158 ,,

Calling furthermore the corresponding capacities of the excited system

$C_{s1} \ldots C_{s7}$

all of them can be at once determined from C_{s1} since

$$C_{s2} = \frac{L_2}{L_1} C_{s1}$$

$$C_{s3} = \frac{L_3}{L_1} C_{s1}, \text{ etc.}$$

Now analogous to previous preceedings of this kind

$$C_{s1} = \frac{0.0162 \times L_1}{6 \times 10^6} \text{ mfd. } \text{ or } C_{s1} = \frac{0.0162 \times 14,526 \times 9 \times 10^5}{6 \times 10^6} = \mathbf{35.298 \ cm.}$$

Taking approximately $C_{s1} = \mathbf{35.3 \ cm}$ we have:

$$C_{s2} = \frac{L_2}{L_1} C_{s1} = \frac{14,947}{14,526} C_{s1} = 1.029 \times 35.3 = 36.324 \text{ cm}$$

$$C_{s3} = \frac{L_3}{L_1} C_{s1} = \frac{15,053}{14,526} C_{s1} = 1.0363 \times 35.3 = 36.58 \text{ cm}$$

$$C_{s4} = \frac{L_4}{L_1} C_{s1} = \frac{15,579}{14,526} C_{s1} = 1.0725 \times 35.3 = 37.859 \text{ cm}$$

$$C_{s5} = \frac{L_5}{L_1} C_{s1} = \frac{15,368}{14,526} C_{s1} = 1.058 \times 35.3 = 37.347 \text{ cm}$$

$$C_{s6} = \frac{L_6}{L_1} C_{s1} = \frac{15,473}{14,526} C_{s1} = 1.0652 \times 35.3 = 37.6 \text{ cm}$$

$$C_{s7} = \frac{L_7}{L_1} C_{s1} = \frac{15,158}{14,526} C_{s1} = 1.0435 \times 35.3 = 36.8355 \text{ cm}$$

from these values follow:

Capacity effective of lamp experiment	$2 = C_{s2} - C_{s1} = 36.324 - 35.3 = \mathbf{1.024 \ cm.}$
,, ,, seal broken	$3 = C_{s3} - C_{s1} = 36.58 - 35.3 = \mathbf{1.28 \ cm.}$
,, of my Roentgen tube exp.	$4 = C_{s4} - C_{s1} = 37.859 - 35.3 = \mathbf{2.559 \ cm.}$
,, double focus tube target connected	$5 = C_{s5} - C_{s1} = 37.347 - 35.3 = \mathbf{2.047 \ cm.}$
,, ,, ,, electrode connected	$6 = C_{s6} - C_{s1} = 37.6 - 35.3 = \mathbf{2.3 \ cm.}$
,, ,, Lennard tube described	$7 = C_{s7} - C_{s1} = 36.8355 - 35.3 = \mathbf{1.5355 \ cm.}$

Colorado Springs

Nov. 24, 1899

A test was made with the object of ascertaining how close the table of inductances prepared from measured data agreed with the values determined by resonance method. The procedure was as follows: the coil with 344 turns on drum 14″ diam., 8 feet long was again used as suitable for the test and it was excited in the manner before described. In order to establish a different relation between capacity and self-induction of the primary circuit these constants were in each case varied and the adjustment completed until the maximum rise on the terminal or terminals of the excited coil took place. As the period of the system remained in each case the same the products of the capacity and self-inductance in primary remained constant also. Now the capacities in the primary in the succeeding experiments being known or exactly measurable, the various values of inductance in primary were obtained from the relation:

$$L_1\,C_1 = L_2\,C_2 = L_3\,C_3 = \text{etc.}$$

The period of the secondary system was $T_s = \dfrac{2\pi}{10^3}\sqrt{LC_{s1}}$.

The inductance of coil 344 turns being about 6×10^6 cm. and the average value for C_{s1} from a number of readings with different values of inductance and capacity in primary circuit being 34.9 cm., the period of secondary or excited circuit was thus given.

A reading was now taken at random and resonance was obtained with constants in primary circuit as follows:

Capacity in primary circuit	Inductance in primary
$\dfrac{36}{2} = 18$ bottles$=0.0162$ mfd	$4\dfrac{13}{16}$ turns$+$conn.

from this the period of primary circuit was:

$$T_p = \frac{2\pi}{10^3}\sqrt{0.0162 \times L} \qquad L \text{ being the inductance in primary}$$

Now $T_p = T_s$ or $0.0162 \times \dfrac{L}{10^9} = \dfrac{6\times 10^6}{10^9} \times \dfrac{34.9}{9\times 10^5}$ and from this we get L in centimeters:

$$L = \frac{6\times 10^6 \times 34.9}{9\times 10^5 \times 0.0162} = \frac{34.9 \times 6}{9 \times 0.0162} = \frac{698}{0.0486} = L = \textbf{14,362 cm.}$$

by resonance method.

Now from table prepared:

$$\text{Inductance } L = 4\frac{13}{16} + \text{conn.} = \begin{cases} \text{Inductance } 4\ 1/2 \text{ turns}+\text{conn.} = 13,474 \text{ cm.} \\ \text{Inductance } 5 \quad\ \text{turns}+\text{conn.} = 15,158 \text{ cm.} \\ \hline \text{Ind. of } 1/2 \text{ turns} \qquad\qquad = 1684 \text{ cm.} \\ \text{Ind. of } 1/16 \text{ turns} \qquad\quad = 210.5 \text{ cm.} \end{cases}$$

Ind. of $5/16$ turn$=5\times 210.5 = 1052$ cm approx.

Consequently inductance $L = \text{Ind.}\begin{Bmatrix} 4\ 1/2 + 5/16 \text{ turn} \\ +\text{conn.} \end{Bmatrix} = 4\dfrac{13}{16}$ turn$+$conn.$=$

$$= 13,474 + 1052 = \textbf{14,526 cm.}$$

Agreement fairly close, within about 1%. This shows readings are reliable.

Nov. 25, 1899

Experiment which follows was made to ascertain how the capacity of the same conductor may be altered by different distribution. The experiment was performed in the following manner. Two lengths of wire No. 10 were taken (rubber covered) and one length was bent zigzag fashion so that a piece with three parallel wires was obtained one meter long. The other length was cut in three pieces 1 meter long each and these were connected at the ends. The difference between the two pieces so prepared will appear

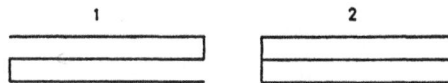

from the sketch in which 1. shows the zigzag wire and 2. the wire cut in three pieces joined in multiple. The distribution in both cases was radically different. These pieces were one after the other placed on the free terminal of a coil with 344 turns and in each case the primary adjusted until resonance was observed. The wires were placed vertically, in the prolongation of the axis of the coil.

An experiment was furthermore made in tuning, not to the real vibration but to a higher harmonic (the next octave) of the primary. The results (with spark gaps slightly changed) were:

Capacity in primary circuit	Inductance primary
$\dfrac{36}{2}=18=0.0162$ mfd **with wires multiple**	1 1/8 turns+conn.
$\dfrac{36}{2}=18=0.0162$ mfd **with wire zigzag**	1 1/16 turns+conn.

Now the primary inductance in the first case was 6736 cm and in the second case 6635 cm. Hence, the inductance remaining practically the same, the capacity (effective) of the zigzag wire was smaller to the extent of nearly **1.6%**.

Nov. 26, 1899

Determination of the capacity of structure of iron pipes by improved method
before described.

Note: Originally the structure, to prevent lateral play, was supported sideways by 8 projecting beams at a height of about 80 feet, each beam having fastened into its end a strong champagne bottle. The necks of the bottles abutted against the iron pole and prevented sideways play. They were wrapped with tape to diminish the danger of the necks being broken. This arrangement was good enough and withstood the storm but it did not allow going beyond a certain pressure as the sparks from the iron pipes would jump to the beams which had the bottles fastened into them. To overcome this defect a plan was adopted, contrived long ago, which consisted in providing a conical roof or hood (made in two parts) rounded on the periphery to reduce loss by leakage and fastening four cords under the roof for the purpose of preventing lateral play and steadying this pole. This arrangement is excellent as the sparks can not jump upon and follow the cords to the ground being fastened under the roof where the electrical pressure was extremely

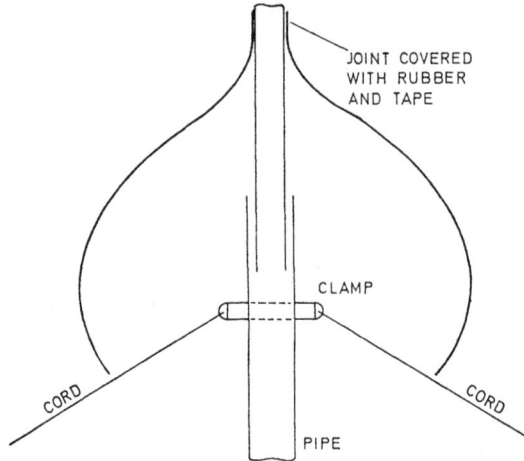

small. The arrangement is indicated in the sketch. The dimensions of the hood were: outside diam. 8 feet., diam. of small circle on periphery 9″. The height of the conical surface about 3 1/2 feet. The cords were led to the four corners of the building and fastened to the same by means of a cable of sheet wires over which was slipped a rubber hose with very thick wall and which was wrapped around three glass insulaters supported on the pole on each corner. It was thought of advantage to insulate the ropes thoroughly and for this purpose they were soaked about a week in linseed oil boiled out and dried in the sun afterward. The cords went down at an angle such that the nearest point of the hood was two feet distant. This arrangement permitted the charging of the pole easily up to a million volts. This is the best arrangement I have found so far for supporting a body to be charged to so high a potential as is necessary for instance in the transmission of messages over great distances. It has been in use since a few weeks ago but the measurements of capacity of the structure before recorded were made without the hood. Presently the readings were taken with the hood with following results:

**Coil 344 turns drum 14″ with iron structure connected
to free terminal.**

Capacity in primary circuit $\qquad$ Inductance in primary circuit

$$\frac{2 \times 36 + 12}{2} = \frac{84}{2} = 42 \text{ bottles} = 0.0378 \text{ mfd}$$ 21 1/8 turns+conn.

**Coil with connecting wire alone shifted away from
structure (4 feet).**

$$\frac{2 \times 36 + 12}{2} = 42 \text{ bottles} = 0.0378 \text{ mfd}$$ 4 turns+conn.

From this follows: the inductance in primary circuit in the first case was 75,548 cm, in the second case 12,319 cm referring to table of inductances before used. Hence capacity of excited system in first

$$C_{s1} = \frac{0.0378 \times 75,548}{6 \times 10^6} \text{ mfd, or } \frac{0.3402 \times 75,548}{60} \text{ cm} = C_{s1}$$

And similarly $C_{s2} = \dfrac{0.3402 \times 12,319}{60}$ cm, C_{s2} being capacity of excited system in second

experiment. This gives for actual or effective capacity of **structure with hood**

$$C_{s1} - C_{s2} = \frac{0.3402}{60} (75,548 - 12,319) \text{ cm} = \frac{0.3402}{60\cdot} \times 63,229 \text{ cm} = \textbf{358.5 cm.}$$

* Small corrections for inductance may have to be taken later.

Effective capacity of structure of iron pipes with new hood again determined by resonance method. The new "extra coil" was used, its inductance being as before 0.02 H. The readings were as follows:

Capacity in primary   Inductance in primary

Coil with structure and connecting wire

$5 \times 36 - 12 = 168$ bottles $= 0.1512$ mfd   20 3/8 turns+conn.

Coil with connecting wire alone placed at 4 ft. distance

$4 \times 36 = 144$ bottles $= 0.1296$ mfd   7   turns+conn.

other readings:

50 bottles $=0.045$ mfd	17 1/4 turns+conn
40 ,, $=0.036$ mfd	20 turns+conn.
38 ,, $=0.0342$ mfd	20 3/4 turns+conn.
39 ,, $=0.0351$ mfd	(approx) 20 3/8 turns+conn.

From these readings follows:

Inductance in first case, with structure, in primary was 20 3/8 turn+conn.$=72,661$ cm.
Inductance in second case, without structure, in primary was 7 turn+conn.$=21,894$ cm.

Calling C_{s1} and C_{s2} the effective capacities of the excited system in the two cases respectively, we have:

$$C_{s1} = \frac{0.1512 \times 72,661}{2 \times 10^7} \text{ mfd, and } C_{s2} = \frac{0.1296 \times 21,894}{2 \times 10^7} \text{ mfd, hence}$$

$$C_{s1} - C_{s2} = \frac{0.1512 \times 72,661 - 0.1296 \times 21,894}{2 \times 10^7} =$$

$$= \frac{10,986.3432 - 2837.4624}{2 \times 10^7} = \frac{8148.8808}{2 \times 10^7} \text{ mfd,}$$

or

$$\frac{9 \times 8148.8808}{2 \times 10^2} \text{ cm} = \textbf{366.7 cm.}$$

effective capacity of structure **with hood.**

Capacity of structure without hood before found with extra coil was 311.2 cm, hence for hood alone we get 366.7—311.2=**55.5 cm.** From first and last reading it appears that the secondary capacities in the two cases were as $\dfrac{168}{39}$. Now the capacity of excited system in last reading was

$$C'_{s2} = \frac{0.0351 \times 72,661}{2 \times 10^7} \text{ mfd, or } C'_{s2} = 144.77 \text{ cm.}$$

From this would follow value

$$C'_{s1} = \frac{168}{39} \times 114.77 = 494.39 \text{ cm.}$$

Hence
$$C'_{s1} - C'_{s2} = 494.39 - 114.77 = \textbf{379.62 cm.}$$

This value does not agree quite closely with that before found but the tuning was not quite exact in last case.

Note: It now seems that the tuning in all previous cases when structure was determined was made to the first octave instead of to the fundamental tone. **This is to be ascertained.** If this be so then capacity would be much greater.

Colorado Springs Notes

Dec. 1—31, 1899

To be completed:

Dec. 11, 12, 13 Application for Page for separation of gaseous mixtures by high tension discharge of oscillator.

Dec. 17 Description of phenomenon on Pike's Peak in the few days of eclipse of moon.

Patent note

Increase of capacity uses in the arts and scientific measurements.

Colorado Springs Nov. 10. 1899.

is barely visible when the slide z is on the point giving the maximum rise on the exciter system. By a little experience it becomes easy to thus locate the maximum within 1/2 of one per cent. By resorting however to ordinary experimental resources it is impracticable to read greater precision still. Of course the greater the resonance of the exciter system the better it is. There are however instruments or detectors of all kinds which allow the method to be refined to any degree desired. A very simple improvement, effective and readily on hand is however to provide a source of energy for bringing the filament or wire just to a point when its luminosity can be detected by the observer. I connect the lamp to a battery of constant e.m.f. through two checking coils graduating the turns of the latter so that the filament is brought preliminarily to the required temperature. A small amount of surplus energy supplied from the secondary loop is then sufficient to make the filament bright. Thus, less energy is taken from the exciter system and the location of the maximum is rendered much more easy. Diagram below illustrates this arrangement in its simplest form. The high frequency current can not of course pass through the checking coils. The method is also very suitable for tuning circuits for many purposes as in telegraphy.

In the present experiments the coil before described: 1314 turns wire No. 18 — diam. 14" diam. 8 feet long was near the object

3

Dec. 1, 1899

Some particulars about the *apparatus* used in the *experiments here.*

The connections of the bottles in the two tanks frequently referred to as the "old tanks", to which the primary cables and the regulating coil were usually connected and by means of which the finer adjustments of capacity in the primary circuit were made, are as shown in the sketch. The top brass plate has 16 plugs disposed at equal distances in a square and the 36 bottles in the tank being connected by copper springs as indicated, each plug enabled the cutting out or in of two bottles, with the exception of the four central plugs which cut in or out three bottles each. Thus the smallest variation of the capacity on one side was one bottle or 0.0009 mfd, approx. But with the tanks in two sets in series as usually employed it was one half of one bottle. Considering the large number of bottles the variation was small enough for most purposes. The bottles in the new tanks were divided in three sets, twelve bottles in each.

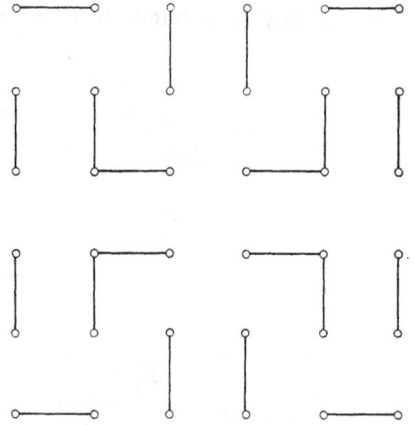

Measured length of all connections on the *condenser sets in primary:*

From top of right condenser to break .... $2' 5\frac{1}{2}'' = 29.5''$

Through break wheel and sp. rod .... $2' 4\frac{1}{2}'' = 28.5''$

From back to regulating coil............................... $3' 10'' = 46''$

Up through second break and rod .... $1' 10'' = 22''$

Connection to left condenser .... $3' 7'' = 43''$

To bottom of left condenser .... $1' 8\frac{1}{2}'' = 20.5''$

Connection on bottom between condensers .... $3' 6'' = 42''$

Up to top of right condenser .... $1' 8\frac{1}{2}'' = 20.5''$

Total length of connections is thus $252'' = 641$ cm, nearly or 21 feet. The section of the connections is partially that equal to the cable used in primary 1 cm radius and partially less. The inductance determined by resonance analysis is about *5000 cm total.* The calculated value a trifle more.

Various ways of tuning circuits or *determining the maximum of resonant rise.*

In the course of the experiments conducted here a number of such ways have been resorted to most of which have been described or at least referred to, but it may be useful to record here once more those which have been found most satisfactory. They are diagramatically illustrated in Figs. 1—5. below:

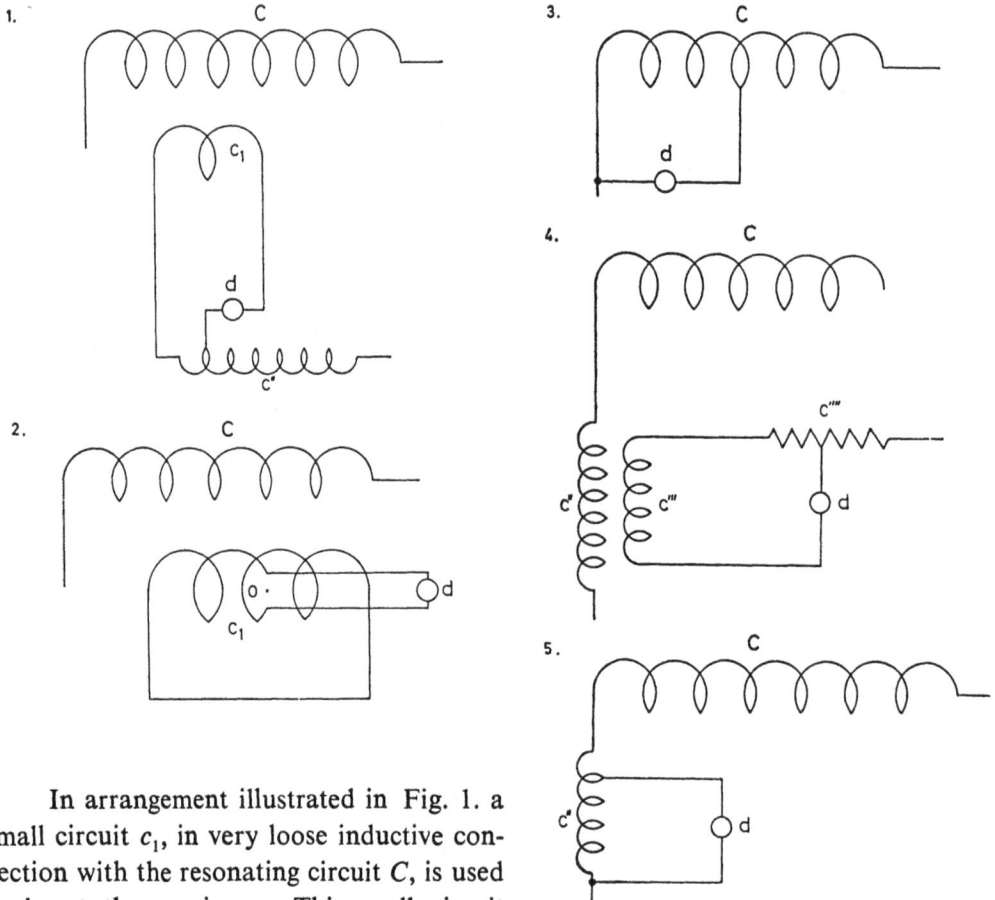

In arrangement illustrated in Fig. 1. a small circuit c_1, in very loose inductive connection with the resonating circuit C, is used to detect the maximum. This small circuit c_1 which exercises no appreciable reaction upon circuit C contains another coil c'' or else a resistance which is noninductive for the purpose of adjusting the effect to suit the indicating device d which is most generally a microscopic spark gap, vacuum tube or any hot wire instrument, as a minute lamp specially made to suit the purpose, a Cardew voltmeter or any other instrument. In Fig. 2. the adjustment of the effect upon d is effected by turning the circuit around a point o or else by approaching small circuit c_1 to or receding with same from circuit C. In Fig. 3. a number of turns of the resonating circuit C are spanned by device d, this number being adjustable. In Fig. 4. again a small coil c'' in series with coil C is placed inductively in connection with a small circuit c''' which again may be adjustable in the manner shown or in any other way. Finally in Fig. 5. a small coil c'' in series with C is spanned by device d. Coil c'' has its turns adjustable. *This method seems best.*

Determination of free vibration of *new "extra coil". No. 10 wire* wound on frame 8′ 3″ diam. and 8′ length modified by taking off five turns on top and placing the last five turns two grooves apart instead of one groove as the rest of turns. In this modification there are just 100 turns in the coil. The excitation was effected from the secondary the connection being made to a point *o* 3/4 turns from the ground plate as indicated in the

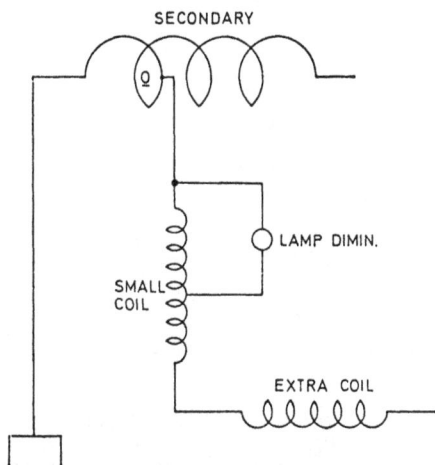

diagram. The maximum rise was determined by means of a diminutive lamp shunting a few turns of an adjustable small coil in series with the extra coil, the small coil being of an inductance entirely negligible as compared with that of the extra coil. The readings for resonating condition were as follows:

Capacity in the primary or exciting circuit

Inductance in the pr. or exc. circuit

$\dfrac{2 \times 36}{2}$ =36 bottles=0.0324 mfd

16 1/16 turns regulating coil+1 primary turn, that is two primary cables in multiple *as modified by reaction of secondary.*

$\dfrac{4 \times 36}{2}$ =72 bottles=0.0648 mfd

5 5/8 turns of coil+primary as above.

From first reading, taking the inductance of primary turn as modified by the secondary 41,000 cm, the period of the oscillation was $T_1 = \dfrac{2\pi}{10^3} \sqrt{0.0324\,L}$, L the inductance of primary being that of primary turn+16 1/16 of regul. coil=41,000+56,000=97,000 cm. This gives

$$T_1 = \frac{2\pi}{10^3} \sqrt{0.0324 \times \frac{97}{10^6}} = \frac{2\pi}{10^6} \sqrt{0.0324 \times 97} =$$

$$\frac{2\pi}{10^6} \times 1.773 = \frac{11.13444}{10^6} \quad \text{and} \quad \mathbf{n = 90,000 \ nearly.}$$

From second reading, the inductance of 5 5/8 turns of regulating coil being 17,300 cm and total inductance of primary $41,000+17,300=58,300$ cm. We have the period

$$T_2 = \frac{2\pi}{10^3} \sqrt{0.0648 \times \frac{583}{10^7}} = \frac{2\pi}{10^6} \sqrt{0.0648 \times 58.3} =$$

$$= \frac{6.28 \times 1.944}{10^6} = \frac{12.2}{10^6} \text{ and } \mathbf{n = 82{,}000 \text{ approx.}}$$

Note: The tuning was very sharp in both cases, but in the second case the secondary reaction *was smaller*, i.e., inductance of the primary greater.

Colorado Springs

Dec. 4, 1899

Experiments to *establish equivalence* between *the inductance of the primary loop* comprising two primary cables in multiple (*with secondary reacting*) and *that of turns of the regulating coil* in primary circuit.

Resonance analysis was resorted to and the coil of 344 turns on drum 14″ diam. 8 ft. long, wire No. 10 was again used. The coil employed had a very small regulating coil in series, and a minute lamp placed across the few turns of this latter coil was employed to ascertain the maximum resonant rise in the coil excited. The results were as follows:

Capacity in the primary circuit	Inductance in the primary circuit
	1.
8 jars on each side, that is 4 jars= =0.0036 mfd	One primary loop as above+6 3/4 turns reg. coil.
	* Small coil in series with coil of 344 turns had 9 turns in circuit *entirely negligible*.
	2.
Capacity same as above	15 1/4 turns regulating p. coil *only*
	* Small coil had 22 turns inserted, still entirely negligible against 344 turns coil

Since resonance was obtained in both cases and the primary capacity in both tests remained the same we have the primary inductances in both cases equal and from this follows that *under the conditions* of these tests the inductance of the primary *as modified* by the *secondary* was equivalent to 15 1/4—6 3/4 turns of the primary regulating coil. Now from the table of inductances the value for the 15 1/4 turns is 52,930 cm and that for the 6 3/4 turns, inserted in the first case, is 21,052 cm. Hence the inductance of primary cables in this instance was

$$L_p = 52,930 - 21,052 = \mathbf{31,878 \ cm.}$$

Thus the secondary, though "open", diminished in this case the primary inductance from 56,400 cm to the above value or about 43.48%.

Colorado Springs

Dec. 5, 1899

Experiments to ascertain *equivalence of inductance of primary loop* (two primary cables in multiple) with *secondary reacting*, and *inductance* of *turns of regulating primary coil* under modified conditions.

Again the coil with 344 turns was used and the maximum resonant rise in same determined in the manner described before. A greater capacity was used this time in the primary so as to come closer to the free vibration of the secondary and thus cause a stronger reaction upon the primary loop. The conditions for resonance were satisfied with the following values:

1.

Capacity in primary circuit	Inductance in primary circuit
$\dfrac{2 \times 36}{2} = 36$ bottles $= 0.0324$ mfd	21 3/4 turns pr. regulating coil *only*

2.

Capacity in primary circuit	Inductance in primary circuit
$\dfrac{2 \times 36}{2} = 36$ bottles $= 0.0324$ mfd	one primary loop as above$+15$ 1/2 turns reg. coil.

Since in both tests the primary capacity was not changed in the least the inductance of the primary loop in this instance was equivalent to that of 21 3/4 turns, less that of 15 1/2 turns of the regulating coil in the primary. This means the inductance of the primary loop was equivalent to that of $21.75 - 15.5 = 6.25$ turns of the primary regulating coil or it was $6.25 \times 3850 = \mathbf{24,063 \ cm}$ only. Induct. diminished 57.33%. This is a still smaller value than found yesterday but is very probably near the maximum as the secondary showed evidence of a resonating condition. These experiments show the danger of allowing for the amount of secondary reaction. This is to be borne in mind.

Further experiments to ascertain *equivalence of inductance of the primary loop* (two primary cables in multiple) with *secondary reacting*, and that of *turns of primary regulating coil* under conditions again modified.

In the present tests the "new extra coil" modified so as to have only 100 turns was used (this is to be explained). The coil was used exactly as that of 344 turns before and the maximum rise was determined in the same way and by the same means as in preceding cases. The results of the two comparative tests were as follows:

1.

Capacity in the primary circuit

$$\frac{5 \times 36 - 24}{2} = \frac{156}{2} = 78 \text{ bottles} = 0.0702 \text{ mfd}$$

Inductance in primary circuit

11 5/16 turns of primary regul. coil *only*

2.

Capacity in primary circuit
same as above=0.0702 mfd

Inductance in primary circuit
one primary turn+3 1/2 turns reg. pr. coil.

From these tests it follows that under present conditions the inductance of the primary loop with secondary reacting was equivalent to that of 11 5/16 turns less that of 3 1/2 turns of the primary regulating coil. Now the table of inductance gives inductance of the 11 5/16 turns=37,775 cm, and that of the 3 1/2 turns=11,164 cm. Hence the inductance of the primary loop under these conditions was

$$L_p=37,775—11,164=\textbf{26,611 cm.}$$

The secondary was resonating to some extent but not as well as in experiments recorded yesterday. The diminution of inductance due to secondary reaction was here 52.82 %.

Note: These tests make it advisable to break the secondary on *more* points when the primary is used for inductance.

Experiments to ascertain to what extent *induced currents in the ground* might affect the *primary and secondary oscillating systems* now used in the laboratory.

Resonance analysis was resorted to and the experiments were performed in the following manner: a *square* frame 11' 1/4" each side and 3 3/4" high, wound with 14 turns wire No. 10 was supported horizontally above the ground, its elevation above the latter being made adjustable. This coil was tuned in various positions and the effect of the ground observed on the change of the period of oscillation in the various positions. In order however to get a better range of reading and also to reduce as much as practicable the effects of the distributed capacity in the coil wound on the square frame the same was joined in series with another coil of 344 turns wound on a 14" diameter drum, which was before used, but so that the coil on the square frame was between the ground connection and the coil of 344 turns, this for the purpose of maintaining between the turns of the square coil a comparatively small difference of potential thus reducing the effect of distributed capacity of the same. By using the additional coil the vibration was rendered slower and a better range of reading was secured. The two coils were excited in the manner repeatedly referred to and described, the maximum resonant rise being determined by means of a minute lamp shunting a few adjustable turns of a very small coil of entirely negligible inductance which was in series with both the coils used.

The results of the tests are indicated in the following table:

Capacity in primary exciting circuit	Distance of square coil from ground	Inductance in primary circuit
1. $\frac{4 \times 36}{2}$ =72 bottles=0.0648 mfd	1' 3"+0	4 1/4 less 1/32 turn reg. pr. coil
2. ,, ,,	1' 3"+2'	4 1/16 turns reg. pr. coil
3. ,, ,,	1' 3"+4'	4 less 1/32 ,,
4. ,, ,,	1' 3"+6'	3 7/8+1/32 ,,

It is to be stated by way of explanation that the *floor* was 1' 3" above ground. A simple inspection of the above results shows that induced currents do not prominently assert themselves. This might have been expected considering the extreme dryness of the ground which, as before pointed out, made it too difficult at the outset of these experiments to get a good ground connection. But it is evident from the above table that the electrostatic capacity is an important element. This is quite evident since the vibration of the system comprising the square coil is *quickened by elevation* of the frame. If there were *induced currents* generated in the ground it would *be just the opposite*. Since the capacity in the primary exciting circuit remained the same in all four experiments the capacity of the excited circuit, the inductance of which remained the same, was varied exactly as the inductance of the primary circuit. Now with reference to the table of inductances the inductances in experiments 1, 2, 3, and 4, respectively, were: 12,824, 12,463, 12,247 and 12,103 cm. The increase of distance from 1 to 2 was 160%; from 2 to 3 61.5%; from 3 to 4 38.2%. Putting results together we have:

Increase of distance		Amount of the diminution of capacity	The numbers indicate a general proportionality but now the capacity of the excited system should be known.
2′	160%	361	
2′	61.5%	216	
2′	38.2%	144	*This will be followed up.*

* Capacity of primary cables with reference to earth should be determined.

Colorado Springs

Dec. 8, 1899

In response to note from I. Hawthorne forwarded to "Northamerican" Philadelphia, dispatch as follows:

"In response to a request from friend I. Hawthorne the following statement:

"Confining myself to my chosen sphere, I believe that the mastery of electrical forces was as great and beneficial an achievement as the Century has witnessed. As to the immediate future my thoughts are dominated by two ideas, one of which is already realized, while the other is on the eve of accomplishment. The art of governing the movements and performances of distant automatons, enabling machines to act as creatures endowed with a mind, will demonstrate to the leading nations the futility of armaments and impracticability of the present life destroying implements of war and will lead to more permanent peaceful relations, in harmony with the humanitarian spirit and enlightenment of the age; while the art of transmitting electrical energy through the natural media, without the use of leading wires, over vast distances, from great centers, as Niagara, will open up inexhaustible resourses of wealth and power and, by rendering immense amounts of the energy of the sun available to the wants of man, will perhaps make it possible for him to produce similar wonderful changes and transformations on the surface of our globe as are, to all evidence, now being wrought by intelligent beings on a neighbouring planet."

N. Tesla

Experiment to determine the inductance of coil wound on the *square frame* before used by means of resonance analysis and *another known coil:*

The square coil was connected in series with coil of 344 turns as before and maximum rise was determined in the same way as described in preceding experiments. The resonant condition was obtained with

Capacity in primary or exciting circuit

Inductance in primary circuit:

$$\frac{36}{2} = 18 \text{ bottles} = 0.0162 \text{ mfd}$$

5 less 1/32 turns reg. coil=15,000 cm approx.

With coil 344 turns alone

$$\frac{36}{2} = 0.0162 \text{ mfd}$$

15 less 1/16 turns reg. coil=51,700 cm

Assuming now that the addition of capacity when coil on square frame is connected in series be negligible, as will be the case when the coil which is to be measured is very small in comparison to the other we may assume, since the capacity in the primary circuit was the same in both experiments, that the inductances in the primary or exciting circuit are as the inductances of the excited circuit in both the successive tests. Calling inductance of square coil L and remembering that inductance of the coil with 344 turns was before recorded to be 0.006 henry, we have:

$$\frac{L + 0.006}{0.006} = \frac{51,700}{15,000} = \frac{517}{150} \text{ and } \frac{517}{150} \times 0.006 - 0.006 = L =$$

$$= 0.006 \times \left(\frac{517}{150} - 1\right) = 0.006 \times \frac{367}{150} = L = \textbf{0.01468 henry.}$$

The correctness of this estimate is of course based on the assumption that the capacity added to the excited system by the square coil is very small compared with the total capacity.

Colorado Springs

Dec. 9, 1899

Results obtained Dec. 7, with square coil in various positions reconsidered.

The results referred to indicate that the induced currents generated in the ground are of little or no moment but that the capacity owing to nearness of ground may be of great importance. To get a better idea of the variation of the capacity and also to settle any doubt as to the influence of the induced currents it is necessary to estimate the capacity of the system which was excited. For the present purpose it is thought the consideration of the "ideal" capacity of the coils will lead to results not far from the truth. Now on a previous occasion the "ideal" capacity of a coil with 344 turns was found to be 34.6 cm. Since this "ideal" capacity increases with the square of the length of the turns, owing to the fact that it increases with the energy stored in simple proportion, and the energy again in proportion to the square of the difference of pressure between each two adjacent turns, the difference of pressure is simply proportionate to the length. Now the length of one turn on square coil is $4 \times 11'$ $1/4'' = 4 \times 132.5'' = 530''$. The coil of 344 turns being wound on a drum 14'' diam., the length of each of its turns is $3.1416 \times 14'' = 44''$ nearly. The "ideal" capacity of the square coil will therefore be, considering also turns,

$$\frac{14}{344} \times \left(\frac{530}{44}\right)^2 \times \text{ideal cap. of coil 344 turns} = \left(\frac{265}{22}\right)^2 \times \frac{14}{344} \times 34.6 = 145.1 \times 1.4 = 203.14 \text{ cm.}$$

Hence total "ideal" capacity will be the *sum* of these, that is approximately **238 cm.** This value obtained at a given height above ground being known, we can now calculate the increments of capacity, or decrements of same as the coil is approached to or removed from the ground and it ought to be found that these increments or decrements are proportionate to the distances of coil from ground in the successive positions of the coil. I think that it is unnecessary to go through the trouble at present because it is very probably so.

Note: Such proportionality would be destroyed if there would be any appreciable currents generated in the ground.

Colorado Springs

Dec. 10, 1899

Estimate of turns of "Extra coil" to be used with the structure of iron pipes as capacity on free terminal.

In a previous test resonance was obtained with new extra coil and elevated structure as capacity with

Capacity in primary exciting circuit	Inductance in primary circuit
168 bottles = 0.1512 mfd	20 3/8 turns reg. coil

In this test the extra coil had 105 turns. The turns being now reduced to 100, the inductance of the coil will be smaller by a ratio of $\left(\frac{100}{105}\right)^2 = 0.907$. Hence instead of 20,000,000 cm as before, the inductance of the coil will be 20×0.907 million cm or 18,140,000 cm = 0.018 henry approx. Had the above test been performed with the coil so modified, the capacity in the primary exciting circuit — assuming the inductance the same — would have been smaller by a ratio of $\frac{18.14}{20}$ or we would have had instead of 168 bottles about $\frac{18.14}{20} \times 168 = 153$ bottles approximately $= 0.0009 \times 153 = 0.13759$ mfd capacity. We have therefore as basis for computation:

Extra coil with structure: 1.

Capacity-primary: 0.13759 mfd

Inductance-primary: 20 3/8 turns reg.
coil.+conn.

Now in another test of extra coil but without structure results were

Extra coil alone: 2.

Capacity-primary: 0.0648 mfd

Inductance-primary: 5 5/8 turns reg.
coil.+conn.

These data enable us to determine how many turns are to be left off on bottom of coil in order that it be in resonance with the primary exciting system under the *best working conditions of the latter;* the structure being connected to the coil.

Now the secondary, to be at best, should work with all jars available and the regulating coil should be *all cut out* in the primary. This is namely the condition corresponding to full output and highest economy. Under such working conditions the secondary will modify greatly the inductance of the primary and for the present approximate estimate it will be close enough to assume a diminution of about 50% of the primary inductance so that the latter may be put at 28,000 cm. This would then give in Test 2:

Extra coil alone

Capacity-primary
0.0648 mfd

Inductance-primary
28,000+5 5/8 turns reg. coil+conn=
=28,000+17,263=45,300 cm. approx.

But capacity in primary with *all jars* as would be required *for best working* would be $\dfrac{8 \times 36}{2}$ =144 bottles=0.1296 mfd. Reduced to this capacity the inductance in primary would have been in reading 2. instead of 45,300 cm only $\dfrac{0.0648}{0.1296} \times 45,300 = $ **22,650 cm.**

Taking now reading 1. when structure was connected to the free terminal of coil the inductance in primary was 20 3/8 turns+conn.=72,660 cm from table. But in this case the capacity was 0.13759 mfd. Reducing this to the capacity when secondary works best, that is, to 0.1296 mfd the inductance with this capacity would have been larger by a ratio of $\dfrac{0.13759}{0.1296} = 1.048$ or it would have been 72,660×1.048=76,150 cm, approx.

Thus we have the following data:

Extra coil with structure connected: a)

Capacity in primary 0.1296 mfd

Inductance in primary =76,150 cm and

Extra coil alone without structure: b)

Capacity in primary 0.1296 mfd

Inductance in primary 22,650 cm.

Now to secure resonance with structure attached and under best working conditions of the exciting system with possibly only a very few turns of the regulating coil in primary, we have from above to reduce inductance of extra coil by a ratio of $\dfrac{22,650}{76,150}$. For same length turns will be

$$\sqrt{\frac{2265}{7615}} \times 100 = \mathbf{54\text{—}55} \ \mathbf{t.}$$

Colorado Springs

Dec. 14, 1899

Test to ascertain free vibration of new *"Extra Coil System"* (latest)

The new "extra coil" latest type is wound on same frame 8' 3" diam. and 8 feet in length, the wire being this time No. 6 instead of wire No. 10 as before. There are presently just 100 turns. A coil of No. 6 wire wound on drum 14" diam. and 8 feet long, which was used repeatedly before, was connected in series with this latest extra coil and resonance determined by means of the adjustable small inductance as usual and a miniature lamp shunting the turns. The results with *no* capacity on the terminal which was "free" were:

Capacity in primary circuit

Inductance in primary circuit

$\dfrac{5 \times 36}{2} = \dfrac{180}{2} = 90$ bottles $= 0.081$ mfd

One primary turn (two cables in multiple)+ +3 turns regul. coil

When a ball 30" diam. with 4 feet of wire was connected to the free terminal the readings were:

Capacity-primary: 0.081 mfd

Inductance-primary: as above, only 7 1/2 turns in regul. coil.

In both cases the excitation was effected through the secondary of the oscillator from a point of the same 3/4 turns from ground plate. In the first case the inductance of primary, taking into consideration the secondary reaction, is estimated at 41,000 cm. Adding to this the inductance of connections and 3 turns of regulating coil=10,000 cm the total inductance was about 51,000 cm. This with a capacity of 0.081 mfd would give the period of the system comprising the two coils in series

$$T_s = \frac{2\pi}{10^3} \sqrt{0.081 \times \frac{51}{10^6}} = \frac{2\pi}{10^6} \times 2.03 = \frac{12.75}{10^6} \text{ approx. and } n = 78,500$$

approx. per second. When the capacity of a ball of 30" diam. is associated with the system we have the inductance of the primary circuit, as before, 41,000 cm for the primary turn with secondary reacting plus 7 1/2 turns regulating coil and connections=23,580 cm, with reference to the table, that is, total=64,580 cm. The period in the second case was therefore

$$T_s = \frac{2\pi}{10^3} \sqrt{0.081 \times \frac{6458}{10^8}} = \frac{2\pi}{10^7} \sqrt{523.098} = \frac{6.28 \times 22.9}{10^7} = \frac{143.8}{10^7}$$

and $n=70,000$ approx. The coil connected in series with the extra coil was wound on a drum used before. In one case this drum was wound with wire No. 10, 346 turns and in this case it had an inductance of 6,040,000 cm. The new "extra coil" latest having 18,000,000 cm approximately, the total inductance may be placed closely enough at $18,000,000+$ $+\left(\frac{283}{346}\right)^2 \times 6,000,000$ cm, the coil in series with the extra coil here used having 283 turns. This gives inductance total $18,000,000+4,000,000$ approx.$=22,000,000$ cm. Now if the latter coil of 283 turns would have been omitted then the system would have vibrated quicker in proportion

$$\sqrt{\frac{22 \times 10^6}{18 \times 10^6}} = \sqrt{\frac{22}{18}} = \frac{\sqrt{11}}{3} = 1.106$$

times or since n in the first case was$=78,500$, the extra coil alone will vibrate $1.106 \times 78,500 = 86,800$ approx. If the inductance in primary be left the same as in the first case then the number of bottles will be reduced by a ratio of $\frac{18}{22} = \frac{9}{11}$ or instead of 90 bottles total we shall want $\frac{90 \times 9}{11} = \frac{810}{11} = 74$ bottles or nearly so; this means 148 bottles on each side or about 4 tanks. This is only an approximate estimate for first guidance. Perhaps the reaction of the secondary is somewhat overestimated in this case as it was not resonating in these tests or near the resonating condition.

* With figures taken from second test the results are very nearly the same.

Colorado Springs

Dec. 15, 1899

List, for future reference, of some coils used in the experiments up to present.

Coil wound on drum 25.25″ diam.		cord No. 20	404 turns	Length of
Coil	,,	,,	cord No. 10	259 ,,	this drum
Coil	,,	,,	wire No. 10	274 ,,	71 1/2″ appr.
Coil	,,	24″ diam.	wire No. 6	207 ,,	70 3/4″ long
Coil	,,	10 5/16 ,,	Bell wire No.18	550 ,,	
Coil	,,	14″ ,,	cord No. 10	346 ,,	Length of
Coil	,,	14″ ,,	Bell wire No.18	1314 ,,	this drum
Coil	,,	14″ ,,	No. 6 wire	283 ,,	8 feet appr.
Coil	,,	5″ ,,	No. 2 ,,	91 ,,	38 1/4″ long
Coil	,,	5″ ,,	No. 6 ,,	129 ,,	38 3/4″ ,,
Coil	,,	4″ ,,	No. 10 ,,	185 ,,	4 ft. long.
Coil	,,	4″ ,,	No. 10 ,,	141 ,,	3 ft. ,,
Coil	,,	30 1/2″ ,,	No. 25 ,,	132 1/2 turns	

* Exp. resonance

Coil on drum   26″ diam.
         upright stand   No. 6 wire   136 turns
Coil on spool   12″ diam.   2 1/8″ long   28 layers Bell wire No. 18

$$\text{turns} \left\{ \begin{array}{l} \text{to 20 lay. 28} \\ \text{,, 28 lay. 27} \end{array} \right\} = 20 \times 28 + 8 \times 27 = 560 + 216 = 776 \text{ turns}$$

* Exp. to determine periods of vibration.

New extra coil as first wound 8′ 3″ diam., 8 ft. long, 105 turns No. 6 wire, between grooves 1/2″. A number of these coils have been before described and their inductances measured. Since the same drums were repeatedly wound with different wires it is easy to determine the inductances on the bases of data before recorded with some of them. It is proposed to make later a complete table of the coils with the inductances and other particulars worked out.

Colorado Springs

Dec. 16, 1899

In carrying on some experiments to ascertain the effects of induction from the primary circuit of the oscillator in the laboratory, at a distance from the same, a square frame described on a previous occasion was used. This frame was 11′ 1/4″=132.5″= =336.55 cm long and as much wide and 3 3/4″=9.525 cm deep. It was wound with 14 turns of wire No. 10. To ascertain the period of vibration of this coil a formula before arrived at and frequently used was again employed. This formula applies to coils of circular cross section but it was thought that the results would be close enough also for the square coil with due allowances. To apply the rule the square surface was converted into a circular one equal to it and the diameter of the latter was calculated. Finally an allowance was made for the diminished length of the wire on the circular coil for the same number of turns. Calling D the diameter of the circle we have $\dfrac{D^2 \pi}{4} = 336.55^2$ from which follows $D=380$ cm. approximately. The formula referred to is

$$p = \frac{3 \times 10^{10} \sqrt{(\tau + d)\,\tau}}{D^2 \, N \sqrt{\pi^3} \, K \sqrt{d}}$$

Here τ=distance of wires=0.46 cm; d=diam. of wire=0.254 cm;

$$(\tau + d) = 0.714; \quad \sqrt{\tau\,(\tau + d)} = 0.573;$$

K as found $= \dfrac{52}{10^6}$; $\sqrt{d}=0.5$; N= number of turns=14; p=natural frequency of coil to be found.

This gives in the present case:

$$p = \frac{3 \times 10^{10} \times 0.573}{3.1416 \times 380^2 \times 14 \times \sqrt{\dfrac{3.1416 \times 52}{10^6} \times 0.5}} =$$

$$= \frac{6 \times 10^{10} \times 573}{3.1416 \times 144{,}400 \times 14 \times \sqrt{3.1416 \times 52}} = \frac{6 \times 10^{8} \times 573}{44 \times 1444 \times \sqrt{3.1416 \times 52}} =$$

$$= \frac{3438 \times 10^{8}}{44 \times 1444 \times 12.8} = \frac{3438 \times 10^{8}}{813{,}300} = \frac{3438 \times 10^{6}}{8133} = 423{,}000 \text{ approximately}$$

$$\text{and} \quad n = \frac{423{,}000}{2\pi} = 67{,}360 \text{ nearly.}$$

This would be for a cylindrical coil of 14 turns, but the turns of the square coil are longer and, since the inductance is proportionate to the square of the length of wire, and the frequency again proportionate to the square root of the inductance, n will be smaller in proportion $\dfrac{4 \times 336.55}{3.1416 \times 380} = \dfrac{1193}{1346}$ or the system will vibrate about $\dfrac{1193}{1346} \times 67{,}360$ or roughly 60,000 per sec. The *experiments showed* it vibrated *nearly this number* of times.

Colorado Springs

Dec. 31, 1899

Of the photographs taken here from Dec. 17 to Dec. 31, 1899 by Mr. Alley the following were forwarded through him to my friends of the Century:

I. Front view of laboratory from Pike's Peak side. Isochromatic 11″ × 14″ plate. Time — afternoon before sunset. This is a very fine photograph showing well the advantage of the pure atmosphere here. Such sharpness of outlines and amount of detail could not be obtained in New York, for instance. I conclude that the high quality of photographs obtainable in these parts is not so much due to the skill of the professionals as to the pure atmosphere and abundance of light.

II. View of interior showing half of circle of oscillator frame with several coils grouped inside; Westinghouse transformer, lightning arresters in background, also part of central "extra coil" latest pattern and a 30″ ball on stand. The photograph was taken late in the afternoon. Light was rather feebly diffused. The plate was as before, 11″ × 14″ isochromatic.

III. View of interior, chiefly showing condensers, break motor and regulating coil in primary of oscillator. Westinghouse high tension transformer and supply transformers in background, also arresters. Plate same as before, the photograph was taken at the same time, practically as that described under II. The diffused daylight was very feeble. Both of these plates are excellent.

IV. View of laboratory from the rear by moonlight; *1 h 20 min.* exposure. Moon about 2/3 full. Showing Pike's Peak Range and all details of building very sharply. Such photographs by moonlight could be secured only in a few places. Plate same as before, 11″ × 14″ isochromatic.

Phot. I. Front view of Laboratory from Pike's Peak side.

Phot. II. View of interior showing half of circle of oscillator frame with several coils inside.

Phot. III. View of interior, chiefly showing condensers, break motor and regulating coil in primary of oscillator. Westinghouse high tension transformer, supply transformers and arresters in background.

Phot. IV. View of laboratory from the rear by moonlight.

Phot. V. View of interior showing a number of coils differently attuned and responding to vibration transmitted to them from an electrical oscillator[41].

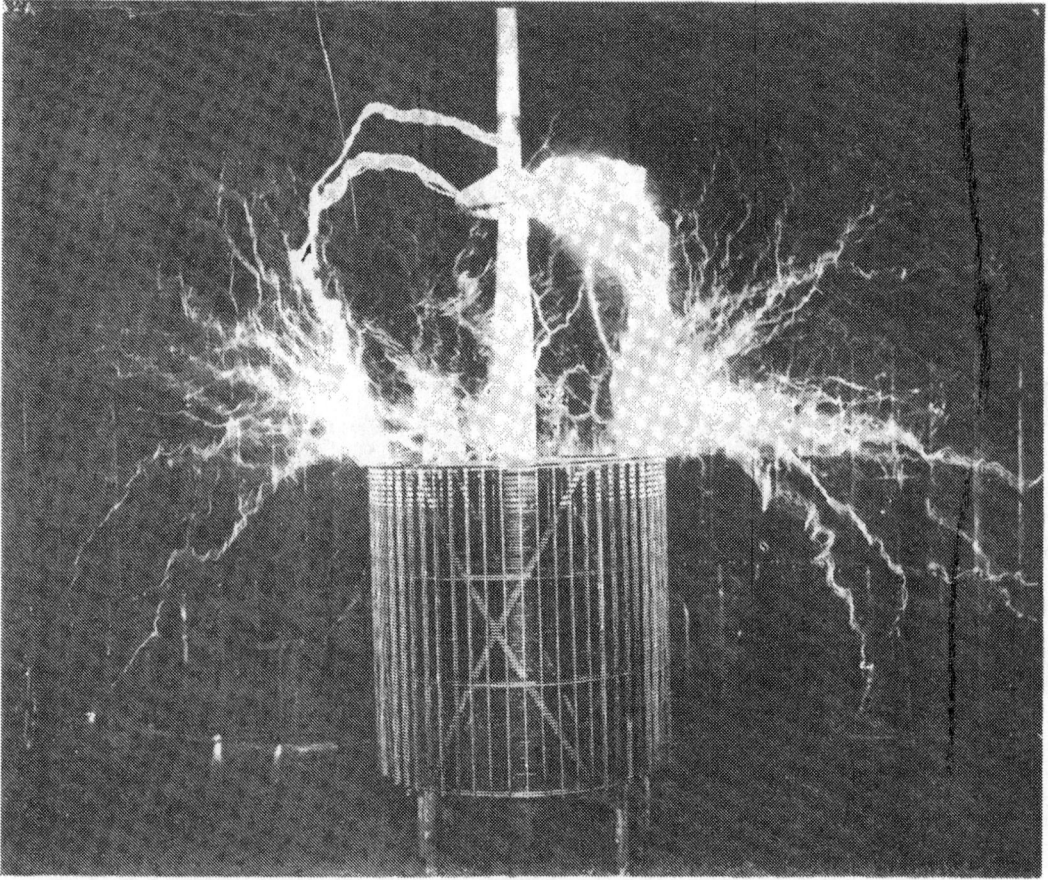

Phot. VI. Normally excited "extra coil".

Phot. VII. View of "extra coil" in action.

V. View of interior much the same as in II. The secondary and various coils placed inside, particularly the central "extra coil", are resonating. The latter is connected to a point of the first secondary turn about 3/4 turns from ground connection, nevertheless the streamers are powerful. Other coils are connected to the same point while a ball 30" on a stand and a coil on a stand are connected to the last turn of secondary. Strong sparks were passing from top to bottom of extra coil and the secondary last turn shows strong streamers. About 100 short flashes or throws of switch and afterwards an exposure of 15 minutes to ordinary arc lamp placed in corner of building for the purpose of photography. The arc light is much preferable to flash light as the time can be closely determined in each case. The isochromatic plate is *decidedly better*. In this plate there is a red dye used in coloring the plate, otherwise it is the same as the Cramer "Crown". The plate used was "instantaneous Cramer isochromatic", same size as those before. This observation suggests a line of experiment which might lead to useful results. It would consist in using plates each dyed with a different color to bring out specific effects. The vibrations of the system was the normal or nearly so as recorded in previous notes.

VI. View of "extra coil" excited as normally. A bare brass ring formed of tube 3/4" diam. is placed on top. The switch is thrown in once and held about 3 seconds. The roof has been slightly opened and also the door in front to create a draught. The effect of the latter is noted easily on the separation of the individual discharges. This feature is particularly noticeable on sparks passing to the hood carried by the structure of iron pipes repeatedly described. These sparks pass in curious ways preferring frequently a long path to a short one. This is peculiar to these discharges of high tension emanating from a single free terminal. Some streamers are broken in places to continue stronger afterwards. From end to end of the longest streamers about 50—55 feet in a *straight* line. This photograph is extremely beautiful on account of the character as well as arrangement of the streamers. The extra coil is excited as normally but not quite full power. The vibration is as previously recorded.

The same kind of plate as before was used and the exposure to arc light was 15 minutes, about half light cut off.

VII. View of "extra coil" in action. Wires slightly inclined to the ground, to prevent discharge from going to roof, were fastened at small distances to brass ring on top of coil. Thus a great amount of streamers was produced and they were necessarily weaker individually. The plate illustrates well this feature as is evident from the hairlike appearance of the streamers. Individual sparks passing to ground sometimes are strong. It is peculiar that discharge will break out more strongly on some wire and then keep on the same place until broken by draught created or otherwise. The path is, however, evidently accidental depending probably on the arrangement of particles floating in the air. This photograph is very beautiful and symmetrical. The length of the streamers is about the same as in preceding case. The vibration of "extra coil" system about normal only slightly modified by the wires attached to brass ring. Some streamers, curious to note when striking the ground and thus becoming *sparks* or spark discharges more brilliant in color, appeared thicker on bottom than on the point of origin. I believe I have recorded this phenomenon some time before. It may, on the plate, appear that the streamer or spark is thicker farther away from the origin without necessarily being so. This is simply caused by the end or lower part of the streamer being closer to the camera than the origin. But the eye is not deceived in this respect and the phenomenon may be frequently noted so that its existence is beyond doubt. This may be explained by assuming a volatilization of the

329

material where the spark strikes, whereby a flame may be produced increasing the brightness of the part of the luminous path nearer to this point. In fact I observe that whenever a powerful spark strikes an object which is of a material easily disintegrated by the heat, such as wood, there is a momentary small flame produced on that point and often one may see the spark bound back as it were or splash over the object like a jet of molten metal. It will be noted on the negative that the points where the sparks hit the floor are always darker, this showing the increased development of light at these spots. The plate used is the same as before.

VIII. This shows a central view of "extra coil" normally excited with streamers issuing from a disk turned towards the camera. In this instance the streamers were longer than before as they were issuing from fewer points, chiefly from the rim of the disk which was a little more than 10″ in diam. Some of the streamers darting towards the corners of the building were 30 feet in a straight line, but considering their curved path they were in all probability 50 feet long. Some of them show interesting bifurcations or splitting up in many branches while others, carried by the draught they create themselves, again show clearly the individual discharges. The brilliant sparks to the hood above the coil are also curious and interesting often passing far above the hood and then returning to the same in paths much longer than the straight route from the top of the coil to the hood which they should follow. Some of the streamers striking the floor show the increased brightness, before referred to, very clearly. It is noticeable that the streamers above the coil are of finer texture and split up, the draught being of course stronger near the opening of the roof. On some of the very long streamers one can see occasionally more brilliant points. The plate in this experiment was the same as before and the vibration of the "extra coil" system likewise normal or nearly so.

IX. This photograph illustrates again the extra coil with streamers and sparks from a pointed wire placed towards the camera. The wire or terminal was turned slightly downwards to cause the streamers or sparks to go more downward, as there was a considerable danger in this experiment of inflaming the roof of the building by the discharge taking an upward course. The streamers, when made to issue from a point as in this instance, are namely very long and in fact it was found impossible to work the apparatus to its full capacity in this account. The excitation of the "extra coil" system was pushed as far as could be done without great risk. The longest streamers reached the side of the building and even the corners sometimes. One of them reached the photographer Mr. Alley in the corner of the building, while another one struck me as I was operating the switch in another corner. They were so feeble at that distance, however, that they did not cause any injury or pain. Another one struck the camera but, as subsequently found, did not spoil the plate. These streamers were about the longest produceable in the present building, with the roof closed, measuring from 31—32 feet in a *straight line* from origin to end. Taking into account the curiously curved path the length was probably more than twice this, so that taking the discharge from tip to tip of these longest streamers, the *actual* path of the discharge through the air was from, say, 124—128 feet! If the building would permit I think that with the present apparatus, by putting about two to three times the copper in the oscillator a discharge extending through approximately twice this distance would be obtained, and by overcoming some defects of the present type of oscillator a further gain of about 50% could still be effected, so that I can certainly expect to reach, measured in this way, a length from 372 to 384 feet from end to end. In an industrial plant

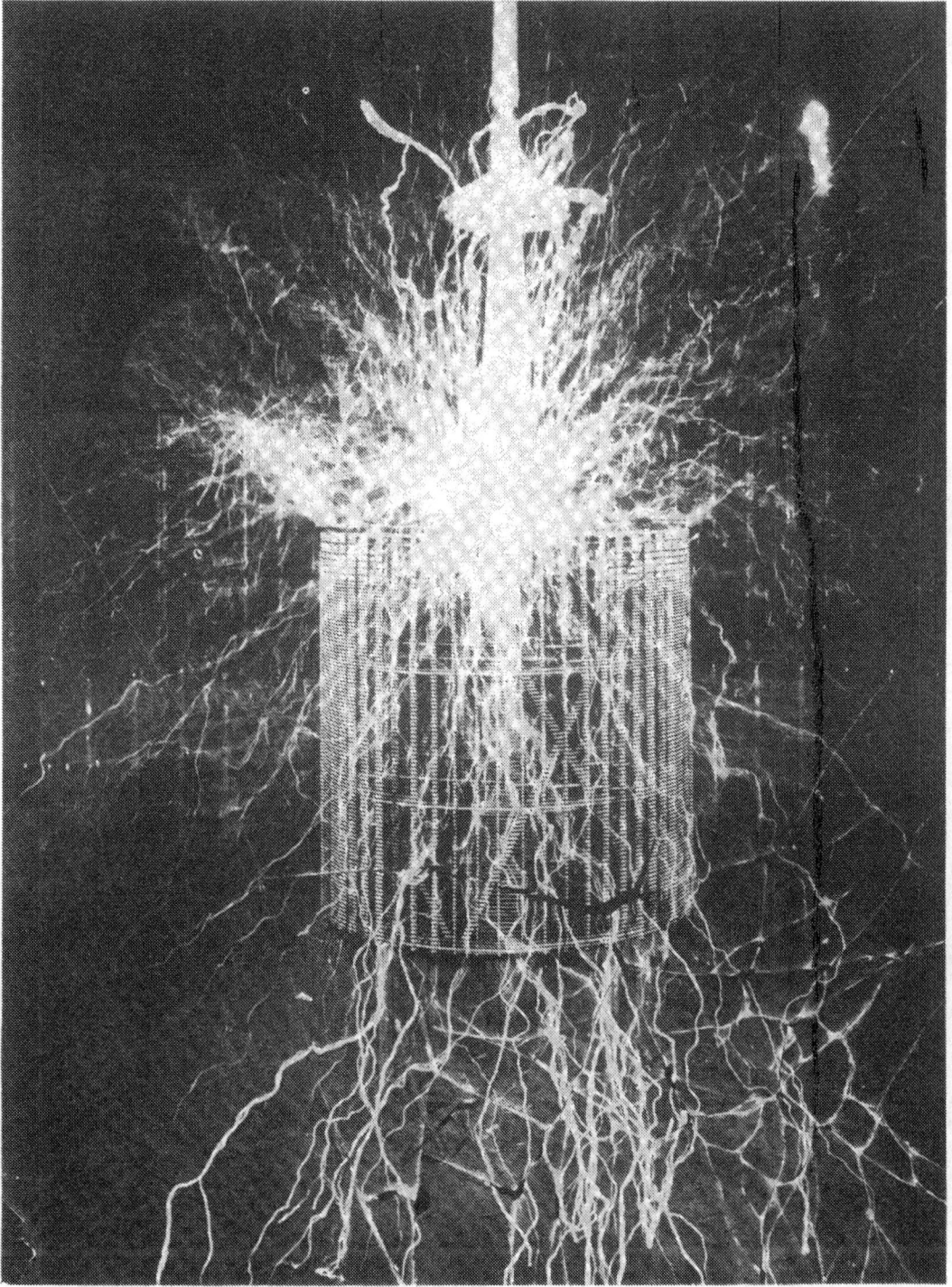

Phot. VIII. Central view of "extra coil" normally excited with streamers issuing from a disk turned towards the camera.

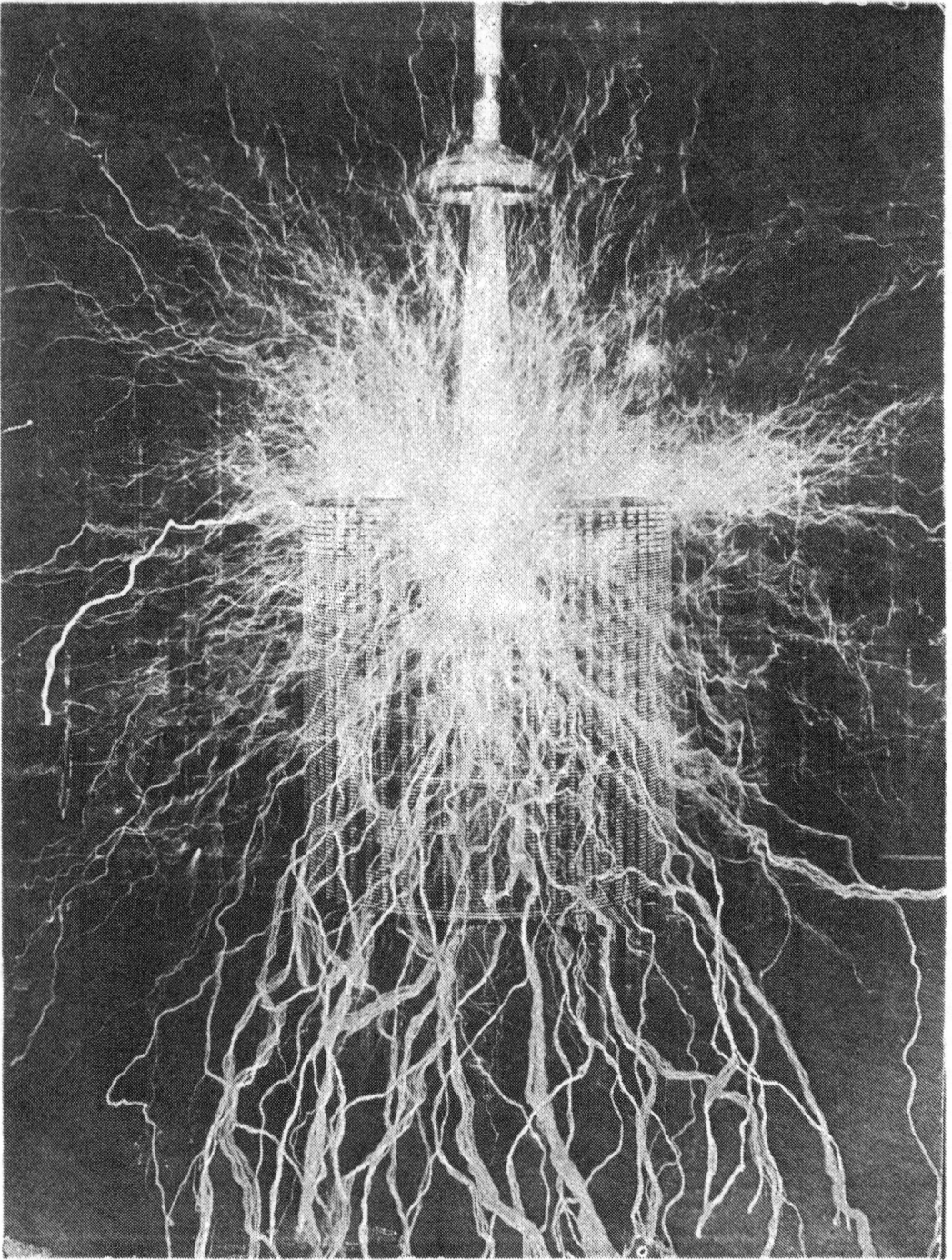

Phot. IX. "Extra coil" with streamers and sparks from a pointed wire placed towards the camera.

it seems to me advisable to push the pressure still further and the difficulties in this respect do not seem to me now to be very great. In the present photograph certain features before commented upon are even better shown than in the preceding plates. For instance, the high luminosity on the bottom where a spark strikes the floor, the "splashing" of the discharge, the branching out and the interruption of a streamer are all well shown. But the most curious feature is the appearance of "fire balls". As already noted in a previous instance a streamer even when not as strong as these here described, will show sometimes one or more points of greater luminosity than the rest. On a plate an effect of this kind may be produced by a streamer suddenly bending or turning, but the actual appearance of these luminous spots or points is unmistakable. In the instance here described the streamers were very powerful and the spots when they appeared were about an inch or possibly more in diameter, actual "fire balls" as they appeared to the eye. Now what is the cause of their formation? I attribute them to the presence of some material in the air at that particular spot which is of such a nature that when heated it increases the luminosity. It is possible that sodium is concerned in the production of the phenomenon. But the luminous "ball" must be extremely short lived as it does not impress itself upon the plate sufficiently despite its high luminosity. One can barely note a small luminous patch on the streamer, the impression of the central portion of the "ball". It is not improbable that the evolution of the "fire ball" may be connected with a process akin to explosion or sudden volatilization. Again, in this instance the same kind of plate was used and the vibration of the extra coil was but slightly quicker than normal.

X. This plate illustrates the discharge issuing from the top of the "extra coil", from the brass ring mentioned before, all over its surface. To produce a more beautiful effect in this instance the switch was thrown in 200 times, but the closure of the circuit was as short as was practicable, only a small fraction of a second, possibly 1/4 or 1/5 of a second. In plates VII, VIII and IX one hundred closures of the switch were made. This photograph is extremely beautiful although the streamers were not so large as in some previous instances. The sparks darting occasionally to the hood rather heighten the effect. The streamers are of fine texture but not quite so as in the plate described under VII. The reason is that the brass ring before mentioned, being of a large diameter, does not permit the streamers to issue from it as readily as the thin wires do and therefore the streamers partake more or less of the nature of sparks, being thicker and more brilliant and bluish white in color at the origin, while the streamers issuing from pointed terminals are of a reddish hue, sometimes quite purple and also less noisy and of feebler luminosity. Owing to their color and small luminosity they do not impress themselves upon the plate as powerfully as the streamers coming from surfaces of a relatively large radius of curvature, which are more or less of a disruptive character, ressembling sparks at the point of issue from the terminal. In this experiment again the vibration of the extra coil was nearly normal, the plate was of the same kind and the exposure to the arc light, as in the previous cases described, about 15 minutes, about half of the light cut off on lens.

XI. Plate illustrates discharge passing laterally to the camera from the central coil across the shop to another coil on a vertical stand. The length of the sparks I estimate about 15 feet in this instance, possibly more. The sparks are very powerful and brilliant. The discharge was produced by 50 successive short closures of the switch, the experiment being, of course, performed in the dark. Then an exposure — with the arc lamp 10 min. lens almost open, with an experimenter (Mr. Alley) sitting in the chair — was made. The picture of a human figure was introduced to give an idea of the magnitude of the discharge.

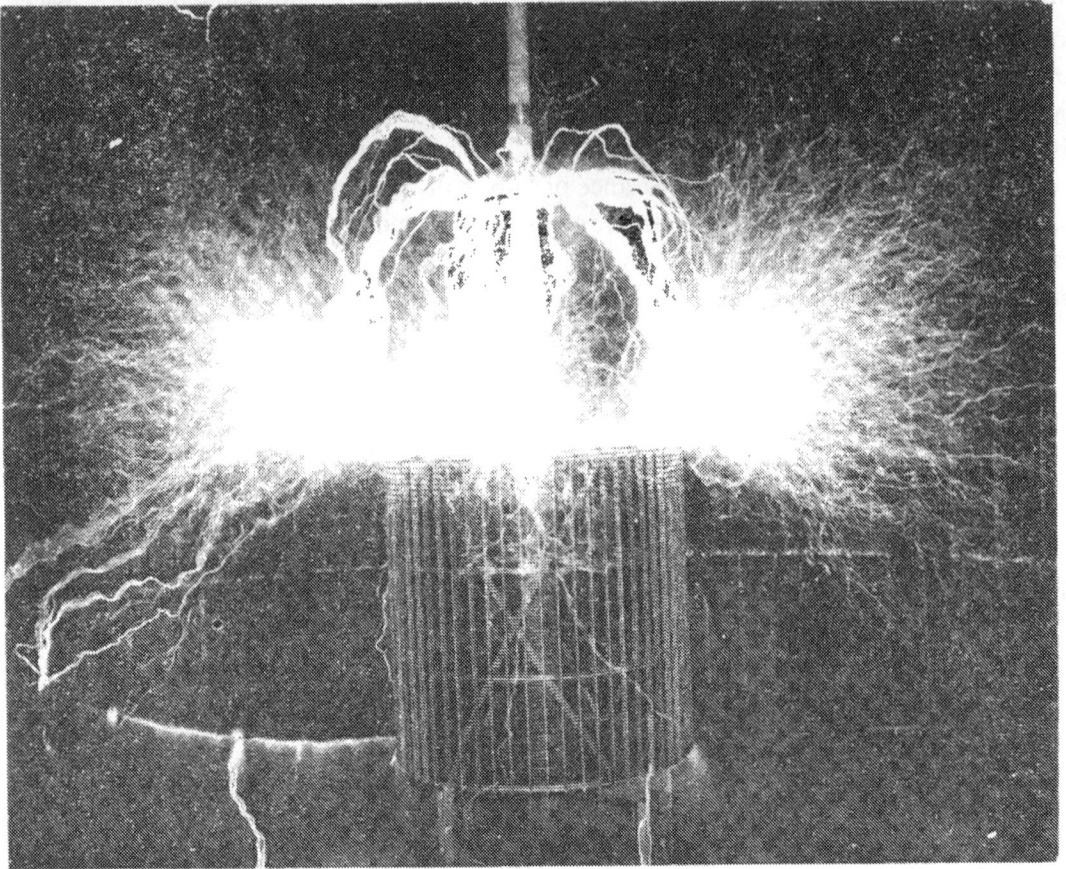

Phot. X. The discharge from the brass ring on the top of "extra coil".

Phot. XI. Discharge passing laterally to the camera from the central coil to another coil on a stand. Mr. Alley is sitting in the chair.

Phot. XII. Repeated experiment shown in phot. XI, but with Tesla sitting in the chair.

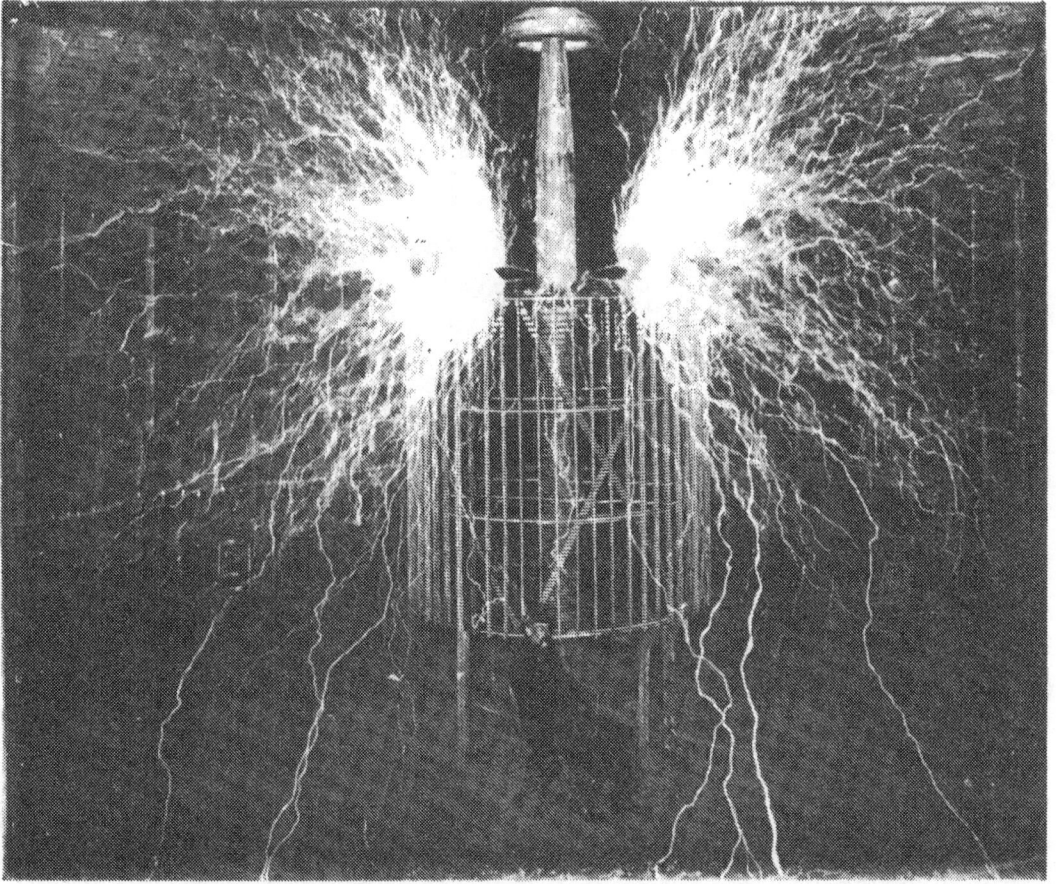

Phot. XIII. Streamers issuing symmetrically from ball with two wires on each side. Mr. Alley is sitting.

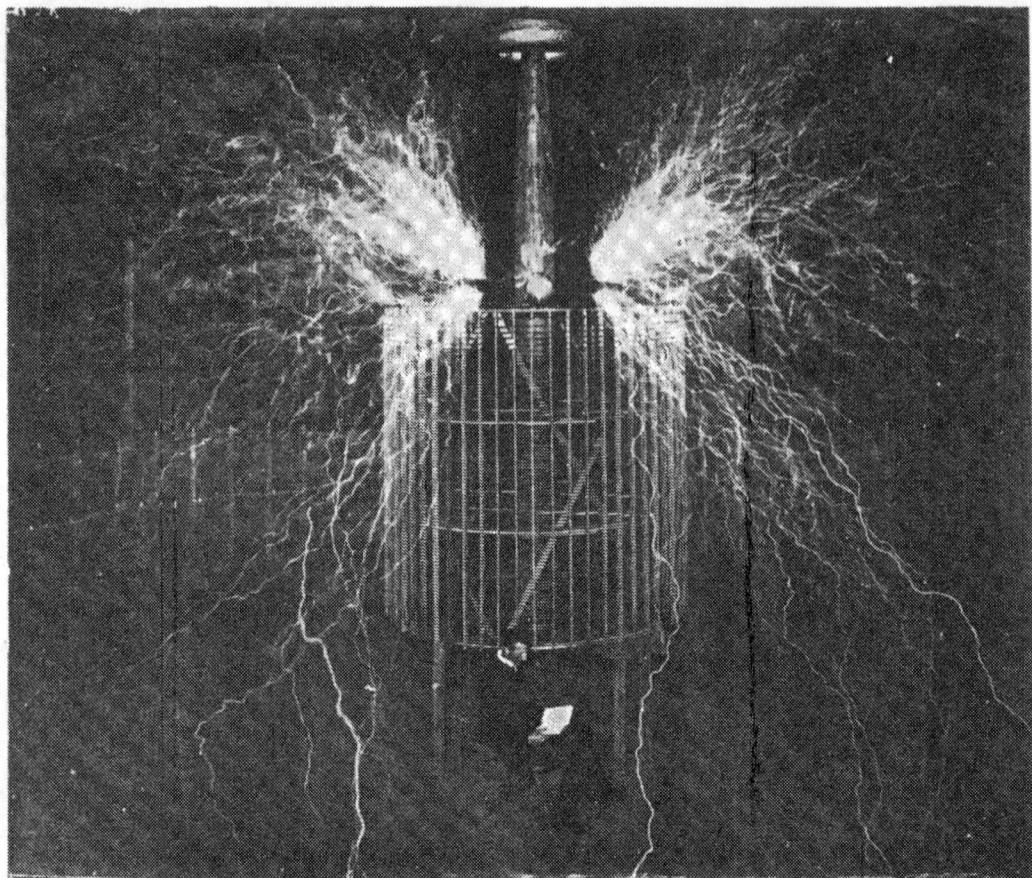

Phot. XIV. Repeated experiment shown in phot. XIII with Tesla sitting.

XII. The same experiment was repeated with the coil on stand being placed farther from the central coil. The sparks were about 20 feet long in a straight line. Fifty throws of the switch were used as before. The procedure was slightly modified by setting off about 1/3 of an Eastman flash powder to bring out my features.

XIII.
XIV. } In these two plates the effect was improved by making the streamers issue

symmetrically from ball 8″ diam. with two wires on each side: ⟩○⟨ . Fifty throws of the switch were used and the procedure was as in XII. In all the later instances described the plates and other particulars were the same as before.

Note: *open lens, arc light 5 min.*

Note: to preceding description of photographs taken with Mr. Alley:

As before stated, in most cases when the "new extra coil" was shown in action, the vibration of the system including it was about the normal, only slightly modified by the attachment of terminals of comparatively very small capacity. The "extra coil" system was namely adjusted to the vibration of the primary system which took place with *one primary turn* (the two primary cables being in multiple) *with all the jars being connected in two sets in series* and with the *regulating coil all but 1/2 turn out.* Under these conditions the secondary was very strongly reacting upon the primary modifying the inductance of the latter so that the inductance of the primary turn instead of being as previously determined about 57,000 cm was diminished to 41,000 cm approx. Now the inductance of the connections to the condensers, and 1/2 of one turn of regulating coil, was as will be seen with reference to the table, often referred to — nearly 6000 cm. So that the total inductance of the primary circuit under these conditions, was owing to the resonating secondary, was only 41,000 + +6000=47,000 cm, or $\dfrac{47}{10^6}$ henry. Now the capacity in the primary was two sets of eight tanks in series or four tanks total, that is, $4 \times 36 = 144$ bottles or $144 \times 0.009 = 0.1296$ mfd, approx. Hence the period was

$$T_p = \frac{2\pi}{10^3}\sqrt{0.1296 \times \frac{47}{10^6}} = \frac{2\pi}{10^6}\sqrt{6.0912} = \frac{2\pi}{10^6} \times 2.47 = \frac{15.512}{10^6}$$

or **n=64,466** per sec. or nearly so. This is a considerably quicker vibration than when the secondary is not active. As the "extra coil" did not have sufficient inductance in itself, another coil was inserted in series with the same to insure the resonating condition with the above primary vibration. This latter coil was wound with wire No. 6 on an old drum 2 feet diam. and 6 feet long. Particulars about this coil will be recorded on another page.

Colorado Springs Nov. 14. 1899

In some experiments with coil wearing 1314 turns wound on drum 14" diam. & feet long the coil was cut in the middle and the two parts 657 turns each connected in multiple. The self-induction was then practically $\frac{1}{4}$ of the self-induction which it had when ordinary. Readings were taken to determine the inductance when the two parts were connected as stated. These readings were:

$$E.m.f. \begin{cases} 214 \\ 212 \\ 210 \end{cases} \qquad C \begin{cases} 10.7 \\ 10.6 \\ 10.5 \end{cases} \quad \omega = 880$$

Average values:

$$\begin{array}{ccc} E & C & \omega \\ 212 & 10.6 & 880 \end{array}$$

$$R = 7.9 \text{ ohm.}$$

from this $\left(\dfrac{E}{C}\right) = 20$, $\left(\dfrac{E}{C}\right)^2 = 400$ $R^2 = 62.41$

$$\left(\frac{E}{C}\right)^2 - R^2 = 337.59 \qquad \sqrt{\left(\frac{E}{C}\right)^2 - R^2} = 8.375$$

$$L = \frac{18.375 \times 10^4}{880} \quad C.m.$$

$$= 20880682 \ C.m \qquad approx. \ 20,881000 \ C.m.$$

The inductance of the coil as ordinarily used would then be approx.

$$= \underline{\underline{83,524,000 \ C.m}}$$

Colorado Springs Notes
Jan. 1—7, 1900

Colorado Springs July 11. 1899.

Now $A = \frac{\pi}{4} D^2$ $l_1 = N(\tau + d)$ hence the inductance will be

$L = \frac{4\pi \cdot \frac{\pi}{4} D^2 N^2}{N(\tau + d) 10^9} = \frac{\pi^2 D^2 N}{(\tau + d) 10^9}$ Henry. Taking these values for L and C

we have with reference to above $\frac{1}{p^2} = \frac{\pi^2 D^2 N}{(\tau + d) 10^9} \cdot \frac{d \ell D}{\tau_n \, 9 \times 10^4} \cdot K$ or

$p^2 = \frac{(\tau + d) \overline{\tau} \times 9 \times 10^{20}}{\pi^2 D^3 \, d \, \ell \, N \, K}$. Since in the preceding the diameter of

the drum is assumed from practical considerations is will be

convenient to find the number of turns N. The quantities D and

$\overline{\tau}$ are of course interconnected since by assuming D and deciding

on the pressure to be obtained beforehand $\overline{\tau}$ is practically given. The

diameter of the wire will in most cases also be selected beforehand

so that then merely N is to be determined to satisfy the condition

of resonance for any frequency specified. Now $l = \pi D N$.

hence substituting this we have from above .

$p^2 = \frac{(\tau + d) \overline{\tau} \times 9 \times 10^{20}}{\pi^2 D^3 \, d \, N \, \pi D N \, K}$ or $p^2 = \frac{(\tau + d) \overline{\tau} \times 9 \times 10^{20}}{\pi^3 D^4 \, d \, N^2 \, K}$ and from this

we get $N^2 = \frac{(\tau + d) \overline{\tau} \times 9 \times 10^{20}}{\pi^3 D^4 \, d \, p^2 \, K}$ or $N = \frac{3 \times 10^{10} \sqrt{(\tau + d) \overline{\tau}}}{D^2 \, p \, \sqrt{\pi^3 \, K} \cdot \sqrt{d}}$ ..... This

formula may serve to give an approximate idea how many turns

are to be wound in all cases when the length of the wire, owing

to the capacity in the excited circuit is smaller than $\frac{\lambda}{4}$, (or

respectively smaller than $\frac{\lambda}{2}$, if the circuit is not one of the kind illustrated in diagrams

above, that is one, in which the potential on one terminal is many

times higher than on the other, but an ordinary circuit, in which there

is a symmetrical rise and fall of pressure at both the terminals), but

the equation assumes that k the dielectric constant is = 1 or nearly so.

11.

Photographs taken Dec. 22. and 23; with Mr. Alley from Dec. 17 to Dec. 31, 1899 and *particulars* about the same:

XV. Shows an incandescent lamp 16 c.p., 100 V placed out on the field with two wires leading to it. Snow on the ground. The lamp is lighted.

XVI. Illustrates the same with three lamps 16 c.p., 100 V placed on the snow and lighted. The lamps are connected in multiple arc.

XVII. Photograph showing once more the same with three lamps as before, the lamps being placed on black cloth to improve effect. These photographs were taken under the following conditions:

A cord of section equal to that of wire No. 10 was laid on the field in the form of a square of 62' 5"=749"=*1902.5 cm* side, the center of the square being from the center of the primary loop of the oscillator in the laboratory a little over 60 feet=720"=*1830 cm*. The ends of the square were connected to two small condensers joined in multiple and each having a little less than 1/20 mfd. Neglecting capacity of the cord against that of the condensers the total effective capacity of this system was, with fair approximation 1/10 mfd, or *90,000 cm*. The inductance of the square, taking it as consisting of two pairs of parallel wires, was

$$L_s = 2 \times 2\,l \left(\log_e \frac{d^2}{r\,r'} + 1/2 \right).$$

In the present instance

$$l = 1902.5 \text{ cm}, \qquad d = 1902.5 \text{ cm}, \qquad r = r' = 0.254 \text{ cm}.$$

$$\frac{d^2}{rr'} = \frac{d^2}{r^2} = \left(\frac{d}{r}\right)^2 = \left(\frac{1902.5}{0.254}\right)^2 = (7500)^2 \text{ and } L_s = 4 \times 1902.5 \times$$

$$\times (\log_e 7500^2 + 1/2) = 7610 \times (17.825 + 0.5) = 7610 \times 18.325 = 761 \times 183.25$$

log 7500=3.875061 $\qquad\qquad\qquad\qquad$ $L_s = 139,450$ cm, or $\dfrac{13,945}{10^8}$ henry

2 log 7500=7.750122

2 $\log_e$ 7500=7.75×2.3=17.825 approx.

From the foregoing we have for the period of the secondary system:

$$T_s = \frac{2\pi}{10^3} \sqrt{0.1 \times \frac{13,945}{10^8}} = \frac{2\pi}{10^7} \sqrt{1394.5} = \frac{2\pi \times 37.34}{10^7} = \frac{234.5}{10^7}$$

and this would give *n=42,640* approx. per second.

Now resonance in this circuit was obtained with all jars being connected as usual and two primary turns in multiple, there being besides in the primary circuit 18 1/2 turns of the regulating coil. This gives approximately the capacity of primary or exciting circuit $\dfrac{8 \times 36}{2}$ jars=144 jars or bottles=0.0009×144=0.1296 mfd. Neglecting for the moment

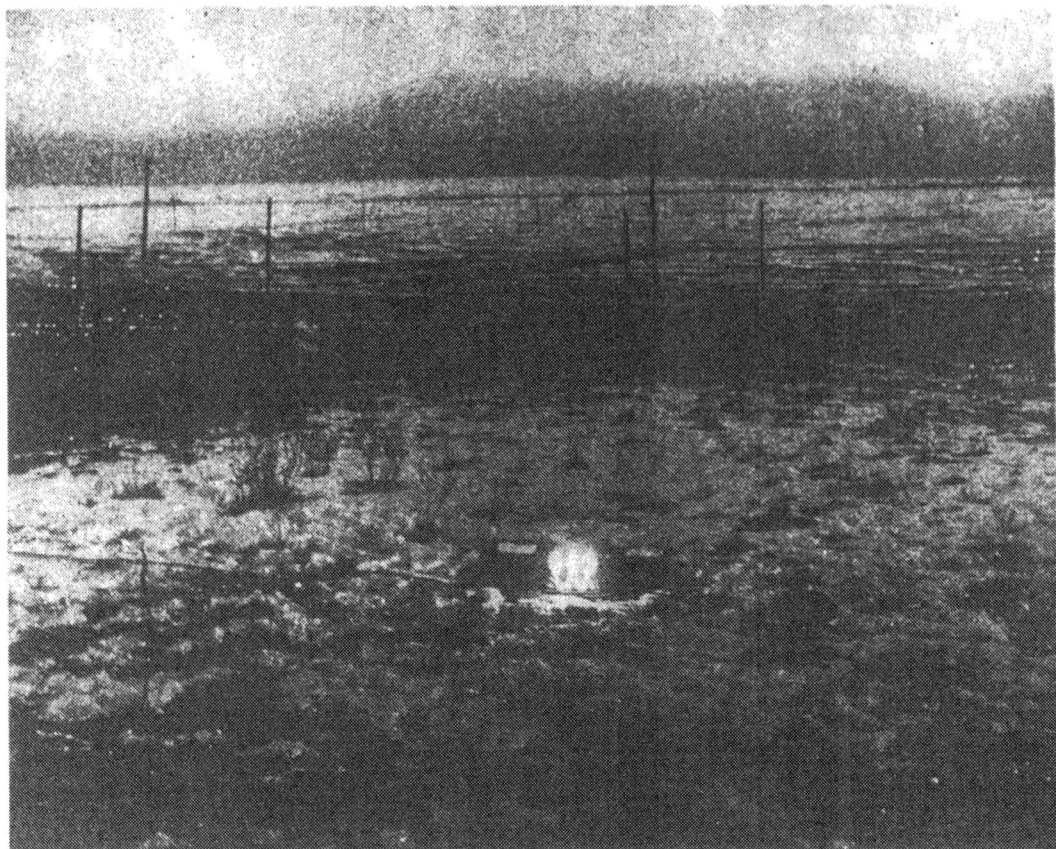

Phot. XVII (?) Experiment to illustrate an inductive effect of an electrical oscillator of great power. The photograph shown is reproduced from Tesla's article[41] in which somewhat different circuit dimensions and position are quoted.

the secondary reaction (that is the action of the secondary of the oscillator upon the primary, the former being thrown out of step by connecting some inductance in series with it), the inductance of the primary circuit with reference to previous data for the primary turns and regulating coil was 122,000 cm. But the inductance calculated from equation

$$\frac{2\pi}{10^3} \sqrt{0.1296 \times L_p} = \frac{2\pi}{10^3} \sqrt{0.1 \times \frac{13,945}{10^8}} = \frac{234.5}{10^7}$$

is smaller, being in fact only about *108,000 cm*. The smaller value is evidently the correct one, being the actual value of the inductance of the primary circuit, as modified by the secondary of the oscillator and the secondary circuit lighting the lamps. A small streamer was seen to issue from the free terminal of the secondary, hence it certainly affected the primary, reducing the inductance. The reaction of the secondary circuit lighting the lamps was comparatively small. In order not to burst the lamps the current of the supply transformers was very much cut down. I think about 30 lamps could have been lighted in this manner by pushing to the normal limit the excitation of the primary loop.

Four photographs were made of the discharge of the secondary of latest type, 20 turns of two wires No. 10. The primary consisted of one turn, the two primary cables being connected in multiple arc. The ratio of conversion being thus 1 : 20, the e.m.f. at the terminals of secondary, with the primary excited to full power, was about 400,000 volts. Owing to the low resistance of the secondary — extremely low when considering the high e.m.f. — the discharge was very powerful, of a dazzling brilliancy, literally blinding, and caused a deafening noise. The candle power of the arc is equivalent to that of five or six arc lights of normal strength. At least I conclude so from comparative tests. Photographs were made by exposing the objects to the light of the secondary discharge and it took only a small fraction of the time needed for an arc lamp to impress the plate. The secondary discharge of this apparatus is so powerful that it was always more or less dangerous for the safety of the laboratory and machinery in the same, and elsewhere, to let it play. A number of times the shop caught fire by sparks passing from some nail, wire or any kind of conductor. When the discharge was playing sparks were seen to fly almost everywhere through the laboratory, from one to another object and it was evident that it was more or less risky to let the sparks from the free terminal pass to the ground, because short waves were produced in the conductors and these were only too apt to rupture the insulation of any apparatus in the circuit or circuits connected with the oscillator or in the neighbourhood of the same. The danger resides chiefly in the *short* waves and the risk was considerably diminished when the secondary, instead of discharging directly into the short ground connection, was made to discharge through a coil or inductance, slowing down the vibration and preventing the formation of *very* high harmonics. When the discharge was effected as in the experiments photographed, a continuous and brilliant arcing took place over the lightning arresters and the dynamo at the supply station was short-circuited in rapid succession. When considering that the arc on the arresters is almost continuous one must admit that these arresters work extremely well. The discharge, owing to the terrible noise it creates is highly irritating and I think also dangerous to the timpanum of the ear. Often pain is experienced in the ears afterward and the buzzing in the ears continues for hours. If signalling with *very short* waves were desirable, nothing better could be used than such a secondary discharge. Although I have not tested its effects at extremely great distances, I conclude, from comparison with other induction apparatus experimented with, that it would affect a sensitive device certainly at one thousand miles

Phot. XIX. Discharge between a ball of 30″ diam. and the ground plate.

and very likely at a greater distance, even on land. The great brilliancy of the discharge is in part due to the comparatively large capacity of the secondary which, as before shown, is without further provision inseparable from such a coil of very large diameter. When looking at the arc for a moment one can clearly perceive the arc proper forming the central and comparatively narrow part of the luminous path, around which there is on each side a brilliant band of 1/2″ to 1″ width apparently. When the discharge is playing, generally sparks pass on some places between the secondary top turns, this showing that they are by no means too far apart.

Of the four plates: XVIII., shows the secondary discharging from the free end or terminal formed by a wire to the ground plate visible on the bottom.

Plate XIX. illustrates the discharge taking place between a ball 30″ diam. and the ground plate. In the former case the discharge is about 3 1/2 feet, in the latter about 3 feet long. These photographs were taken through about half of the full lense opening, nevertheless the discharge is not sharp for although the focusing was carefully effected the wide luminous band on each side of the arc proper, which was referred to, blurs the image. To improve the photographs two more plates were exposed, one the same as plate XVIII., marked XX., showing discharge between a wire and ground plate, and the other, marked XXI., illustrating discharge playing between a ball 18″ diam. and the ground plate. In these instances a small opening was used and the images are sharper. To be quite sharp a pin-hole diaphragm should be used. The vibration in these four instances was the normal as before determined, all jars and one turn in primary, 25 throws of switch, very short, flash afterwards.

Colorado Springs

Jan. 2, 1900

Photographs taken with Mr. Alley from Dec. 17 to Dec. 31, 1899 and *particulars* about the same.

XXII. This photograph shows the "new extra coil" as last modified, having 98 turns wire No. 6 and on top two turns or nearly so of wire No. 10 covered with a thickness of 3/8″ rubber. This wire was repeatedly referred to in previous notes. It was necessary to use it in many of the experiments recorded for the purpose of preventing or at least reducing leaks. In many cases despite the excellency and great thickness of the insulation the latter was found inadequate to withstand the strain, as is evident from a number of photographs showing the coil in action energized to full power. The picture illustrates five incandescent lamps lighted — and to much more than normal candle power — on a table in front of the coil. The lamps are in series, one end of the series being connected to the ground by a wire seen on the bottom while the other end of the lamp series is joined to the lower end of the coil, the upper end being entirely insulated and remote from objects which might act upon it inductively. The connection will be best understood from the diagram below. In the experiment illustrated there is no appreciable induction exerted

Phot. XXII. Five incandescent lamps lighted by current passing from "extra coil" to the ground plate.

upon the extra coil as the wires of the oscillator proper, wound on the wooden structure seen in the back behind the coil, are *short-circuited*. One of the terminals of the condensers is grounded so that when they are discharging through the circuit, chiefly composed of a number of turns of the regulating coil, there is a strong vibration propagated through the ground which through the ground wire w reaches the "extra coil". Now, generally,

the energy which can thus be transmitted to the coil would be minute, but when the oscillations passing through the ground are exactly of the frequency of the "extra coil" system itself, a considerable current passes into the coil which then acts just as a hole would in a pipe through which a fluid is pumped by means of a pulsating piston. As the magnifying factor of the coil is very large the feeble impulses reaching the ground wire and lamps magnify the impressed e.m.f. and create considerable movement of electricity through the lamps which are thus brilliantly lighted, as shown in the photograph. In the experiment the capacity in the exciting oscillating circuit, impressing the vibrations upon

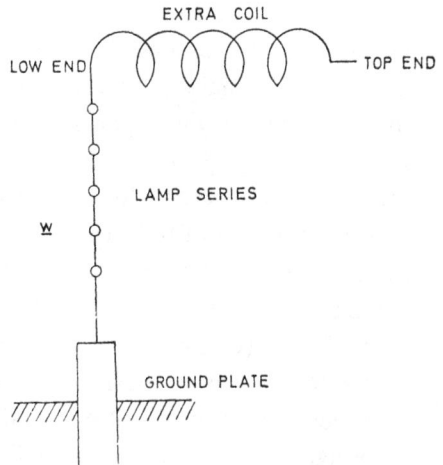

the ground and wire w, was 3 tanks on each side or 1 1/2 tanks total, that is 54 bottles or $0.0009 \times 54 = 0.0486$ mfd, approx. The total inductance was 41,000 cm+ind. of 6 1/4 turns of regulating coil $= 41,000 + 19,368 = 60,000$ cm, approx. or 0.00006 henry. From this the approximate period of the vibration impressed upon the ground would be

$$T_p = \frac{2\pi}{10^3} \sqrt{0.0486 \times \frac{6}{10^5}} = \frac{2\pi}{10^5} \sqrt{0.02916} = \frac{2\pi}{10^5} \times 0.1708 = \frac{1.074}{10^5}$$

approx. and $n = 93,110$ and $p = 585,000$ approx. λ would be very nearly 2 miles and $\frac{\lambda}{4} = 1/2$ mile or about 2640 feet. In reality the length of the wire in the excited system — that is extra coil and ground wire, was found by measurement to be *2660 feet* (98 turns, wire No. 6, 25' 11" each turn $= 2540' + 3/4$ turn of cable inside of secondary frame $= = 112' +$ continuation of ground cable outside of circle to ground plate $= 28' +$ wire $w = = 20' +$ rubber covered wire on top of coil $= 50'$ that is, total $2450' + 112' + 28' + 20' + 50' = = 2660$ feet. From above data and taking resistance of extra coil at *1 ohm* (in reality a little less) we get magnifying factor for coil alone $\frac{pL}{R} = \frac{585,000 \times 0.018}{1} = 10,530$.

Taking, however, into consideration that the resistance of the lamps was about 1000 ohms roughly, when including the latter in the system the factor would be only about $\frac{1}{1000}$ of this, or approximately only *10.5*. But I believe that the resistance of the lamps when operated by currents of such extreme frequencies is much smaller than the measured resistance according to the usual methods. The currents, namely, when produced in such ways as these here employed, have very high maximum values and the carbon is brought periodically to a much higher temperature than when operated with steady currents or currents of ordinary frequencies. I have observed this repeatedly. Furthermore when such currents as these here are used some part of the discharge also passes through the rarefied gas in the bulb and there is a corresponding diminution of the effective resistance of the lamp on this account. In fact, I think that it is chiefly owing to this that the resistance becomes very small, it being a fact that such currents pass with the greatest freedom through the rarefied gas particularly when it is maintained at a high temperature, as in the case here considered. The heating of the gas has the effect of increasing the incandescence of the carbon and it is well demonstrated that an incandescent lamp takes, for a given luminosity, less energy when operated with currents of such extreme frequencies. Owing to these reasons the magnifying factor of the excited system even with the lamps included must have been very much larger than the figure last mentioned. It was astonishing to note, in the experiment recorded on the plate, how much energy can be in this manner conveyed to such a carefully synchronized coil through the ground. The supply transformers were cut down by a regulating coil in the primary to less than one half, in fact to about 1/3 of full capacity and inasmuch as only 3 tanks on each side in the primary or exciting circuit were used while 8 tanks were available, it is evident that, if the excited system would have been designed to work with *full output* of the exciting apparatus it would have been quite practicable to light, say $3 \times 8/3 = 8$ times as many lamps, or about 40 lamps. However, inasmuch as the five lamps were far above candle power it would have been, in all probability, possible to light 60 lamps or so to normal candle power by a specially designed coil, with a liberal allowance of copper, vibrating in unison with the system exciting the portion of the ground containing the ground plate. Nothing could convey a better idea of the tremendous activity of this apparatus and a simple comparison with well ascertained data, obtained with other induction apparatus, shows that one of the problems followed up here, that is the establishment of communication with any point of the globe irrespective of distance, is very near its practical solution. The existance of stationary waves proves the feasibility of the project almost beyond any doubt. The great amount of energy which can be conveyed to such a synchronized circuit, by conduction through the ground, makes it appear possible, that the necessity of elevating terminals in my system of energy transmission to a distance may be dispensed with in many instances and that, with a very moderate elevation of, say, a few hundred feet enough energy may be conveyed to a circuit to serve for one or another useful purpose beyond mere signalling, or such uses of the system in which a minute amount of energy is required. Certainly, the amount of energy conveyed in this manner was, in some experiments with this apparatus, surprising at first. An interesting consideration in this connection may be the following: As before stated the period of the exciting or primary circuit, when resonance

with the extra coil was attained, was $T_p = \dfrac{2\pi}{10^3}\sqrt{0.0486 \times \dfrac{6}{10^5}}$. Now the period of

the excited system was $T_s = \dfrac{2\pi}{10^3}\sqrt{0.018 \times C_s}$, in which C_s is the "ideal" capacity as

designated in previous instances, that is the capacity which would have to be joined to the free end of the "extra coil" of inductance of 0.018 henry but devoid of all distributed

capacity. Since $T_p = T_s$ we find $C_s = \dfrac{0.0486 \times \dfrac{6}{10^5}}{0.018}$ mfd, or

$$C_s = \dfrac{9 \times 10^5 \times 0.0486 \times \dfrac{6}{10^5}}{0.018} = \dfrac{54 \times 0.0486}{0.018} = \dfrac{54 \times 486}{180} = \dfrac{3 \times 486}{10} = \textbf{145.8 cm.}$$

Suppose an ideal system of this kind excited in the manner described, so that capacity on the free end is charged each time as the current alternates to a potential P. Then, since as before stated, the system in experiment illustrated was vibrating about 93,000 times per second, the total energy set in movement in the system would be

$$2 \times 93,000 \times \frac{P^2 \times 145.8}{2 \times 9 \times 10^{11}} \text{ watts.}$$

Let it be further assumed that 1% of the total energy set in movement is frittered down in the lamps and that the number of these were *60*, as might have been the case in the presently described experiment. Suppose each lamp to take 50 watts, the total energy consumed in the lamps would be 3000 watts hence, under the above assumptions, the total energy set in movement in the excited system would have to be 100 times this amount or 300,000 watts. To satisfy this condition we would have

$$2 \times 93,000 \times \frac{P^2 \times 145.8}{2 \times 9 \times 10^{11}} = 300,000$$

and

$$P^2 = \frac{54 \times 10^{13}}{186 \times 145.8} = 10^{12} \times \frac{540}{186 \times 145.8} \text{ or } P = 10^6 \sqrt{\frac{540}{186 \times 145.8}} =$$

$$= 10^6 \sqrt{\frac{540}{27,118.8}} = 10^6 \sqrt{0.02} \text{ approx. or } = 10^6 \sqrt{\frac{2}{100}} = 10^5 \sqrt{2} =$$

$$= 10^5 \times 1.414 \text{ or } \textbf{\textit{P} = 141,400} \text{ volts.}$$

Not very much, as will be seen, for such a pressure is extremely small with apparatus of the kind used here. Taking P roughly as 140,000 volts and assuming that the ground plate be at such a distance that only 1000 volts are impressed upon the same then the magnifying factor would have to be only 140. Of course, this is merely an example to support the above statement that considerable energy may in this way, and by such apparatus, be conveyed to a distant circuit which is connected to the ground at only one point directly or, if desired, through a condenser.

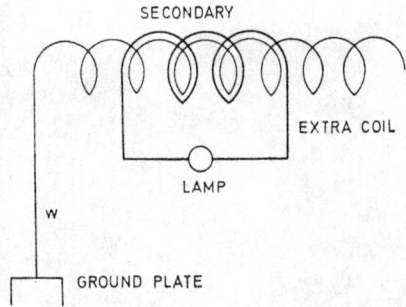

In another photograph, marked Plate XXIII., taken with same apparatus a similar experiment is illustrated. Here the extra coil as indicated in the diagram above is connected directly through the wire w to the ground plate and another coil designated in the diagram as "secondary" is placed in inductive relation to the extra coil excited in the same manner as in experiment before described, and a lamp is lighted by the currents generated in this "secondary". Only one lamp was used as it was the object of the photograph merely to illustrate a novel experiment, but with reference to the above it will be understood that as many as *60* lamps, or nearly so, might have been lighted with the apparatus used in this manner. All the particulars were practically the same as before. The "secondary" had four turns, the excitation of the extra coil being reduced so that the lamp was somewhat above normal candle power. In this, as well as in the preceding experiment, the switch on the Westinghouse high tension transformer was thrown in and out

Phot. XXV. A coil outside laboratory with the lower end connected to the ground and the upper end free. The lamps is lighted by the current induced in the three turns of wire wound around the lower end of the coil.

about 50 times and after this, as usual, an exposure to arc light for the detail of the apparatus was made, the time being 10 minutes with a small opening.

In order to make these two experiments still more interesting they were performed outside with a smaller receiving coil and photographs were taken, which are numbered and of which Plate XXIV. shows a coil standing outside on a table to which a groundwire leads, which is connected to one of the terminals of a small lamp, while the other terminal of the same is joined to the lower end of the coil. The upper end of the coil is free, a metal tube being connected to the same, as shown clearly in the picture. This tube is placed axially and serves to increase slightly the capacity of the excited system.

In another experiment, with the same coil illustrated in Plate XXV., the coil is placed on the ground away from the laboratory and the lower end is connected to the nearest ground, while the upper end or terminal is free. Three turns of wire are wound around the lower end of the coil and the ends of this wire are connected to a lamp socket with its lamp which is, as shown, lighted by the currents induced in the three turns of wire through the oscillations transmitted through the ground to the coil.

One more experiment of this kind was photographed, the same coil being again used and placed far out into the field, this being shown in Plate XXVI., giving a clear view of the Pike's Peak Mountain Range in the background. The diagrams and several remarks before made apply to some extent also to these three photographs which were taken after sunset when the dark began to set in, as it was impracticable to take them at another hour of the day. They might have been taken by moonlight but the time was otherwise occupied. During the hours when the light was strong it would have been necessary to exclude first the daylight in some way, flash the lamp in the dark and finally make a short exposure to the full daylight to get the detail of the apparatus. It was found impracticable to get a good photograph by flashing the lamp in full daylight as the latter was too strong and the lamp did not have enough time to impress the plate as strongly as was desirable, even if it was pushed to much higher candle power than the normal. In getting the photographs, generally about 100 throws of the switch were sufficient with the lamp being pushed considerably above the normal. When the daylight was still deemed too strong Mr. Alley helped himself by covering the lens during the short interval when the lamp was not lighted and thus regulated the effect of the daylight, keeping it down to the required value.

The *particulars* were as follows: The coil used in these three experiments was wound on a drum before referred to of 25.25″ diam. and had 274 turns of wire No. 10, rubber covered. Since another coil wound on the same drum had 404 turns and an inductance of approximately 40,000,000 cm. or 0.04 henry the inductance of the present coil was with fair approximation $\left(\frac{274}{404}\right)^2 \times 0.04$ henry or $\left(\frac{137}{202}\right)^2 \times 0.04$ henry. The wire leading from the ground plate to the lower end of the coil placed on the table or ground consisted of two pieces of cord No. 10, one 308 feet and the other 84 feet long. The inductance of these two pieces of wire was estimated at 113,000 cm and compared with the inductance of the coil itself was very small, almost negligible. Calling the total inductance of the excited circuit comprising the two pieces of wire and the coil used L_1 we have for this inductance value $L_1 = \left(\frac{137}{202}\right)^2 \times 40,000,000 + 113,000 = 18,400,000 + 113,000 = 18,513,000$ cm, or $L_1 = $ $= \frac{185}{10^4}$ henry. This inductance, with its distributed capacity, gave a system responding

Phot. XXVI. Experiment to illustrate the transmission of electrical energy without wire. The photograph shown is reproduced from Tesla's article[41].

to the primary vibration when the capacity in the primary or exciting circuit was 1 2/3 tanks or 60 bottles on each side, or 30 bottles total, that is, $30 \times 0.0009 = 0.027$ mfd, total.

The vibrations were impressed on the ground plate by the oscillator with normal connection, that is, two primary cables in multiple or *one* primary turn, the approximate inductance of which was 56,400 cm or, say, 56,000 cm, which is close enough for the present consideration. This inductance may have been modified by the secondary, but the effect of the latter must have been very slight as, with the capacity used, it was "out of tune" and the current through it was necessarily very small. Taking then the inductance of the primary exciting circuit at 56,000 cm, the period of this circuit was

$$T_p = \frac{2\pi}{10^3} \sqrt{0.027 \times \frac{56}{10^6}}.$$

Now calling C_s the "ideal" capacity of the excited circuit, the period of the same was

$$T_s = \frac{2\pi}{10^3} \sqrt{\frac{185}{10^4} \times C_s} \text{ and equating we have } C_s = \frac{10^4}{185} \times 0.027 \times \frac{56}{10^6} = \frac{56 \times 0.027}{185 \times 10^2} \text{ mfd,}$$

or $C_s = \dfrac{9 \times 10^5 \times 56 \times 0.027}{185 \times 10^2} = \dfrac{243 \times 56}{185} =$ **75.2 cm,** approx. From above

$$T_p = \frac{2\pi}{10^3} \sqrt{0.027 \times \frac{56}{10^6}} = \frac{2\pi}{10^6} \sqrt{1.512} = \frac{6.28}{10^6} \times 1.23 = \frac{7.7244}{10^6}$$

and $n = 129,500$ per second nearly.

The theoretical wave length would thus be $\lambda = \dfrac{186,000}{130,000} = \dfrac{186}{130} = 1.43$ miles approx.

or $\dfrac{\lambda}{4} = \dfrac{1.43}{4} = 0.3575$ miles or $0.3575 \times 5280 = 1888$ feet $= \dfrac{\lambda}{4}$.

The actual length of wire in the experiment was: 274 turns of the coil, each $79.29'' = 1810$ feet $+$ one piece of wire 304 feet $+$ one piece of wire 84 feet $= 1810 + 304 + 84 = 2198$ feet or nearly 15% more than the theoretical value. The fact is, the adjustment for resonance was not quite close as the lamp lighted could not withstand the current by closer adjustment. Two of these lamps were broken. The energy transmitted through the ground to the coil was, of course, small in this instance, since only a small part of the available primary capacity was used, that is, $\dfrac{1.66}{8}$ of the available capacity and the current of the supply transformers was reduced as far as practicable. If a coil especially adapted for the full output of the oscillator would have been used it would have been practicable to transmit many times the amount of energy needed for lighting the lamp. The lamps used in this experiment were special ones each taking, under the conditions of the experiment, perhaps 10 watts or nearly so. Assuming again a circuit under ideal conditions with the capacity of 75.2 cm on the free end of a coil without distributed capacity, and calling the potential to which this capacity would be charged P, the total energy set

23*

in movement in the excited system would be $2 \times 129{,}500 \times \dfrac{P^2 \times 75.2}{2 \times 9 \times 10^{11}}$ watts. If we assume that, as before, 1% of the total energy of the system is frittered down in the lamp we would have, in conformity with what was stated before for determining P, the equation

$$2 \times 129{,}500 \times \frac{P^2 \times 75.2}{2 \times 9 \times 10^{11}} = 1000 \quad \text{or} \quad P^2 = \frac{18 \times 10^{11}}{259}$$

or

$$P = 10^5 \sqrt{\frac{180}{259}} = 10^5 \sqrt{0.695} = 10^5 \times 0.834$$

or **83,400 V** nearly, which is a small e.m.f. The length of wire in excited circuit was as before stated 2198 feet; the wire being No. 10, with a resistance of 1 ohm per one thousand feet, the resistance of the circuit was about 2.2 ohm. From above $p = 2\pi n$ was $= 6.28 \times 129{,}500 = {} = 813{,}260$ or, say, $813{,}000 = p$. The inductance being, as shown, $\dfrac{185}{10^4}$ henry, the magnifying factor in the coil was $\dfrac{\frac{185}{10^4} \times 813 \times 10^3}{2.2} = 6840$ nearly. The lamp was one with a very short filament and its resistance may have been possibly 6 ohms. Thus with the lamp comprised the magnifying factor was still very considerable, that is, $\dfrac{185 \times 813}{82} = {} = 1830$ or nearly so. Taking it at 1800 we see that it was necessary, under the conditions assumed, to impress upon the ground plate, or near portions of the ground an electromotive force of only $\dfrac{83{,}400}{1800} = \dfrac{834}{18} = 52$ volts or nearly so! This seems very little indeed, it can be scarcely believed, but the figures seem to be not far from truth. These remarks refer particularly to the experiment illustrated on the plate marked XXIV. in which the connections were the same as in the diagram shown when discussing Plate XXII., the lamp or lamps being in series with the excited coil or system.

In the experiments illustrated in the Plates marked XXV. and XXVI., the connections were schematically the same as in diagram shown a propos Plate XXIII. and the vibrations and other particulars were practically the same as in experiment shown in Plate XXIV. just described. It is to be stated that when a secondary circuit is used, as in experiments described under XXIII., XXV. and XXVI., in connection with the excited coil, this secondary should for maximum effect be placed near the lower end of the coil; the exact position may be determined by experiment or approximately calculated. Namely, if the coil which is excited were devoid of capacity and the necessary capacity were all on the upper or free end of the coil, then the secondary circuit should, for maximum effect, be just at the center of the coil. But in the experiment as shown, the capacity is distributed and the current is strongest in the first or lowest turn, diminishing towards the top of the coil in each turn. The resultant maximum effect is thus always found near the lower end of the coil, but not quite at the end, since the upper turns also effect the secondary circuit, though proportionately less than the lower ones. The calculation of the maximum position of the secondary is complicated by the fact that generally neither the capacity nor the potential is uniformly distributed, the distribution being greatly varied by very slight irregularities in the dimension of the individual turns or their position, or the position of the nodal point or points on the wire and many other causes. It is proposed to investigate this subject specially when time permits.

XXVII. This photograph shows an incandescent lamp 16 c.p., 100 V connected with *one* of its terminals to the top or free end of the extra coil, the lower end of the latter being connected to the ground plate. The carbon filament is brought to incandescence by the currents transmitted from the ground plate and the rarefied gas is also glowing as evident from the photograph. It is instructive to note the great actinic power of the glowing gas which, though appearing to the eye of feeble luminosity as compared with the lamp filament, nevertheless impresses the plate at least as strongly, if not more so, than the incandescent filament.

In many experiments made a few years ago I observed this and also that certain gases are particularly adapted to impress the plate. This fact, again observed, impresses me more and more with the value of powerfully excited vacuum tubes for purposes of photography. Ultimately, by perfecting the apparatus and selecting properly the gas in the tube, we must make the photographer independent of sunlight and enable him to repeat his operations under exactly the same conditions, which is almost indispensable in order to attain the best results. Such tubes will, however, enable him to regulate the conditions and adjust the light effects at will. Such a facility would offer a very great advantage to the artist as with the sunlight, particularly in large cities where the air is not very pure, he has to rely much on chance and where it is almost impossible for him to perform two successive operations under the same conditions or to adjust beforehand the light effects. The photograph described shows also the high actinic power of streamers when they are of a bluish or violet color as frequently, which is the case in this instance. The red streamers are comparatively very slow in their action, but it should be stated that apart from the color the action on the plate is determined much by the power or intensity of the streamers. Thus with the powerful apparatus which I have perfected here the actinic rays are much more powerful than with the New York apparatus. In the experiment illustrated the results would have been the same if *both* of the terminals of the lamp would have been connected to the free end of the "extra coil" instead of only one, as illustrated in the diagram, which is added for the purpose of showing more clearly how the connections in this case were made.

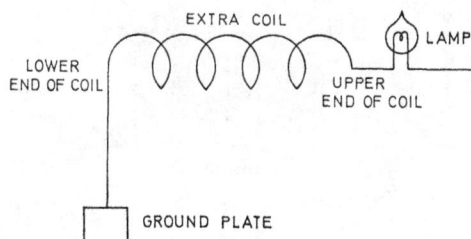

Now, as to the other particulars of the experiment, the capacity in the primary or exciting circuit was 3 tanks on each side or 1 1/2 tanks total, that is 54 bottles or $54 \times \times 0.0009 = 0.0486$ mfd. As to inductance of the exciting circuit there was the primary turn with closed secondary = 41,000 cm. and 5 3/4 turns of regulating coil, including connections, making an inductance, according to the table, of 17,684 cm, that is, the

Phot. XXVII. An incandescent lamp connected with one of its terminals to the top or free end of "extra coil", the lower end of the latter being connected to the ground plate.

total inductance of the exciting circuit was 58,684 cm, or, with fair approximation, 59,000 cm $= \dfrac{59}{10^6}$ henry. These data give

$$T = \frac{2\pi}{10^3} \sqrt{0.0486 \times \frac{59}{10^6}} = \frac{2\pi}{10^6} \sqrt{2.8674} = \frac{6.28}{10^6} \times 1.7 = \frac{10.676}{10^6}$$

and $n = 93,700$ *approx.* and $588,000 = p$ nearly.

If we assume, as in previous cases, that one percent of the total energy of the vibrating system is used up in the lamp in unrecoverable form, the energy of the system would then be 100 wats per each watt used in the lamp, or taking the lamp at 16 c.p. consuming, say, 50 watts the energy of the system would have to be 5000 watts. From this we get the potential P on the free end of the coil approximately by equating:

$$5000 = \frac{c}{9 \times 10^{11}} \times \frac{P^2 \times 93,700 \times 2}{2}$$

Here c would be the capacity of the lamp or other object, in centimeters, and taken in the same way as before explained. In previous measurements by resonance analysis the capacity of such a lamp was found to be only *1 cm*, approximately. Taking this value for c we have:

$$P^2 = \frac{5 \times 10^3 \times 9 \times 10^{11}}{937 \times 10^2} = \frac{45 \times 10^{12}}{937}$$

and

$$P = 10^6 \sqrt{\frac{45}{937}} = 10^6 \sqrt{0.048} = 10^6 \times 0.22$$

or $= 220,000\ volts = P$. This shows that P would be rather large on the assumption of but 1% frictional work in the lamp. In reality, as I know by experience, there will be generally much more energy, of the total energy of the system, used up in the lamp so that the potential P will be found in practice much smaller. But it is to be stated that *owing to the small capacity* of the lamp it will be relatively very high. By providing capacity in any way, so as to enable the lamp to take more energy, the potential required may be reduced at will.

This will be apparent from another photograph which is marked **XXVIII**. In this instance the same connections were used as before, only the free terminal of the lamp was connected by a wire to the structure of iron pipes above. The wire can be scarcely distinguished. In this experiment resonance with the exciting circuit was obtained with 6 tanks on each side, this would mean a capacity of three tanks or 108 bottles $= 108 \times 0.0009 = 0.0972$ mfd.

The inductance of the exciting circuit comprised a primary

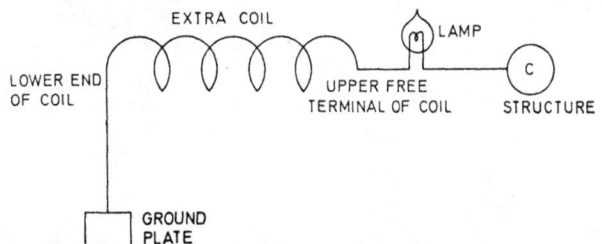

EXTRA COIL LAMP
LOWER END OF COIL UPPER FREE TERMINAL OF COIL c STRUCTURE
GROUND PLATE

Phot. XXVIII. An incandescent lamp lighted by the current passing from "extra coil" to the structure of iron pipes. The lower end of the coil is connected to the ground plate.

turn with secondary short-circuited=41,000 cm, as before, and *all* the turns of the regulating coil with connections, that is about 85,000 cm, giving total inductance at $41,000+85,000=126,000$ cm or $=\dfrac{126}{10^6}$ henry. From this follows

$$T=\frac{2\pi}{10^3}\sqrt{0.0972\times\frac{126}{10^6}}=\frac{2\pi}{10^6}\sqrt{12.2472}=\frac{6.28\times 3.5}{10^6}=\frac{21.98}{10^6}$$

and

$$n=45,496 \text{ or approx. } \mathbf{45,500=n.}$$

As will be seen from the consideration of the diagram above, in this experiment the *entire* energy supplied to the structure or capacity C had to pass through the lamp. Suppose the latter to consume 50 watts, and taking the capacity C with reference to previous estimates roughly at 500 cm. with the protecting hood, and designating with P again the potential to which the structure is charged, we have

$$50=\frac{500}{9\times 10^{11}}\times\frac{P^2\times 45,500\times 2}{2} \text{ or } P^2=\frac{9\times 10^8}{455}$$

and

$$P=\frac{3\times 10^4}{\sqrt{455}}=\frac{3\times 10^4}{21.3}$$

approx. or $=\mathbf{1409\ volts=P}$, which is a very small pressure indeed. This pressure could have been, of course, still further reduced by using a resonating circuit of still higher frequency. In the manner just described and illustrated a great many lamps could have been lighted by the vibrations transmitted to the ground and some of the facts pointed out before will enable one to make an approximate estimate in this respect. However, by connecting the "extra coil" as in normal operation, that is, to the free end of the secondary instead of to the ground, a number of lamps might have been lighted corresponding to the full output of the oscillator, say, one thousand lamps or more. Even through the ground, as in the experiment described, the action of the exciting circuit was so intense that the current of the supply transformers had to be cut down to but a very small fraction of the current taken by full output.

It should be remarked that also in the experiment described before (Plate XXVII.) a considerable number of lamps might have been lighted, but not nearly as many as in the case presently described, owing to the very small capacity of the lamps, as before stated.

Coming now to the other experiments, a lamp was lighted in the manner illustrated in Plate XXIX. in which case the lamp was connected in a *shunt* to the extra coil instead of in series with the same, as in the previous case. The diagram illustrates the connections clearly. In this experiment resonance with the exciting circuit was obtained with 22 bottles on each side, that is, 11 bottles total capacity or $0.0009\times 11=$ $=0.0099$ mfd total capacity. There were 4 1/2 turns in the regulating coil of an inductance according to

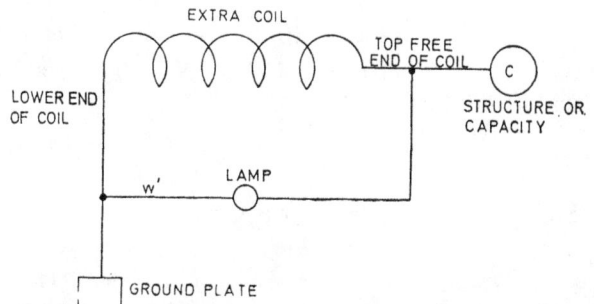

the table of 13,474 cm. This, with the 41,000 cm as before, would give an approximate inductance of 54,500 cm for the exciting circuit and from this

$$T_p = \frac{2\pi}{10^3}\sqrt{0.0099 \times \frac{545}{10^7}} = \frac{2\pi}{10^6}\sqrt{0.00099 \times 545} =$$

$$= \frac{2\pi}{10^6}\sqrt{0.53955} = \frac{2\pi}{10^6} \times 0.735 = \frac{4.6158}{10^6}$$

and $n = 217,000$ per sec. approx.

As in this case the capacity may be put approximately at 500 cm, we may roughly estimate the effective inductance in the excited circuit. Namely, calling this inductance L_s we have, for the condition of resonance,

$$T_p = T_s \text{ or } \frac{2\pi}{10^3}\sqrt{0.0099 \times \frac{545}{10^7}} = \frac{2\pi}{10^3}\sqrt{\frac{500}{9 \times 10^5} L_s}$$

and from this

$$L_s = \frac{0.0099 \times 545 \times 9}{5 \times 10^4} = \frac{48.56}{5 \times 10^4} \text{ henry}$$

or

$$\frac{48.56 \times 10^5}{5} = \frac{4,856,000}{5} = \mathbf{971,200} \text{ cm.}$$

This was the effective or actual inductance, or nearly so, of the combined system of extra coil and lamp as connected in the diagram on p. 361. Evidently, in the experiment the lamp might have been lighted by doing away entirely with the extra coil and in fact I have done so. Time did not permit taking a photograph of the experiment thus modified.

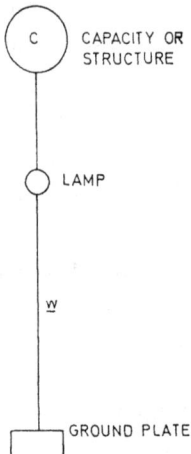
C — CAPACITY OR STRUCTURE
LAMP
w
GROUND PLATE

In such a case, the inductance of the wire w (see diagram) need be very small, hence the frequency of the currents impressing the vibration upon the ground plate will be very high and the potential, to which the capacity C will have to be charged in order to pass enough energy through the lamp or other working circuit, will be comparatively very small. By way of example, suppose wire w, forming practically all the inductance of the excited circuit, had 10,000 cm and C were the same structure as in the experiment last described of a capacity of 500 cm; then calling P_2 the potential to which the capacity is to be charged in order to supply 50 watts to the lamp or working circuit, we would have:

$$50 = \frac{P_2^2 \times 2n}{2 \times 9 \times 10^{11}} \times 500 \text{ or } P^2 = \frac{9 \times 10^{10}}{n}.$$

Now, calling T the period, we have

$$T = \frac{2\pi}{10^3}\sqrt{\frac{10,000}{10^9} \times \frac{500}{9 \times 10^5}} = \frac{2\pi}{3 \times 10^7}\sqrt{5} = \frac{6.28 \times 2.236}{3 \times 10^7} = \frac{4.68}{10^7}$$

and from this

$$n = \frac{10^7}{4.68} = 2,137,000 \text{ per sec.}$$

Substituting this for n we get

$$P^2 = \frac{9 \times 10^{10}}{2,137,000} \text{ or } P^2 = \frac{9 \times 10^{10}}{214 \times 10^4} = \frac{9 \times 10^6}{214}$$

and

$$P = \frac{3000}{\sqrt{214}} = \frac{3000}{14.63} = \textbf{205 volts} \text{ only!}$$

We see that with such an extreme frequency the voltage of the ordinary incandescent lamp supply circuit would be sufficient to pass enough current through the lamp. Under the above conditions the current would be

$$I = EC\omega = \frac{205 \times 500 \times 2\pi \times 214 \times 10^4}{9 \times 10^{11}} = 1.5313 \text{ amp. approx.}$$

In the experiment illustrated in Plate XXIX. it will be seen that while the inductance of the "extra coil" was greatly reduced by shunting the same with the lamp, yet the coil was still effective, as is evidenced from the streamers visible on the wire leading from the lower end of the coil to the lamp, which is the wire marked w' in the corresponding diagram.

Passing now to the experiment illustrated in plate designated number XXX., the connections in this case are shown in diagram below. Here the lamp was lighted only

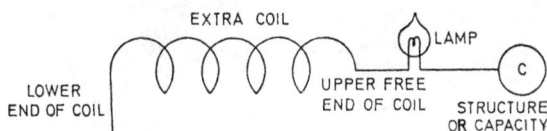

by induction from the primary circuit, the ground connection being omitted. In this case the capacity was 3 tanks on each side or 1 1/2 tanks total=54 bottles=$54 \times 0.0009=$ =0.0486 mfd and the inductance 41,000+3 turns of the regulating coil=41,000+10,000= =51,000 cm, or $\frac{51}{10^6}$ H approx. This gives

$$T_p = \frac{2\pi}{10^3} \sqrt{0.0486 \times \frac{51}{10^6}} = \frac{2\pi \times 1.37}{10^6} = \frac{8.6}{10^6}$$

and $n = 116,300$ approx.

Note: Here the excitation of the primary circuit had to be strong, showing that the inductive action was feeble in comparison with action through the ground. The tuning was not quite exact in all these experiments described as time was pressing. In the last experiment n ought to have been just twice the value found in case XXVIII.

XXXI. This is a Roentgen photograph taken in a peculiar manner. It so happened that a workman who thought he had injured one of his figures desired to have a photograph taken and a tube was inserted between the ground plate and a coil, as illustrated in the diagram under XXII. It was doubtful whether the tube could be energized in this manner but the experiment proved that it could and a photograph was taken by flashing the tube a few times, after the adjustment of the vibration in the primary and excited circuit was completed and resonating condition in the latter secured. Nothing peculiarly interesting beyond the manner of taking the photograph was contemplated, nevertheless an inspection revealed that this photograph, probably owing to the high economy of the oscillating systems used in these experiments, or possibly on account of the frequency, was distinctly different from many others taken with different apparatus. The tube was *not strongly excited* and the exposure was scarcely more than a few seconds, yet much of detail was recorded. It is quite curious how plainly the nails are shown, much as in an ordinary photograph.

Colorado Springs laboratory fully developed.

On this occasion I must point out a peculiar feature about the action of the currents developed by this apparatus upon Roentgen tubes. As might be expected, some experiments were carried on in this line also and possibly with greater pleasure than those in other directions, for my conviction is growing stronger every day that, with apparatus such as the present, wonderful results must be secured provided only that a tube is constructed capable of taking up any amount of energy. On my return this task will be a serious one. Many times tubes have been worked here from the secondary but curiously enough, for a reason which is to me not yet clear, they can only work for a *few seconds at the most* as, almost instantly, they become very highly rarefied and the sparks begin to dart over the glass, the tube becoming useless. No matter how the current was cut down the action took place, unless it was reduced to such an extent that the tube was scarcely excited at all. Already in New York with an apparatus similar to this, though much smaller, I observed that such an action always occurs, *in some degree,* when the vibrating system possesses a very *small resistance* and the electrical movement in the circuit connected to the tube is very large. This will be further investigated.

XXXII. This is a central view through the open entrance spoiled...

Note: Plate XXXIII. is missing. One more photograph is to be made showing a lamp lighted without a coil, merely in series with a plate or object as capacity. If time permits a special coil will be constructed for *full power* of apparatus and all photographs, as far as practicable, taken *outside.* Taken in this manner they would be undoubtedly much more interesting to scientific men.

Colorado Springs

Jan. 3, 1900

Photographs taken with Mr. Alley from Dec. 17 to Dec. 31, 1899 and *particulars* relating to the same.

In the photographs, which will be presently described, 11″×14″ Cramer plates were used, most of which were "instantaneous isochromatic".

XXXIII. View of the Pike's Peak Range taken from room through the window glass by moonlight, the night of an eclipse. A gale was blowing and it being impossible to place the camera outside, the photograph was taken in the above manner. As the building was trembling more or less during the sudden gusts of wind the picture is not as perfect as it might have been. The light was magnificent, the moon almost full and a good photograph might have been taken in 10—15 minutes. There was, of course, considerable loss in the transmission through the window glass though it was thoroughly cleansed, nevertheless the photograph is as good as if it had been taken in daylight. The exposure was much too long, *two hours*, from 9—11. Had the wind not been blowing a perfect picture would have been obtained in a quarter of that time. The wonderful brightness of the moonlight was increased by the snow which was unusually heavy.

XXXIV. This photograph shows the laboratory viewed from behind with the mountain range from Pike's Peak to Cheyenne Mountain in the background. There was

Phot. XXXVIII. View of the laboratory from the Pike's Peak side.

not much snow on the ground and little or none on the roof. The exposure was only 35 minutes, the moon nearly full. As the wind was blowing, the ball on top of the pole does not show well.

XXXV. This is another view of the laboratory from the rear with the Pike's Peak Range as background. Considerable snow on the ground. The exposure was *two hours.* It was again a windy night and the image of the upper part of the pole and ball is marred owing to swaying.

XXXVI. Again a view of the laboratory from the rear with the lower mountain range on the right side of Pike's Peak as background. There was snow on the ground. Extremely cold and very windy. Exposure 40 *minutes.*

XXXVII. Once more a view of the laboratory from the rear with the Pike's Peak Range as background. The moon was waning and a longer exposure — *three hours* — was made. It was found that the plate was somewhat over-exposed. Otherwise the conditions were not unfavourable. The wind was not strong and there was still considerable snow on the ground. The above photographs satisfied fully the novel pleasure of taking pictures by the fascinating moonlight of Colorado and attention was then turned to less agreeable but more useful work.

XXXVIII. This is a view of the laboratory from the Pike's Peak side taken in the morning by sunlight, fresh fallen snow on ground.

XXXIX. A view of the laboratory from the rear taken under the same conditions.

XL. Illustrates discharge of "extra coil" laterally across the field from a pointed wire on the top of the coil to a ball 30″ diam., provided with a point and a coil on a stand which supports the ball. Strong streamers and sparks, the latter passing to the floor; the coil mentioned and hood fastened to structure of iron pipes in the center of the building. Some sparks passed also to the roof causing, as usual, considerable concern, the fireproof paint notwithstanding. Some streamers in the same positions as others relative to the camera are very weak, this is probably due to their red color. Many streamers show clearly the phenomenon of splitting up or ramification and the sparks, where they strike the floor, develop increased luminosity. The feature of "splashing" on the floor upon striking the same is also well illustrated, particularly on one of the streamers which is thinner and sharper than most others. To give an idea of the magnitude of the discharge the experimenter is sitting slightly behind the "extra coil". I did not like this idea but some people find such photographs interesting. Of course, the discharge was not playing when the experimenter was photographed, as might be imagined! The streamers were first impressed upon the plate in dark or feeble light, then the experimenter placed himself on the chair and an exposure to arc light was made and, finally, to bring out the features and other detail, a small flash powder was set off. It was found necessary to sit in the chair during the exposure to arc light as, otherwise, the structure of the chair would show through the body of the person, if the same were exposed merely to the light of the flash powder. As the weather during these experiments, which were carried on late at night, was most generally far below zero, I tried to overcome the above necessity but neither I nor Mr. Alley could device a practical remedy.

The simplest remedy would have been to employ a very powerful light enabling an instantaneous exposure, but this was not practicable with the arc light. On the other hand, the ball being very large, a great quantity of flash powder was required which it was not advisable to use because of danger and impairment of the quality of the photographs. In the instance described, the streamers shown were produced by 50 closures of the circuit energizing the Westinghouse transformer, each closure lasting approximately one half of one second, possibly less. The exposure to arc light lasted 5 minutes, about half of the full lens opening, and in flashing about one third of a large size Eastman powder was set off.

XLI. This is a similar view of the discharge (fifty flashes, very short) of the extra coil from a pointed wire on top, or free terminal. The photograph shows, in an interesting manner, the splitting up of streamers or sparks near the floor. Another curious feature observable is the splitting and again uniting of a streamer or spark. I say "streamer or spark" because the former becomes a "spark" or perhaps stated better, an "arc" when it strikes some object which causes a strong current to pass through the path of the streamer, which then suddenly assumes the brilliancy and color characteristic of "sparks". I have seen at times a very strong discharge dart out from a point or surface in the form of a spark, continue for 6—10 feet or so and then split up in streamers. It is more than likely that this phenomenon will be found recorded in some of the photographs taken on closer inspection, as it is not unfrequent. At times, again, I have observed a small length of a streamer, anywhere along its path, assume a relatively very considerable luminosity and assume in this part the character of a spark. This I have, if I am not mistaken, already pointed out elsewhere. Some such luminous parts or points are to be seen plainly in this and other photographs.

When the action is very energetic, owing to the power of the streamer and other causes, the luminous portion of the same becomes a veritable "fireball". This observation which, to my greatest astonishment, I have frequently observed in experiments with this apparatus, shows now clearly how "fireballs" are produced in lightning discharges and their nature is now quite plain. I have heretofore always been inclined to believe this phenomenon to be merely a visual impression, similar to one which is experienced upon a violent blow on the eye, or some part of the head, or the spine, or which follows upon a sudden and very intense manifestation of light most generally. Although the vision of a moving ball of great luminosity is experienced only in the rarest instances, the person as a rule seeing luminous spots, "stars" or flaming tongues.

With the present experiences I am satisfied that the phenomenon of the "fireball" is produced by the sudden heating, to high incandescence, of a mass of air or other gas as the case may be, by the passage of a powerful discharge. There are many ways or less plausible in which a mass of air might be thus affected by the spark discharge, but I hold the following explanation of the mode of production of the "ball" as being, most likely of all others which I have considered, the true one. When sudden and very powerful discharges pass through the air, the tremendous expansion of some portions of the latter and subsequent rapid cooling and condensation gives rise to the creation of partial vacua in the places of greatest development of heat. These vacuous spaces, owing to the properties of the gas, are most likely to assume the shape of hollow spheres when, upon cooling, the air from all around rushes in to fill the "cavity" created by the explosive dilatation and subsequent contraction. Suppose now that this result would have been produced by one spark or streamer discharge and that now a second discharge, and possible many more, follows in the path of the first. What will happen? Before answering the question we must remember that, contrary to existing popular notions, the currents passing through the air have the strength of many hundreds and even thousands of amperes.

It was a revelation to myself to find that, even with the apparatus used in these experiments, a single powerful streamer, breaking out from a well insulated terminal, may easily convey a current of several hundred amperes! The general impression, if I am not mistaken, is that the current in such a streamer is small but this belief is due to the comparative unfamiliarity of the electrician with such apparatus as I am now using. As a matter of fact it is quite easy to consume in such streamers, as are illustrated in these photographs, most of the energy developed by the apparatus and the currents conveyed through the air may be, by suitable provisions, made as strong as those circulating in the wire or coil itself which produces them. No wonder then, that a small mass of air is "exploded" with an effect similar to that of a bombshell, as noted in many lightning discharges.

But to return now to the explanation of the "fireball", let us now assume that such a powerful streamer or spark discharge, in its passage through the air, happens to come upon a vacuous sphere or space formed in the manner described. This space, containing gas highly rarefied, may be just in the act of contracting, at any rate, the intense current, passing through the rarefied gas suddenly raises the same to an extremely high temperature, all the higher as the mass cf the gas is very small. But although the gas may have been brought to vivid incandescence, yet its pressure may not be very great. If, upon the sudden passage of the discharge, the pressure of the heated air exceeds that of the air around, the luminous ball or space will expand, but most generally it may not do so. For assume, for instance, that the air in the "vacuous" space was at one hundredth say, of its normal pressure, which might well be the case, then, since the pressure in the space would be as the absolute temperature of the gas within, it would require a temperature which seems scarcely realizable, to raise the pressure of the rarefied gas to the normal air pressure. It is therefore reasonable to expect that, despite the high incandescence of the rarefied air, the space filled with the same will continue to contract, and here an important consideration presents itself. When, as before explained, the vacuous space was formed, the spark or streamer passed through the air *disruptively*, therefore the path was necessarily very thin, threadlike, and the minute quantity of the air which served as a conductor for the current was expanded with explosive violence to many thousand times its original volume. Owing to the fact, however, that the quantity of matter through which the current was conveyed was small, a great facility was offered for giving off the heat so that the highly expanded gas-owing to its expansion and to radiation and convection of heat-cooled instantly.

But how is it when the second discharge and possibly many subsequent ones pass through the rarefied gas? These discharges find the gas already expanded and in a condition to take up much more energy by reason of the properties it acquires through rarefaction. Evidently, the energy consumption in any given part of the path of the streamer or spark discharge is, under otherwise the same conditions, proportionate to the resistance of that part of the path; and since, after the gas has once broken down, the resistance of other parts of the path of the discharge is much smaller than that including the vacuous space, a comparatively very great energy consumption must necessarily take place in this portion of the current path. Here, then, is a mass of gas heated to high incandescence suddenly but not, as before, in a condition to give up heat rapidly. It can not cool down rapidly by expansion, as when the vacuous space was being formed, nor can it give off much heat by convection. To some extent even radiation is diminished. On the contrary, despite the high temperature, it is compelled to confinement in a limited space which is continuously shrinking instead of expanding. All these causes cooperate in maintaining, for a comparatively long period of time, the gas confined in this space at an elevated temperature, in a state of high incandescence, in the case under consideration. Thus, it is that the phenomenon of the "ball" is produced and the same made to persist for a perceptible fraction or interval of time. As might be expected, the incandescent mass of gas in a medium violently agitated, could not possibly remain in the same place but will be, as a rule, carried, in some direction or other, by the currents of the air. Upon little reflection, however, we are led to the conclusion that the ball or incandescent mass, of whatever shape it be, will always move from the place where an explosion occured *first*, to some place where such an explosion occured *later*. This will be most generally in the path of the discharge, from its origin to its end, but not necessarily always so. For example, it may so happen that a spark produced in some place strikes an object of a material which is evaporated or volatilized with difficulty, and that *later*, in another place, this same or other spark hits an object of a material more volatile. If so then the explosion on the later place will sooner occur, and the result will be that the current of air, when both the explosions have subsided, will move from the later to the former locality. But I believe that, in most cases, the current of air will take the opposite course, as before stated. In whatever direction the movement may occur, it is plain that the velocity can not be very great. In fact, all observers concur in the opinion that such a "fireball" moves slowly. If we interpret the nature of this wonderful phenomenon in this manner, we shall find it quite natural that when such a ball encounters in its course an object, as a piece of organic matter for instance, it will raise the same to a high temperature, thus liberating suddenly a great quantity of gas by evaporating or volatilizing the substance with the result of being itself dissipated or "exploded". Obviously, also, it may be expected that the conducting mass of the "ball" originated as described, and moving through a highly insulating medium, will be likely to be highly electrified, which accords with many of the observations made. A better knowledge of this phenomenon will be obtained by following up experiments with still more powerful apparatus which is in a large measure already settled upon and will be constructed as soon as time and means will permit. There may be a way, however, of intensifying in this respect, the action of the present machine. A very important matter is to use better means of photographing the streamers exhibiting these phenomena. Much more sensitive plates ought to be prepared and experimented with. The coloring of the films before suggested might also be helpful in leading up to some valuable observation. It being a fact that this phenomenon may now be artificially produced, it will not be difficult to learn more of its nature. Photography will be, of course, the best means to investigate it and the first efforts ought to be in this direction. With the present plates, although the "balls" produced with the apparatus experimented with are probably up to 1 1/2″ diam. and possibly more, they leave only a small dark spot on the plate, only the nucleus or central portion impressing itself.

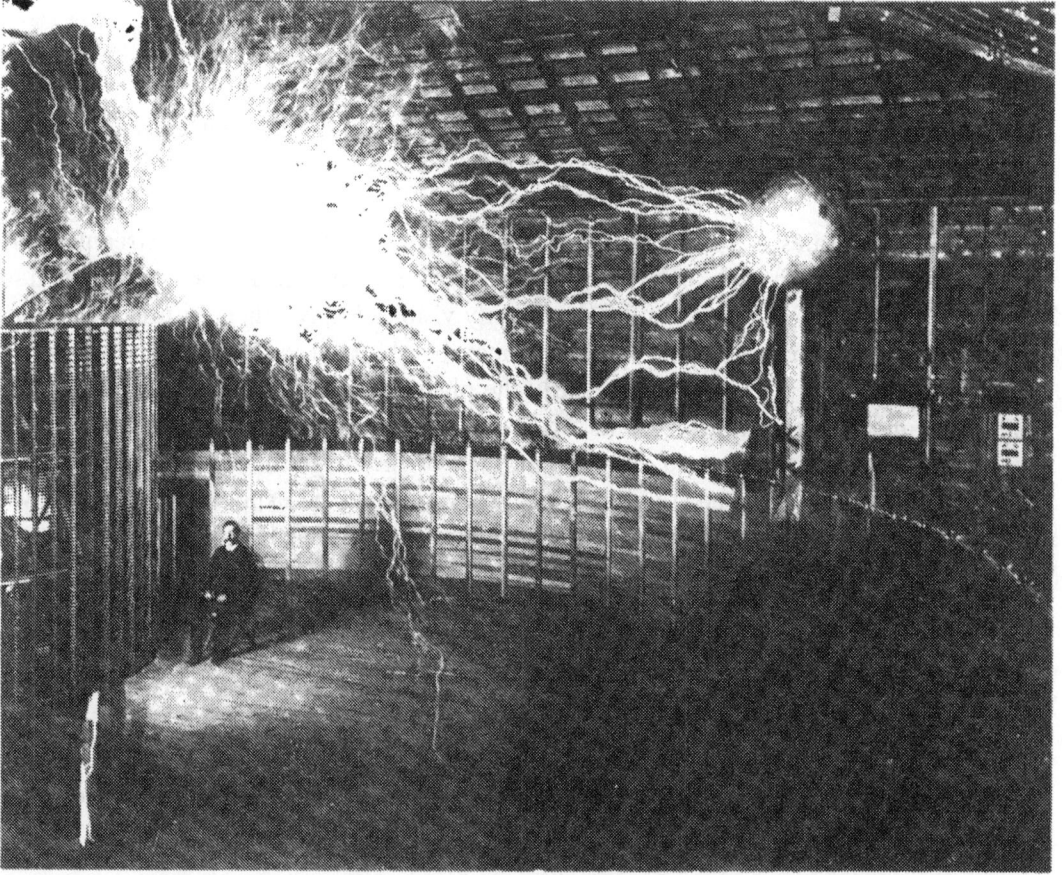

Phot. XLIII. The discharge of "extra coil" laterally across the shop. Mr. Alley is sitting.

Photographs taken with Mr. Alley from Dec. 17 to Dec. 31, 1899 and *particulars* about the same:

XLII. Illustrates a similar view of same apparatus in action. The streamers and sparks are produced by 25 short throws of the switch. The sparks pass to hood, floor and roof. The "splashing" on the floor is plainly visible. Many other features of interest, some of which have been already described, may be observed in the photograph. So, for instance, one or two streamers show clearly the phenomenon, which has been dwelt upon already, namely more luminous spots or "balls". The loss in luminosity of a streamer branching out is also illustrated. To give an idea of the magnitude of the display again a human figure is introduced, this time Mr. Alley sitting. The exposure to arc light was 5 minutes, about half opening of full lens and afterward about one half of the large Eastman powder was set off.

XLIII. This shows again the same discharge laterally across the shop, as before. The sparks are more powerful this time and there are more of them, fifty throws of the switch being made. The sparks and streamers are made to issue this time from a curved wire forming the terminal of the extra coil. The sparks to the hood and to the roof are particularly interesting. The point used before on the ball 30″ diam. was taken off. The extra coil unfortunately broke through some places and also to the floor.

Some of the streamers form actual loops turning back upon themselves. A curious feature is presented by a long streamer striking the wooden support of the ball and splashing from there upon the ground wire leading from the ball. Some streamers are seen to pass along the wire without striking the same, lighting finally on the ball. This plainly illustrates that the path of such a discharge is accidental, dependent on the arrangement of particles floating in the air on the currents in the latter. Luminous points are again observed on some of the streamers, as in previous instances. In this experiment the other particulars were the same as before. Mr. Alley was photographed once more, a flash being used after the arc light exposure as in the preceding case.

XLIV. This plate shows the extra coil discharging laterally across the shop and to the floor, the streamers and sparks issuing from several thin wires tied together and spread apart on the end. In this experiment several discharges to the floor were so powerful as to actually inflame the wood on the spot where they struck the wood. Several instances of "splashing" on the wood are observed. The splitting or branching is nicely shown in one of the streamers. One of the sparks strikes the wood and disappears emerging again at some distance from the spot, having evidently followed a better conducting path through the wood. A powerful spark passes to the coil on the stand, jumps out and strikes the top wire of the secondary, instead of taking the shorter and easier route along the wire to the ground. Some very curious curves and twists are observed on a number of the sparks and streamers. In the experiment, 100 throws of the switch were made and the particulars were in other respects the same as before, the vibration of the "extra coil" being the normal or nearly so.

XLV. This shows a slightly different view, the discharge of the extra coil taking a similar course laterally across the shop. Strong sparks pass to hood, floor, high coil on stand and also to roof. The photograph presents features of interest similar to the foregoing. One of the streamers behaves curiously, going for some distance away from the coil and then turning back upon the same. The feature of the thickening of some sparks and streamers on the lower end near the floor is well shown in a few instances, and unmistakably. The discharges are long as they are made to issue from the tip of a wire, mostly the leaks on other points being comparatively small. The glow of the top wire of the secondary is strong. Fifty throws of the switch were made in this instance, other particulars remaining substantially the same.

XLVI. In this case again the extra coil discharges laterally as in a number of instances just described. As there was a bad leak on a spot of the rubber cable forming the last two or top turns of the extra coil, a ball of 30″ diam. was connected to the cable at that point, the object being to take up the pressure and prevent the leak. It was expected that the discharge from the end of the wire, from where it was intended to issue, would thus be strengthened. This proved to be the case decidedly, but the presence of the ball slowed down the vibration of the extra coil to some extent, thus destroying the best condition for resonance obtainable with the apparatus, for some inductance had to be inserted in series with the primary of the oscillator and this meant a slightly decreased economy. Some streamers, issuing from the large ball, despite the drain on the end of the wire, illustrate clearly the immense electromotive force and great quantity of electrical movement in the system. Many strong sparks pass to the high coil and still many more to the ground through the wooden floor, the latter showing some of the interesting phenomena described before. Particularly clear is the action of some sparks in inflaming or volatilizing the wood on the spot where they strike it. The splashing over the surface of the wood is also well shown in a number of instances. A portion of the floor struck by a number of sparks near together is considerably illuminated, chiefly by the increased light of the sparks on the surface of the floor and near the same. This additional light is, as may be seen during the experiment and also by an inspection of the photograph, due to the momentary ignition of the material to which also the thickening of the sparks near the floor might be ascribed, observable in some of the sparks. A curious screw movement is noticeable on some of the streamers. There were again 100 throws of the switch made in this experiment, the other particulars being, as nearly as practicable, the same as before.

XLVII. Once more a similar view was taken, illustrated on this plate, but the ball was not connected to the top of the extra coil and the vibration of the latter was very nearly the normal. There are some leaks on the cable on top, despite the most careful wrapping with mica and rubber, neverthelles some of the sparks and streamers issuing from the tip of the wire attain great length. A number of the sparks fly in curiously curved paths to the top wire of the secondary. These are about 22 feet or more in a straight line. There is a draught created by the discharge, as is evident from the splitting up of some sparks or streamers in individual discharges effected by the operation of the break wheel. Many interesting formations of the streamers and some phenomena before described are again noted in this case. Some sparks are seen to continue along the floor. The sparks to the hood show clearly and beautifully the effect of the draught created in consequence of the rapid heating of the air. A curious feature is afforded by some streamers actually doubling upon themselves. The secondary resonates very strongly but the strong streamers on the top wire are partially due to the reaction of the excited system. In the present instance also, 100 throws of the switch were made, the other conditions remaining practically the same as before.

XLVIII. This plate shows one of the strongest discharges viewed similarly, as in the previous cases just dwelt upon. The streamers and spark issue again from the tip of the "free terminal" wire of the extra coil. Some of the sparks pass again in winding paths, which are very long, much longer than the straight course which they should take, to the top of the secondary. These paths have easily a length of some fifty feet or so and are of extreme brilliancy. This is partially due to the capacity of the ball of 30″ diam. connected to the top turn of the extra coil. Some very brilliant and thick sparks pass from the ball to the hood above and others much longer, though less brilliant, strike the high coil on the stand across the laboratory. One of the streamers is wonderfully interesting on account of the curiously twisted and curved appearance. It is hard to conceive how a discharge can pass through the air in this way when there exists a strong tendency to make it take the shortest route.

The curiously curved path clearly shows how extremely sensitive discharges of great length, and particularly those passing out into the air from a single terminal, are to currents of the air. This sensitiveness is still further increased when the streamer or spark is not compelled to issue always from the same point or points, as when the terminal is constituted by a pointed wire, but can issue with *equal facility* from other points as, for example, when the terminal is constituted by a large ball or disk. The slightest draught is, in such a case, sufficient to alter the position and shape of the streamer. In such an instance the discharge is also highly sensitive to other influences, as magnetic or electric actions, Roentgen rays, light and other forces or disturbances in the medium.

I now understand better why a "rotating brush", which I have described some years ago, is so wonderfully sensitive, many times more than any other sensitive device of which I have knowledge at present. It seems to me that a sensitive device in telegraphy on this principle will have to be ultimately adopted in preferance to others. The trouble is that it is dificult, or at any rate inconvenient, to produce and maintain the phenomenon but this difficulty may be overcome in time. The photograph described shows very beautifully how a streamer falls apart and spatters after striking a wooden structure, owing to the sudden heat evolution and gas generation at that spot. The photograph conveys well the idea of the fierceness of the discharge produced by 100 throws of the switch, other particulars remain'ng.

Colorado Springs

Jan. 5, 1900

Photographs taken with Mr. Alley from Dec. 17 to Dec. 31, 1899 and *particulars* about the same continued:

XLIX. This plate shows again the extra coil viewed centrally and discharging from the brass ring on top, mentioned before, to which thin wires, pointing upwards, are fastened. As the streamers are made to issue from great many points at once, they are decidedly weaker than when, as in some previous cases, permitted to issue only from a few points. This is evident from the fineness of their texture, as shown in the photograph. But the total amount of energy spent upon the air is very great in this experiment and the coil in action produces the effect of a hot furnace, creating a strong current of air through the open roof which, rushing upward, takes effect upon the streamers as is plainly recorded on the plate, particularly on the upper streamers. A very interesting long spark passes to the wire leading from the secondary. It is evident that many streamers are split up to such an extent that they do not impress the film sufficiently. In the experiment also 100 throws of the switch were made and the other particulars were nearly the same as before. The e.m.f. at the terminals of the extra coil system is necessarily smaller because of the facility afforded for the escape of the streamers and great frictional loss in the air, which causes the free oscillations of the system to die out more quickly than in some of the foregoing experiments.

L. This photograph represents a similar view of the extra coil discharging from the brass ring and wires fastened to same; in this case the wires being directed downwards. The streamers show much the same character as before and many are evidently not strong enough to record their path on the film, on account of the great quantity of them. Most of them produce the effect of a strong glow only. Some strong sparks pass occasionally to the wire leading from the secondary. Again, the photograph was produced by 100 closures of the switch and normal excitation of the extra coil system.

LI. This is once more the extra coil discharging under similar conditions and viewed in the same manner. The wires fastened to the brass ring forming the last turn of extra coil are in this case pointed still more downwards. To strengthen the streamers somewhat the number of the discharging wires is reduced. The streamers on the top part of the picture are very beautiful, their paths being curiously curved. But the fine texture of the discharge again shows the weakening effect of the many wires or points of issue, though the individual discharges are evidently stronger than before. Some sparks passing to the hood and floor are interesting and show that the e.m.f. has been increased by reducing the number of the discharging wires. In fact some sparks and streamers attain great length and there are a number of instances of their splitting up in curious ways illustrated. A strong draught effect is also apparent. The photograph is rendered more beautiful by its symmetry. *All particulars* remained as before.

Colorado Springs

Jan. 6, 1900

Photographs taken with Mr. Alley from Dec. 17 to Dec. 31, 1899 and *particulars* relating to the same continued:

LII. This plate shows a discharge from the top of the extra coil, produced by a single closure of the circuit or throw of the switch, of very short duration. The discharge issues again from the brass ring on the top of the extra coil, which is viewed centrally. Some streamers pass into the air and sparks dart to the hood above the coil. A curious spiral or screw motion in one or two streamers is noticeable. The light of the discharge, though lasting but one fraction of a second, is strong enough to reveal a part of the structure and the top of the coil. The current of air separates the sparks and streamers into the individual discharges produced by the break wheel. This effect is beautifully shown. The vibra-

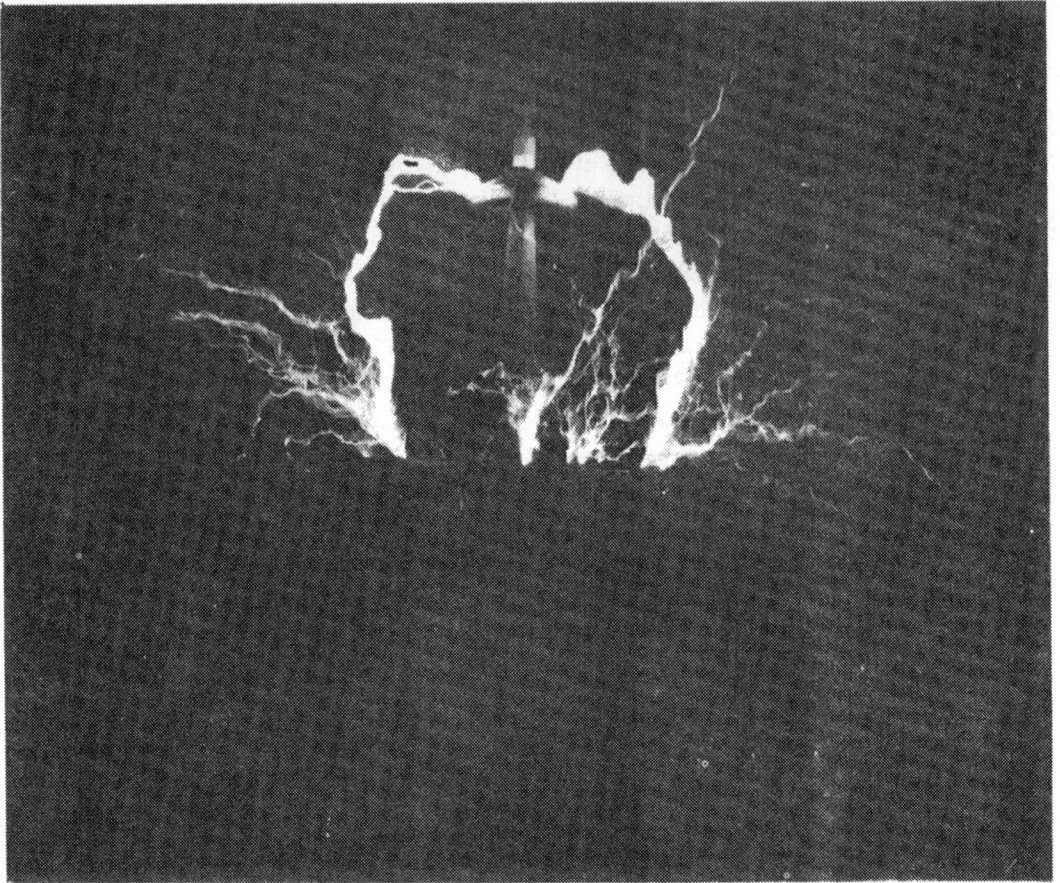

Phot. LII. Discharge from the top of "extra coil", produced by a single closure of the circuit of very short duration.

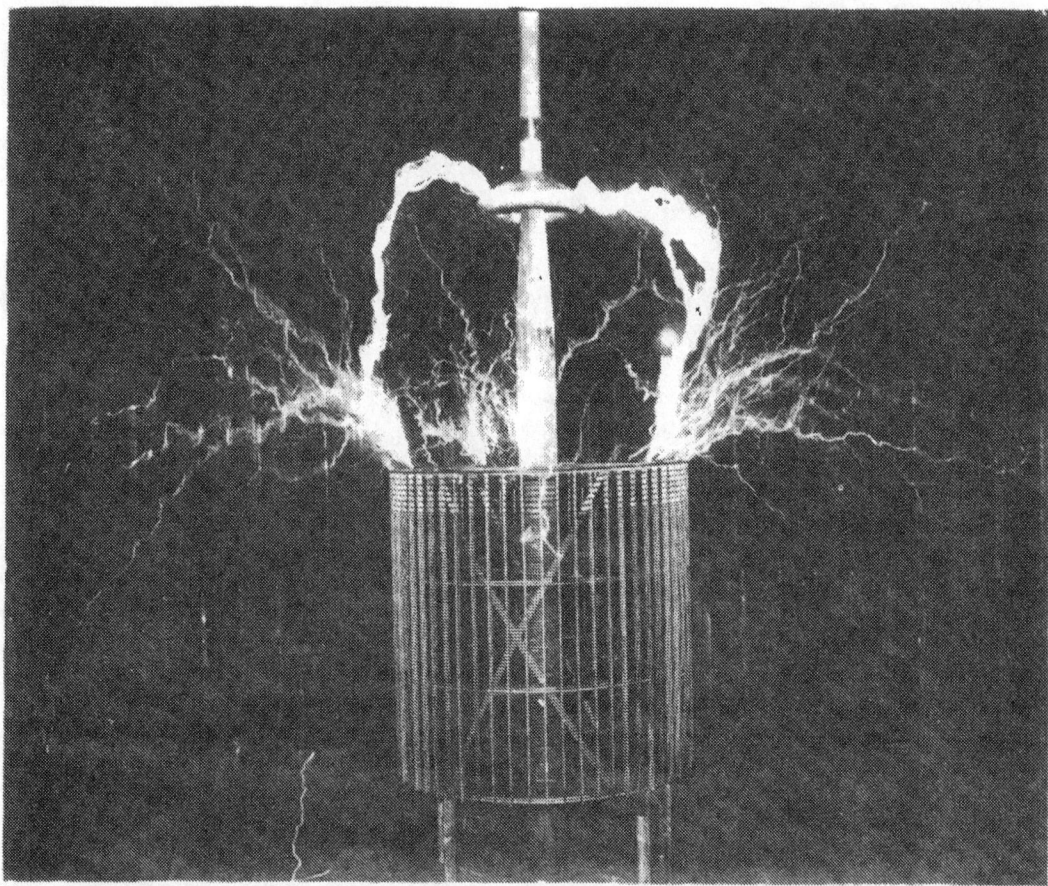

Phot. LIII. Produced under similar conditions as phot. LII by a single closure of the circuit but of longer duration.

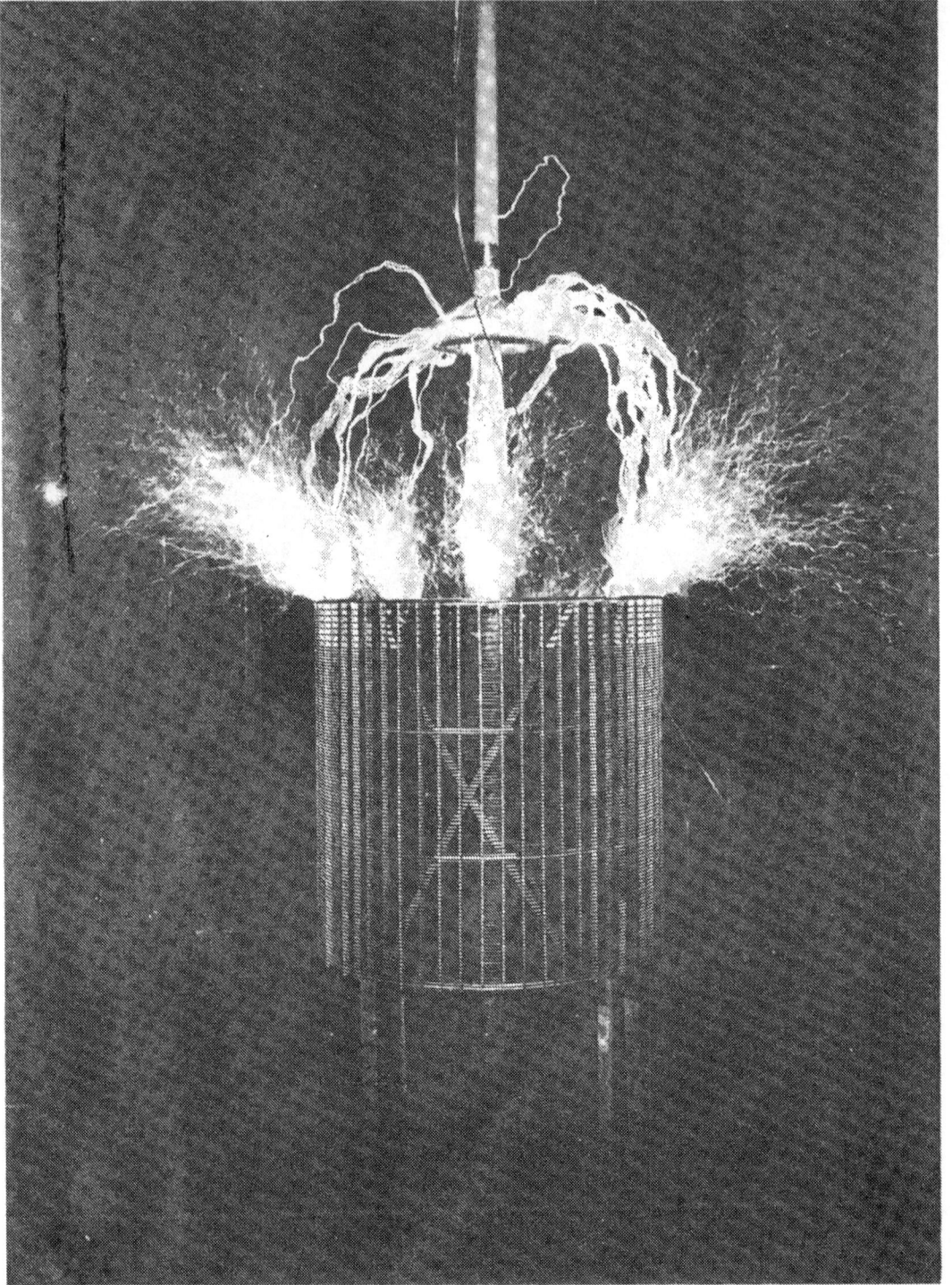

Phot. LIV. "Extra coil" discharging upwards from the brass ring on the top.

tion of the coil and all other particulars are as in most previous instances recently described. *No* after illumination by arc or flash.

LIII. This illustrates again the discharge of the extra coil under similar conditions as produced by a single closure of the circuit through the switch, but of longer duration. This is a side piece to Photograph VI already described, only the door was not open as in that case and the draught was consequently much smaller. Some very interesting bending upon themselves and twisting of the streamers and sparks may be observed in the picture. In places, some streamers are actually interrupted. This singular phenomenon has already been described on a previous occasion. Some streamers again appear exceptionally thin though they ought to appear as thick as the others, judging from their position relative to the camera. This is evidently due to the smaller actinic power of these particular discharges, owing to which only the central and white portion impresses itself upon the film. A spark striking the floor shows the feature of "splashing", dwelt upon before very clearly. A few sparks reach the roof but do not continue along the same, being rather weak at that distance. The photograph is on the whole very beautiful. All the particulars were as normally or nearly so.

LIV. In this photograph the "extra coil" is again shown in central view discharging upwards from the brass ring or turn on the top, under conditions nearly normal. 100 throws of switch were made in the experiment. The plate is very interesting on account of the exceptional fineness of the streamers and perhaps still more so owing to the curious curves of the sparks. One of the latter is seen to pass quite close to the hood without striking it, preferring a point of the iron pole far above the hood. Another feature of interest is afforded by a streamer passing close to the iron pole and escaping through the opening in the roof, seemingly not being affected by the presence of the pole. These two discharges are very long, the latter particularly. The photograph is very successful, the focusing being excellent.

LV. This plate illustrates the discharge passing laterally across the laboratory from the tip of a wire, forming the "free" terminal of the extra coil, chiefly to the floor and top wire of the secondary. The sparks and streamers are very powerful and long, exhibiting many of the phenomena described in some cases before. One spark passing from the terminal of the extra coil across the laboratory is of rare beauty showing the individual discharges effected by the break wheel exceptionally well. A few sparks and streamers show luminous points on some places which may be "fireballs". I have already dwelt at some length on this fascinating phenomenon and have explained what I consider to be its true nature. It may be of interest to state that such luminous points may be produced in other ways without partaking of the nature of the fireballs. One of these ways may be described here. It is well known — at least I assume so — that by causing two or more vibrations of different pitch to pass along a conductor, nodal points and points of maximum effect may be produced and caused to shift slowly along the conductor. Such a result I have frequently obtained with two vibrations of but slightly different period, the period of one of the vibrations being adjustable. If I recollect rightly I have described a result of this kind somewhere before this.

Substituting for a conductor of copper a vacuum tube of great length, I have also produced in the latter more or less luminous strips, striae or portions which would move along the tube with a velocity dependent on the relative wave length of both the vibrations,

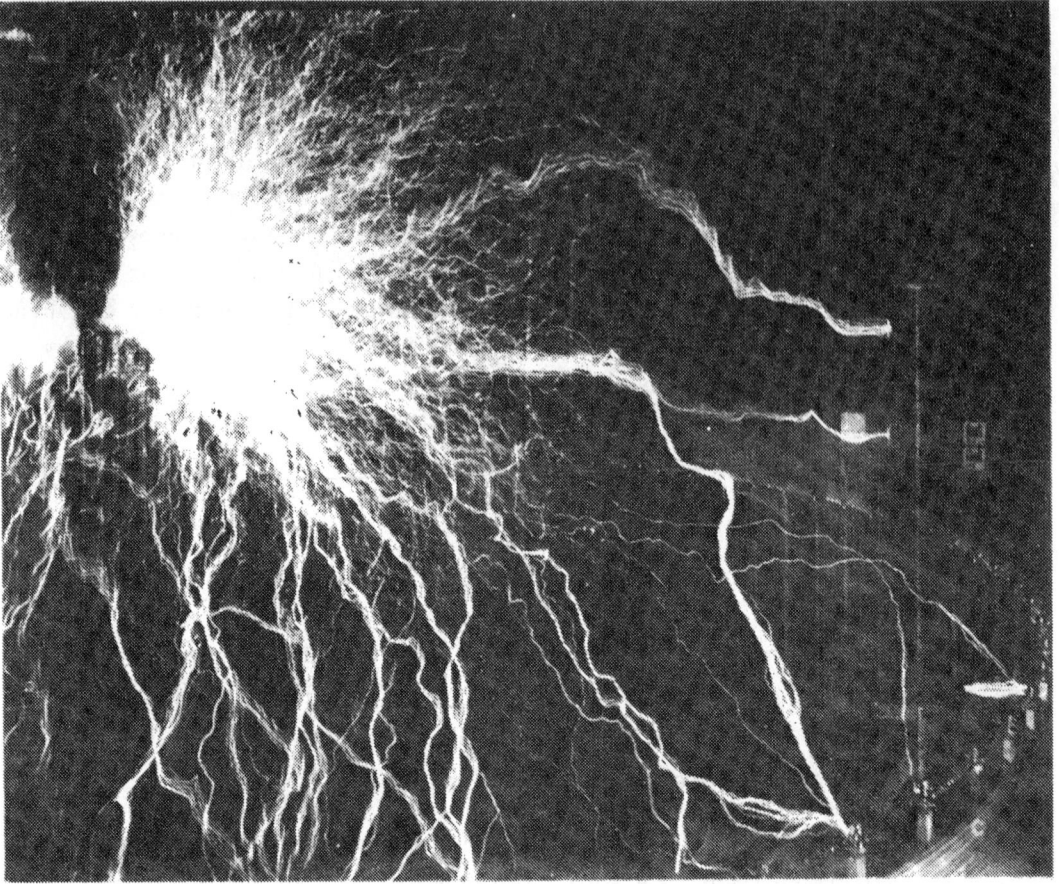

Phot. LV. "Extra coil" discharging laterally across the laboratory from the tip of a wire forming the free terminal of the coil to the floor and top wire of the secondary.

and which was at will adjustable by an adjustment of the wave length of one of the vibrations. The truth is, however, that I observed this phenomenon in vacuum tubes long before without being able to render myself an account of the nature of the same until I obtained the same effect on a wire in the manner stated. Now it is quite clear that, since a streamer is a conducting path comparable to a wire, the same phenomenon may take place on the streamer itself — being the result of two (or possibly more) vibrations of different wave length. This is all the more probable as in such an apparatus two vibrations of this kind may readily occur since the capacity, or the inductance of the circuit or both, may undergo variations as the discharge is playing, thus modifying the period to a slight extent — sufficient to cause the production of this phenomenon. In fact I have frequently observed such variations of the constants of the circuit which is oscillating, these variations being produced in various different ways. Thus it may happen that there is seen on such a long spark or streamer a point or portion of greater or smaller luminousity, or more such points, moving along the path of the discharge with a small velocity, without having anything to do with the occurrence of the fireballs, as before explained.

If the interruption of a discharge in some part of its course, which was noted before, would not have been actually and unmistakably observed, I would think it quite possible that this phenomenon might be due to a dark part or nodal portion formed on the streamer in the manner above set forth, this part being either stationary or slowly shifting along the path of the discharge, as the case may be. I think that I shall be able to settle this point in the following experiment which I propose to carry out. The idea is to provide a streamer which will be preferably straight and will pass continuously through the same path, thus enabling an effect propagated along the streamer to be observed just as on a wire. The streamer should be preferably also of very great length. This I am convinced I can realize as follows:

A glass tube of pieces joined together temporarily, of rather large diameter, and of a length of, say, 30 to 50 feet is to be provided. The end of a well insulated wire, forming the "free" terminal of a coil, as the extra coil here used, is to be led in one end of the glass tube, in the center of the same so that the streamer, when formed on the point of this wire, will have the tendency to pass along the tube on the inside of the same. In order, however, to keep the discharge away from the glass, suction is to be applied on the other end of the tube or else a current of air is to be forced through the tube — from the side of the discharging wire towards the other open end of the former, in any convenient way — so as to compel the streamer to pass along the axis of the glass tube. If the tube is of large diameter I do not think that it will be difficult to carry out the experiment. Now this streamer is to be produced by oscillations of small wave length and under these conditions it will be quite easy to produce stationary or shifting, nodal or maximum points along the path of the streamer.

But, to return to the description of the present photograph, some of the paths described by the discharge are curious in the extreme. Many features dwelt upon before are again and even more clearly shown. So the "splashing", the splitting up and reuniting is plainly visible. Some streamers strike the roof and one particularly was dangerous, the plate showing that after hitting the roof it divided in three parts following the structure. *This will scarcely print.* An ignition of the roof would have been unavoidable had the switch been held on only a fraction of a second longer. But in manipulating the switch I always took care to throw off the handle instantly when, by chance, one of the discharges

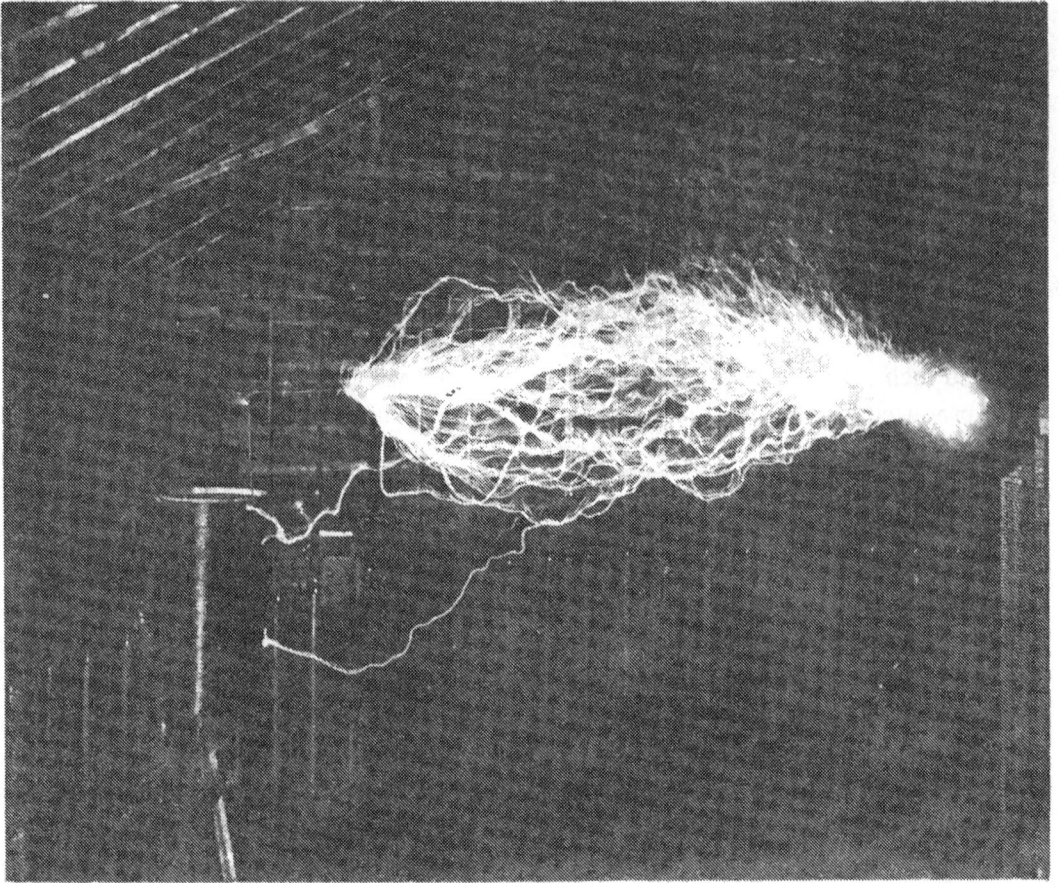

Phot. LVI. Discharge of "extra coil" issuing from a ball of 30″ diam. forming the "free" terminal of the coil towards a coil on a stand.

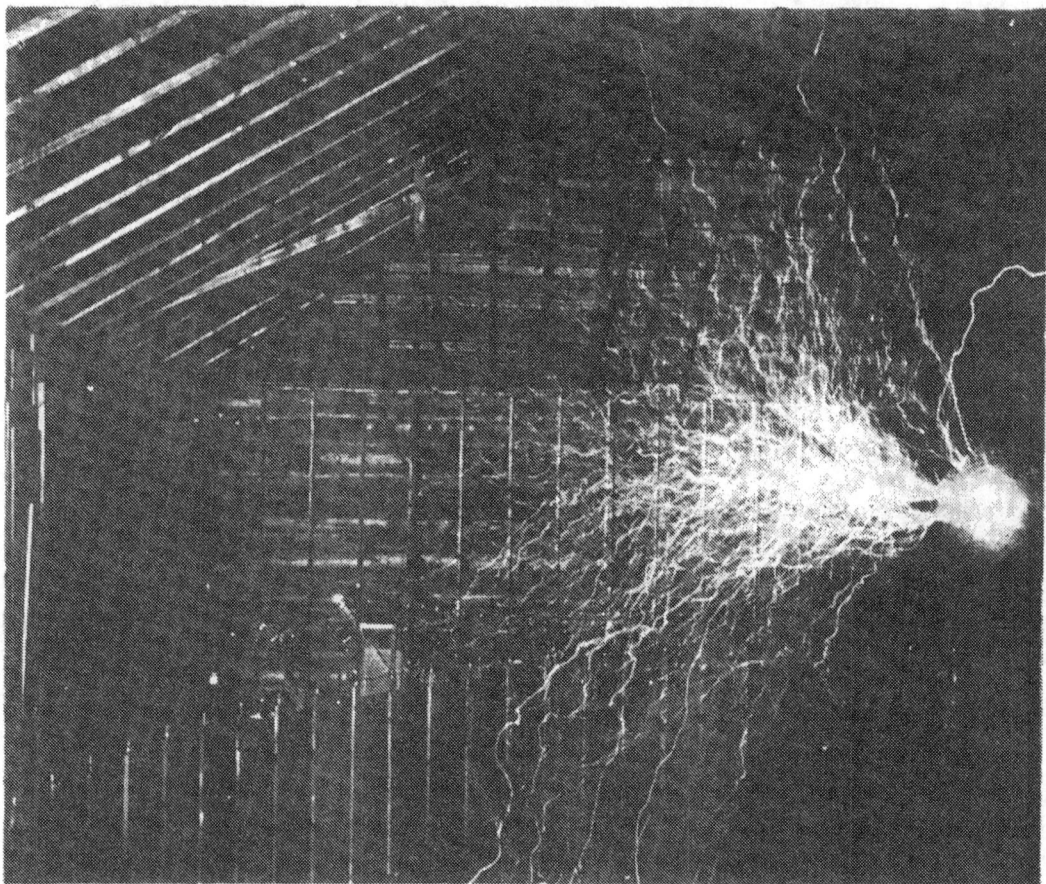

Phot. LVII. Similar as in phot. LVI but discharging across the laboratory into the air.

would dart to the roof. It is to be regretted that the building, although very large for ordinary experiments, did not allow the production of discharges still stronger than those before described, which would have been easily practicable with the present apparatus which — with more copper in the coils and particularly in the "extra coil", and possibly without any change — would have in all probability enabled me to reach twice or three times the length of the actual discharges. In the experiment a great many sparks were seen to pas to the top turn of the secondary. These discharges would, without adequate provision, infallibly injure the condensers and the Westinghouse transformer and also other apparatus connected with the circuits or at a small distance from the same, no matter how well insulated they might be. By grounding the circuits in proper ways this danger is in a large measure reduced. In the experiment just described there were again 100 closures, rather short, effected by the switch and the other particulars not dwelt upon were the same as before.

This statement repeatedly made in the description of these plates should be specified. The truth is, each experiment required a special adjustment as the size and form of the terminals and the character of the discharge affected, to some extent, the constants of the oscillating system or systems. But the departures from the conditions designated as normal were very slight, the inductance of the primary or exciting circuit being varied only by inserting a very few turns of the regulating coil.

LVI. In this photograph the discharge of the extra coil, issuing from a ball of 30″ diameter forming the "free" terminal, passes across the laboratory to a wire turned toward and extending from the top of a coil on a stand. The discharge is made of smaller length for the purpose of heightening its brilliance. The shortest distance in a straight line from the ball to the wire is eleven feet. As the light of the sparks would produce a marring of the images, the photograph is taken through a diaphragm with a very small oponing. The individual discharges corresponding to the closures of the circuit by the break wheel are very clearly shown. It is interesting to observe the curved paths of the sparks which are much longer than the shortest route open to them. Some of the sparks avoid the wire preferring to pass through fully twice the distance through the air. An interesting feature is afforded by one of the longest streamers which in a portion nearer to the wire, extending from the coil on the stand, changes from a *streamer* into a *spark*. It is also curious to observe its path passing far beyond the wire point and returning to the same. In this instance there were 50 throws of the switch made and the vibration was slightly slower than the normal on account of the large ball at the free terminal.

LVII. This plate shows the extra coil, with a ball of 30″ diam. as "free" terminal, discharging laterally across the laboratory into the air. The discharges are much stronger when breaking out from the ball, requiring a much higher e.m.f. Capacity also adds to their volume and fierceness. Many streamers again show luminous points and attain great length, one in particular, which traverses the entire laboratory striking the wall. The end is probably too fine to print clearly. Many discharges again are carried to the roof by the draught they create. Once more 100 throws of the switch were made. The conditions were otherwise the same.

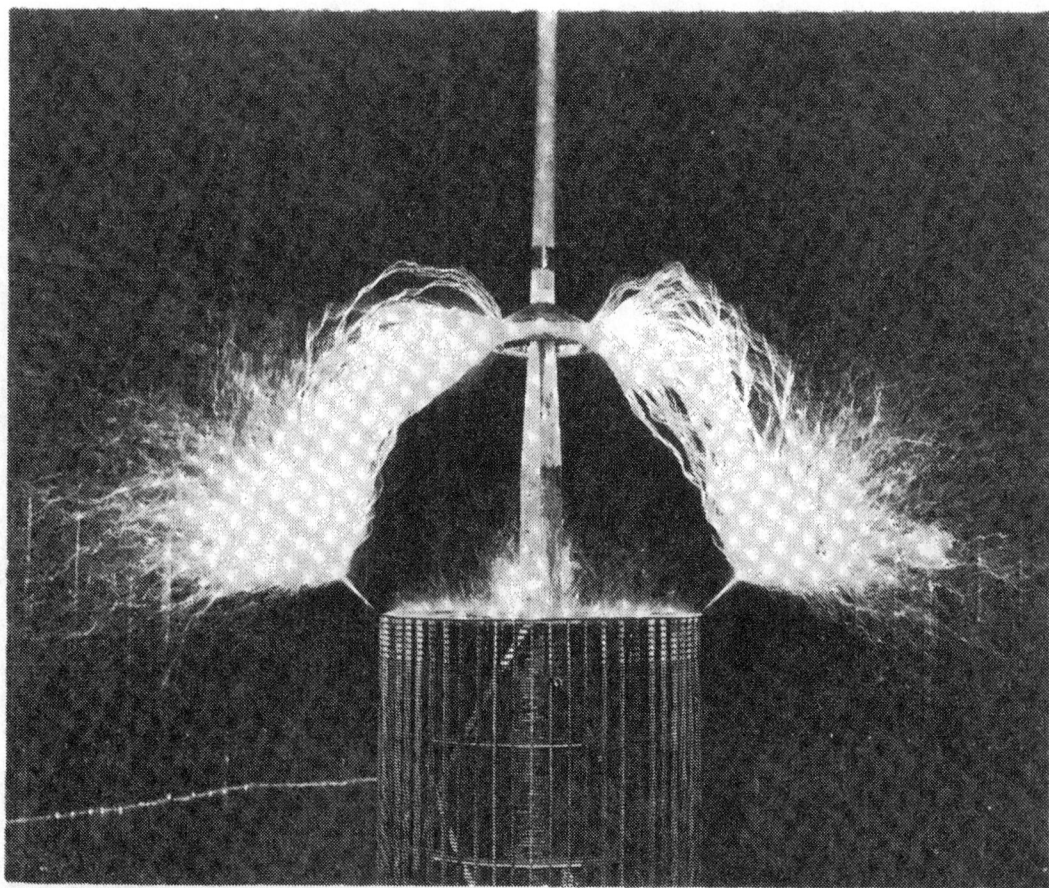

Phot. LVIII. Discharge from "extra coil" issuing from two diametrically opposite wires, pointing upwards and fastened to the brass ring on the top of the coil.

Phot. LIX. Discharge of "extra coil" issuing from many wires fastened to the brass ring as on phot. LVIII.

Photographs taken with Mr. Alley from Dec. 17 to Dec. 31, 1899 and *particulars* about the same continued:

LVIII. This photograph shows the extra coil in central view with the discharge issuing from two diametrically opposite wires, pointing upwards, which are fastened to the brass ring or last turn on the top. The sparks, passing abundantly to the hood above the coil, produce a most beautiful symmetrical figure, which is rendered still more so by the fine texture and sharpness of the discharge paths. On the top of the coil from the brass ring delicate streamers rise producing the effect of a flame. The two wires from the ends of which the sparks and streamers sally forth *glow all along*. This is remarkable and indicates the great quantity of electrical movement. The glow of the top turn of the secondary and also that of the wire leading to the extra coil is strong, that of the latter wire being much stronger. Some exceptionally brilliant sparks occur occasionally. This happens when the discharge breaks out on one wire or point much earlier. In this case the spark continues along this path and practically all the energy supplied goes this way. In the experiment, as in most cases before, 100 throws of the switch were made, other particulars remaining unchanged.

LIX. In this plate the extra coil is again shown in central view, only the discharge is made to issue from many wires instead of only from two as before. The wires are fastened to the brass ring and are pointed upward as in the preceding case. This photograph is also symmetrical and extremely beautiful. Some stronger sparks and streamers occurring occasionally present many features of interest. A number of sparks starting from the back avoid the hood, though passing near the same, circle around and finally strike the hood on the front side. One streamer, extremely long, starts upward following the pole for some distance then turns, passing for a good distance horizontally, and finally darts upward through the opening in the roof, evidently carried by a sudden gust of the draught. Several sparks pass closely to the hood upward and return to the same after traversing a considerable distance. All the wires emitting the streamers glow. Again 100 throws of the switch were made and the vibration, as well as other particulars, remained nearly normal.

LX. This is a very interesting plate, the discharge taking place from a ball of 8″ diam. and two diametrically opposite wires fastened to the same. The extra coil is again viewed centrally. Very strong sparks pass to the floor, some twisting and darting about curiously and exhibiting several phenomena described already. Sparks and streamers also pass to hood, roof and sides of the building. Some of the streamers attain great length being thirty feet, at least, in a straight line, while some sparks measure twenty feet or more. These latter are very brilliant and fierce. The upper streamers indicate the existence of a strong current of air created by the heat developed by the discharge. Again 100 throws of the switch were made, other particulars remaining the same.

LXI. This is another most beautiful photograph showing streamers and sparks issuing from a disk facing the camera. The extra coil is viewed centrally, as before, the disk forming the free terminal being on a point of the vertical axis of the coil. It seems that in some quick moving streamers the individual discharges are recorded, at least the texture appears far too fine for the break wheel period. Assume, for example, 4000 breaks per

25*

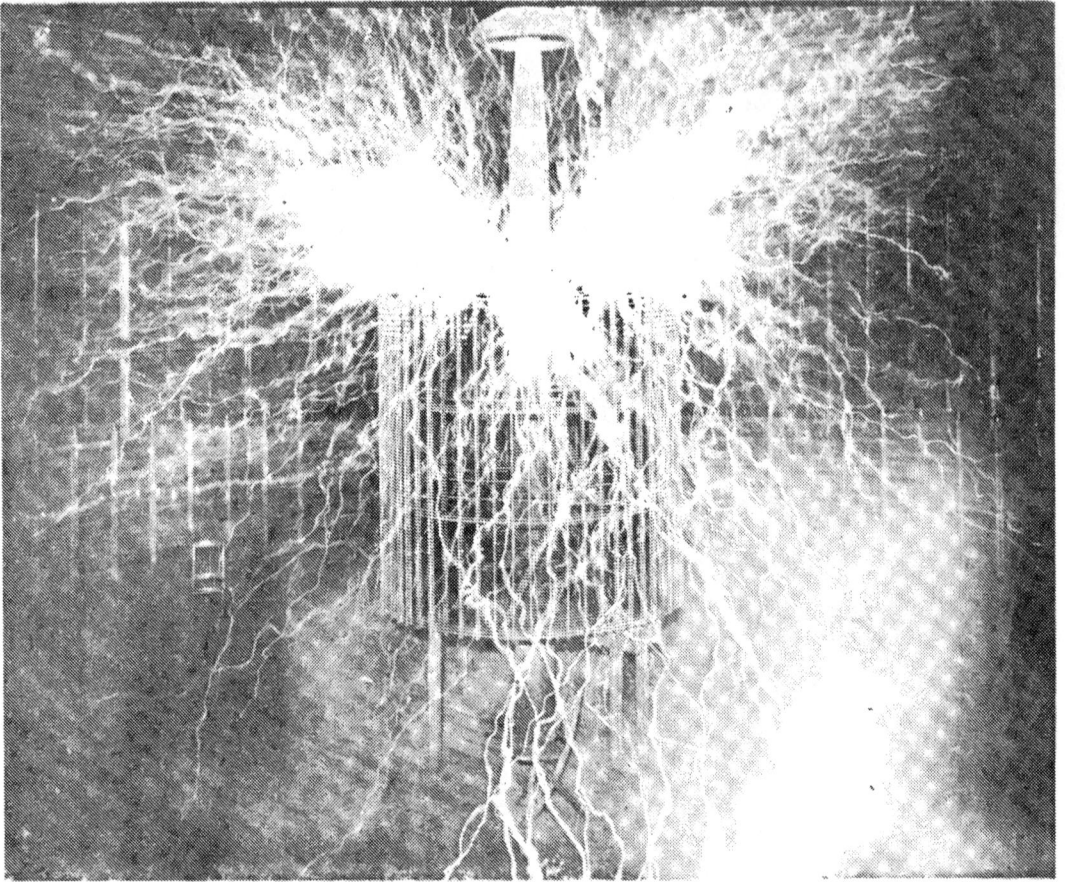

Phot. LX. Discharge from a ball 8'' diam. with two diametrically opposite wires fastened to the same.

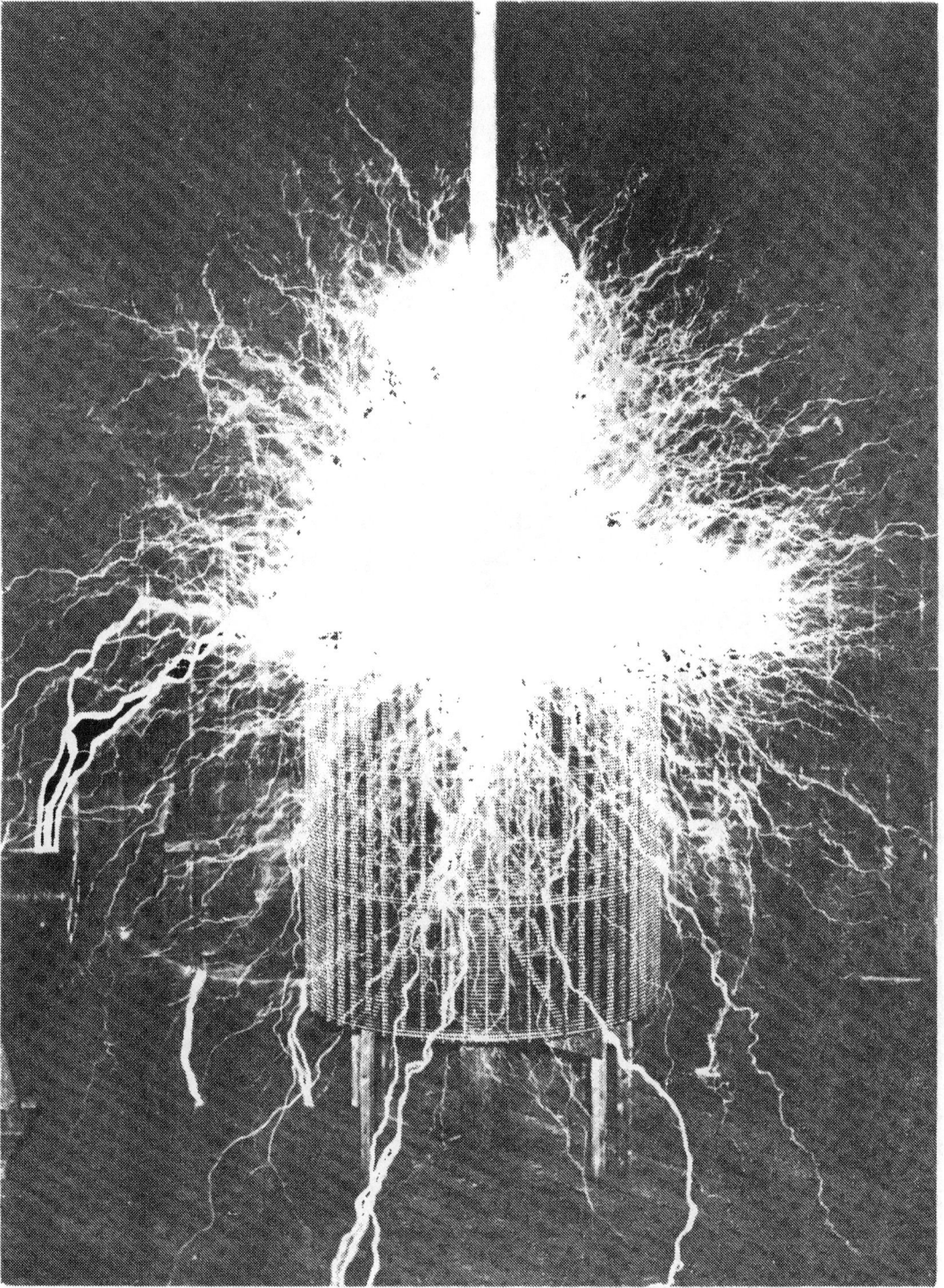

Phot. LXI. Streamers and sparks issuing from a disk facing the camera.

Phot. LXII. Discharge issuing from the base of a cone upon which a ball of 30″ diam. is resting.

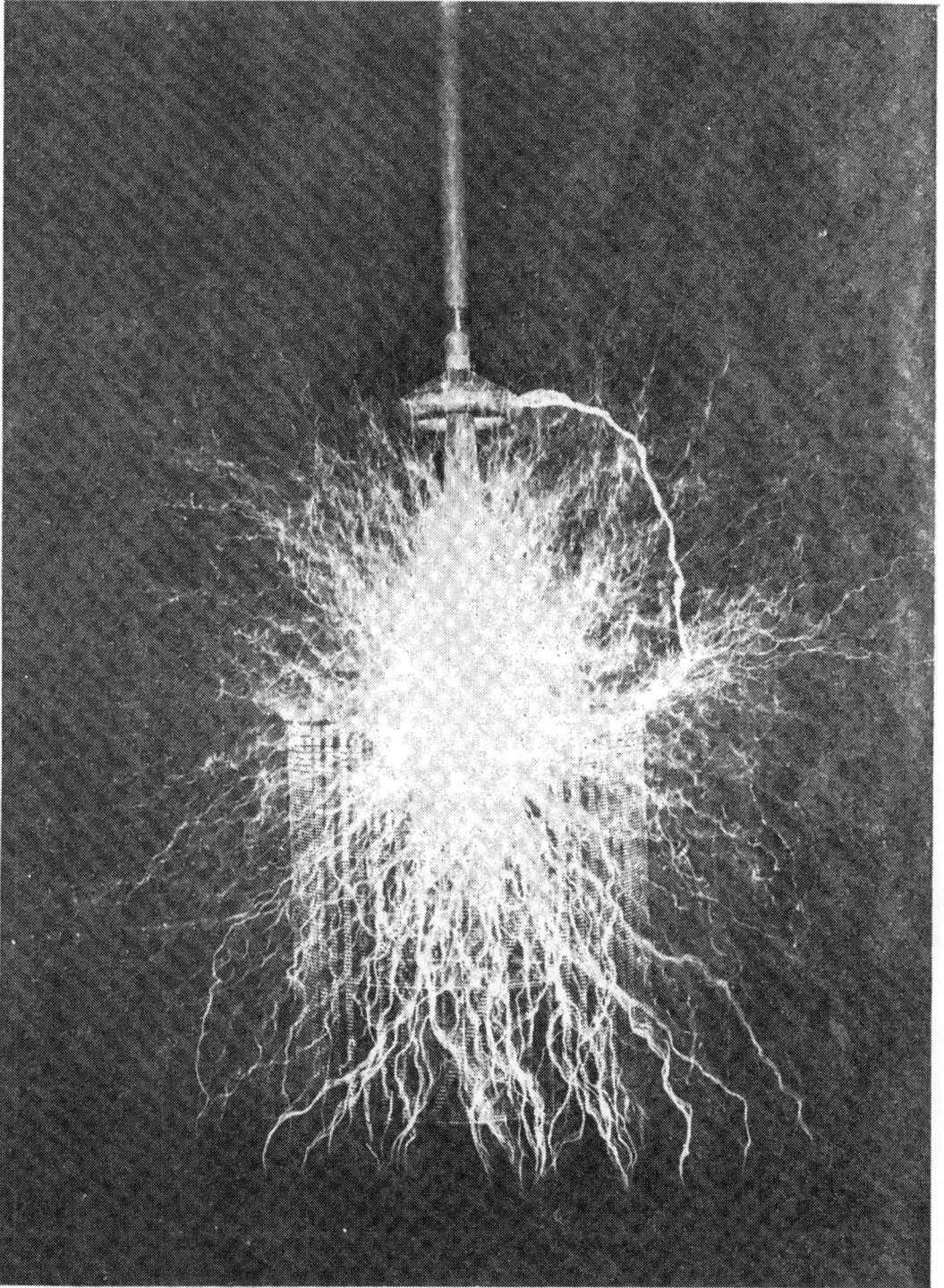

Phot. LXIII. Discharge from a ball 30″ diam. supported on a vertical coil.

second and let the vibration be, say, 60,000 per second — then there would be 30 individual discharges in each break wheel period, understanding by this the time interval from one break to the next, or from one closure to the next. Since the break wheel effect is always seen, even when the streamer moves slowly, it would seem that the finer threads indicate the individual discharges. At any rate, with a stronger draught the latter can be evidently easily recorded and this might be a simple way of exactly determining the vibration of the system, certainly simpler than analysis by a revolving mirror.

This experiment I expect to perform on my return for it is indispensable to determine the vibration quite exactly. Up to the present this necessity was not imperative, the method used being satisfactory so long as the chief purpose was to perfect the apparatus and make general observations. But now quantitative estimates have become important. To return to the description, the strong sparks to the hood are particularly curious. The strong current of air is evident from their behaviour and also from the appearance of the upper streamers. Some very strong sparks pass to the coil in series with the extra coil. One of the streamers striking the floor ignites the wood. Many streamers are carried through the opening in the roof. One, exceptionally long, passes to the photographer in the corner of the building. The shock is but slight as might be expected. A *spark* might be fatal, but there was no possibility of a spark taking that course without being stopped by conducting objects nearer to the origin. Some of the upper streamers are chopped up curiously indicating the presence of small whirls or eddies in the air current passing through the roof opening. The switch was here also closed 100 times; the closures were very short and the other particulars remained as before.

LXII. This is one of the most beautiful plates taken. It shows a discharge issuing from the base of a cone upon which a ball of 30'' diameter is resting. Streamers, though a few only, issue also from the ball giving evidence of the immense electrical pressure and quantity of electric movement. It was indispensable to employ the metallic conical vessel for the purpose of preventing the discharge from following the wooden support, upon which the ball of 30'' was supported, to the ground. This would unavoidable occur even if the support were of the most excellent material as regards insulation, as of glass for example, and no matter how high the support might be.

In fact, I have found from long experience with these discharges of extreme electromotive force, that it is almost impossible to insulate a terminal without some such provision as used in the present instance. The fundamental idea is to provide an arrangement such that the place of support, or that part of the terminal which rests upon the support, is guarded by the conductor projecting beyond. In other words the terminal must be resting on the support on points where there is *no electrical* pressure or, at any rate, a very small pressure. This amounts to screening the support statically. Another way to prevent the current from following the support is to place a coil, through which the current passes, beneath the support passing axially through the coil. The hood, referred to repeatedly in these descriptions, is used for the same purpose as without the same the current would pass along the pole to the ground. But the hood might be dispensed with by using an extra coil of much smaller diameter, axial with the wooden pole supporting the iron structure, but it would be necessary for insuring safety to let the coil finish very close to the bottom of the iron structure where it rests upon the insulating support. This has been one of the great difficulties encountered in the course of this work and it has required much time and trouble to overcome it successfully. For the same purposes also the large hood, above the cords keeping the iron pole in position, is employed. The rims of both these hoods are curved so as to enable a greater pressure to be reached by preventing the streamers to break out into the air easily and at a much smaller pressure. In a new apparatus, now under consideration, these improvements will be carried much further and I am confident to obtain results far beyond anything arrived at so far.

In the present experiment it is rather astonishing to see some streamers breaking out from the surface of the ball of such a large radius of curvature, when they can pass out so easily from the edge or base of the cone underneath. This shows the existence of violent surgings and a great quantity of electricity set in motion in the system. The photograph shows some luminous spots on a very powerful streamer passing to the floor. Many sparks are very curious on account of the curved paths they follow. The streamers here are also mostly of fine texture, this being due to the facility with which they break forth from the edge of the inverted pass and to their abundance. It may be seen from the plate that one of the supports of the extra coil caught fire. The particulars in this experiment were as in most cases before, 100 throws of the switch being made.

LXIII. This plate again illustrates a most beautiful discharge taking place from a ball of 30″ diameter supported on a vertical coil. Owing to the abundance of the streamers the ball can not be seen. The streamers are of a peculiar character, probably due to the manner in which they were produced, which is different from that practiced in the instances before described. It so happened that when the switch was thrown in the discharge from the ball always darted to the floor. Now, in order to make it pass upward also and so to produce a symmetrical figure, I held the switch on longer expecting that the heated air rising upward would carry the discharge towards the roof. Indeed this happened, the discharge always starting towards the floor and then gradually rising until it was vertical and passing out through the opening in the roof. The last and permanent position was attained in 3—4 seconds of time, this showing the great amount of energy spent in heating the air and rapid heating and rising of the latter. The air current exercises such a predominating influence that some streamers pass vertically upwards very close to the iron pole, without apparently being affected by the presence of the pole.

A peculiar feature of the present photograph is that the upper streamers are not sharp although the focusing was carefully done. This is evidently due to their fluttering motion caused by the sudden and changing gusts of the draught. This again forcibly illustrates the great sensitiveness of such streamers to air currents and I am once more impressed with the possibility of turning this property to some good use. The symmetry of the photograph is somewhat destroyed by a spark, which breaking the insulation of the top turn (3/8″ rubber), passes to the hood above the coil. All the particulars in this case remained nearly normal. There were again 100 throws of switch made.

Aleksandar Marinčić
COMMENTARIES

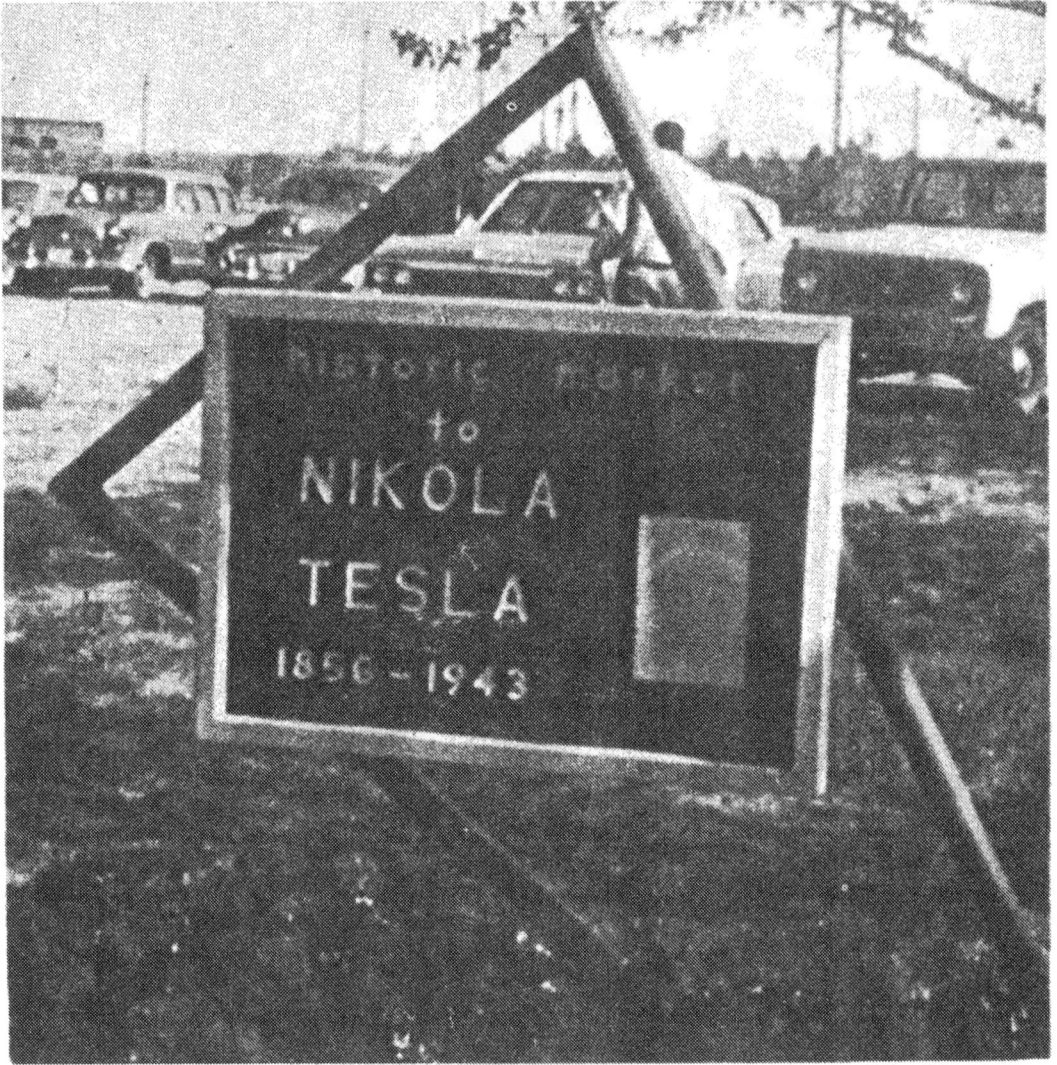

In Colorado Springs' Memorial Park, near site of Tesla's experimental station, an historic marker notes the operation.

1 *June*

Tesla mentioned a similar application of a magnet for extinction of the arc in rarefied gas in his lecture to the IEE in London 1892[5]. The initial idea for this type of detector may date from the time when he was intensively studying phenomena associated with currents in vacuum.

3 *June*

During the year 1899 Tesla filed applications for four patents[8-11] which made use of the principle of "accumulating energy of feeble impulses". It may be seen from these patents that the function of the capacitor was to store energy from the commutated (in fact rectified) HF current. The condenser is connected to the receiver (a relay) which periodically makes contact[8,10] when the condenser charges up enough. Both these patents were filed 24th June 1899[8,10]. The other two[9,11] were filed 1st August 1899. They also refer to a method of accumulating energy but the way the incoming signal controls the charging of the condenser is different: here it causes variations in the resistance of a "sensitive device" which controls the current charging the condenser from a battery. The condenser discharges periodically through the receiver as in the previous case.

Tesla developed the magnet method while he was in Colorado Springs.

5 *June*

Tesla does not state the origin of the formula he uses to calculate M, the power induced in the secondary, receiving coil, by the primary fed with a power of 4 kW (or 4×10^{10} erg/sec, not ergs). Although he himself expresses doubts about the calculation, the conclusion he draws is correct, i.e. that these mode of transmission is greatly inferior to that which he calls the method of "disturbed charge of ground and air", in fact that of electromagnetic radiation.

6 *June*

Working with a darkened Crooke's tube, on the 8th of November 1895 Röntgen noted the luminescence of barium platino-cyanide crystals and discovered that it was due to some unknown radiation which he termed X-rays. Towards the end of that year he held a lecture on his discovery, and in an amazingly short time the whole world knew about his work[66].

Tesla spent some time in intensive research on X-rays, publishing his results in ten articles in the period 11th March 1896 to 11th August 1897[7]. On the 6th of April 1897 he also gave a lecture on his X-ray studies[17] and presented designs of a number of devices for generating powerful rays. During this lecture he reported interesting data from his earlier experiments with Crooke's tubes in 1894. He had then observed that some tubes which produced only feeble visible light had more effect on photographic plates than tubes which were brighter to the eye. The goal of his research was to obtain true phosphorescence ("cold light"), so that he postponed further investigation of this phenomenon, and of the cause of various spots and hazing on photographic plates which had been kept in the laboratory for a time before use. When he finaly did get around to it a fire broke out in the laboratory, destroying practically everything (13th March 1895). It was several months before he could resume his work, and in the meantime Röntgen made his discovery. When Tesla heard about it, it was immediately plain what had been happening in his laboratory. He repeated Röntgen's experiments, which were rather cryptically described, and realized that he had been mistaken in not following up certain chance observations during his work with Crooke's tubes.

During 1896 and 1897 Tesla carried out many experiments with X-rays, also speculating about their nature. He thought "that the effects on the sensitive plate are due to projected particles, or else to vibration far beyond any frequency which we are able to obtain by means of condenser discharges" (Lit. [1], p. A-30). He immediately realized the importance of high voltages for producing powerful rays and suggested using his single-terminal tubes connected to the secondary of the disruptive discharge coil. It is interesting to note that Röntgen too, in a lecture to the Physical Medical Society of Würzburg the same month as Tesla published his first article, also pointed out the great advantage of using Tesla's high-frequency oscillator in generating X-rays.[66]

Tesla measured the reflection and transmission of X-rays for several metals, lead glass, mica and ebonite. It is not clear, however, whether what he measured was true reflected radiation or secondary radiation. He also tried to detect refraction but did not succeed, for reasons which are today obvious. In papers and in a lecture before the New York Academy of Science he described a number of tubes for producing powerful X-rays, most of them resembling Lenard tubes (which he often mentions) but without the anode terminal.

7 June

Descriptions of the high-frequency transformer are to be found in Tesla's publications and patents from 1891 onwards[15,4], but he did not patent it until 1897[26]. The invention protected by this patent is "A transformer for developing or converting currents of high potential, comprising a primary and secondary coil, one terminal of the secondary being electrically connected with the primary, and with earth when the transformer is in use, as set forth". It in particular protects the spiral form of the secondary, and a conical form is also mentioned. For ordinary uses a cylindrical secondary divided into two parts is proposed. A new feature is the specification that the length of the secondary should be "approximately one quarter of the wavelength of the electrical disturbance in the secondary circuit, based on the velocity of propagation of the electrical disturbance through the coil itself", or, in general, "so that at one terminal the potential would be zero and at the other maximum".

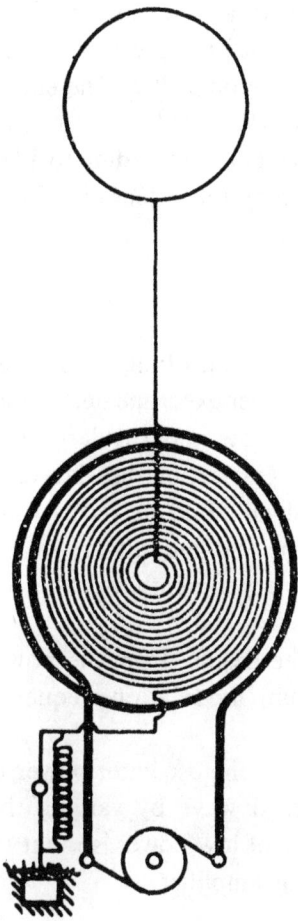

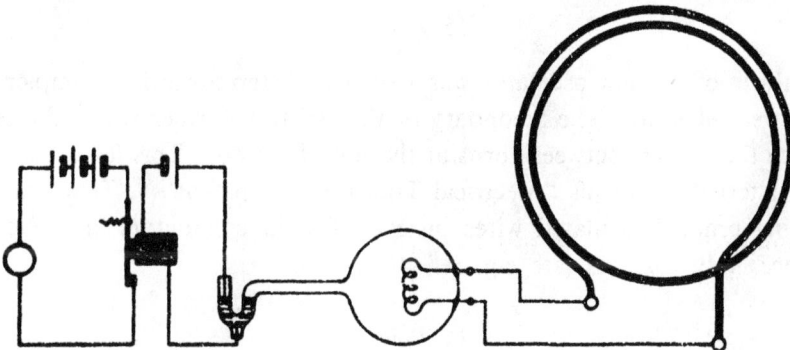

9—12 *June*

In his efforts to construct a sensitive detector for small signals Tesla worked out several designs making use of the thermal effect of high-frequency current. Since the energies involved are very small (according to Tesla of the order of 1 erg), receivers based on this principle would be extremely delicate.

In the archives of the Nikola Tesla Museum, Belgrade, a slide has been found which evidences that Tesla was probably preparing to file a patent on a receiver similar to that which he described in the diary the 9th of June (see drawing on p. 399). The entry for 11th June is stamped on the back "U.S. Patent Office, Nov. 15, 1902."

The basic principle of these detectors is of an earlier date. According to Fleming[33], Gregory carried out measurements of radiation intensity by the extension of a thin wire in 1889, and Rubens and Ritter in 1890 using a bolometer.

13 *and* 14 *June*

From the very start of his work on wireless transmission of signals in 1892—1893 Tesla advocated the use of continuous HF current, while other experimenters were working with damped impulses. The advantage of continuous currents is particularly great in the transmission of continuous signals, such as speech. The entries for the 13th and 14th of June describe two modifications of the HF oscillator which could be used for amplitude modulation. These two circuits were probably in fact the first modulators in the history of radio. It is not known whether Tesla carried out any experiments with this apparatus, but similar ideas were implemented later[19].

Tesla's notes illustrate how carefully he studied the design, from the power supply to theoretical aspects such as the ratio of the maximum modulation frequency to the carrier frequency.

The transmitter using "controlled arc" modulation of the oscillator power described in the entry of June 14th produces amplitude modulated wave by varying the carrier power about a mean value. The modulating signal can be of low power, so that the device as a whole can also be considered a frequency-shifting amplifier.

15 *June*

This trial run of the new oscillator was Tesla's first step towards the implementation of his high-power generator. The secondary of the HF transformer was made conical in order to reduce the voltage between turns at the top of the coil. This feature is described as one of the alternatives in his "Electrical Transformer" patent[26]. Tesla was the first to suggest using braided insulated wires instead of solid conductors in HF circuits in order to reduce eddy currents (see e.g. ref. 46, p. 60).

16 *June*

In these experiments Tesla investigated the influence of grounding* on the HF oscillator. The main point of interest for him was the propagation of electrical waves

---

* At this time there was relatively little experience with grounding. He explains in this entry that grounding was made in "the usual way as here practised", probably referring to lightning conductor grounding. Grounding for single-wire telegraphy dates from 1838, when Stinheil demonstrated that the Earth could be used as the return conductor. In 1893 Tesla described his system for energy transmission

through the Earth. He had already put forward the hypothesis that the Earth could be used as one of the conductors in transmission of energy from a transmitting to a receiving aerial in 1893[6]. He further developed this hypothesis in his patent application "Apparatus for transmission of electrical energy" filed in 1897[13].

18 *June*

The secondary circuit was modified by the addition of another coil, altering its response to the primary and the spectrum of the oscillations. Tesla had already found in experiments in New York that this "extra coil" had a good effect. This coil was not inductively coupled to the transformer (some coupling probably existed, though weak).

19 *June*

Continuing his study of the receiver components referred to between 9th and 11th June, Tesla describes a sensitive detector using the attractive force between the plates of a charged condenser. Descriptions of allied devices are to be found in several of Tesla's patents. In patents[8] and [9] this effect is used to periodically make the receiver circuit when main condenser charges up sufficiently (there is no preexcitation nor quenching, since the circuit quenches itself when all the stored energy gets discharged). A fuller description, where it is noted that the performance of the device is improved under reduced air pressure, may be found in patents[70].

20 *June*

Tesla did not make a strict theoretical analysis of the mode of operation of his oscillator, but determined all the main parameters by tests on a simplified representation of the oscillator circuit. For example, he estimated the power supply drain from the energy in the primary circuit capacity multiplied by the rate of discharge. This involves the assumption that the condenser charges up by the same amount before each discharge (which cannot be the case when it is charged from an AC supply) and that all the energy gets dissipated before the next charging.

The "vibration", i.e. the resonant frequency of the primary circuit is calculated from the measured inductivity of the primary with two turns (see June 17th) and its capacity. Since he was now using one turn, the inductance is divided by 4. The capacitance was somewhat greater than that measured on June 18th with the old jars. Using these L and C values he calculates the resonance period of the primary using Thomson's formula for a lossless circuit.

He then finds the wavelength of the primary oscillations, and hence works out the number of turns the secondary must have so that its length is one quarter of a wavelength (see the commentary to 7th June). It is not surprising that he went astray in trying to set up a representation of the secondary circuit as an oscillatory system since the distributed

---

without wires[6] where the alternating current source is connected "with one of its terminals to earth (conveniently to the water mains) and with the other to a body of large surface P". Popov's receiver of 1895 also used grounding via a water pipe[32]. Around 1895 Marconi did some experiments with a Hertz apparatus grounding one terminal of the inductor and leaving the other connected to an elevated conductor with a terminal capacity. Exhaustive studies of the influence of the form of grounding and the nature of the ground were made around 1905 and later[30].

capacitance was not determined. Tesla's own doubts about this way of determining the secondary in terms of length of wire are best revealed when he refers to checking its resonant frequency treating it as an oscillatory circuit.

21 June

Returning once more to the problem of the conversion of mains power into HF power Tesla calculates the energy in each charging cycle of the condenser (see comments for June 20th). Taking it that all the energy in the charge condenser will at some instant be found in the condenser of the secondary circuit, he in fact works out the peak voltage on the secondary condenser. The energy equation for lossless coupled circuits has the general form

$$\frac{1}{2} C_p U_p^2 = \frac{1}{2} C_s U_s^2$$

where p refers to the primary, s to the secondary circuit, U is peak voltage. It should be noted that Oberbeck's theory[29] yields the same ratio between the voltage on the primary condenser just before discharge begins and the peak voltage on the secondary condenser.

22 June

The circuit with two condensers, one being charged from the power supply and the second via a spark from the first represents a modification of Tesla's classic oscillator*. Theory shows that protraction of the oscillation in the primary circuit lowers the efficiency of the oscillator because energy pulses back and forth between the primary and secondary. However, in this circuit protraction of the spark does not have the same effect because while it lasts the primary capacitance is $C+C_1$, but when it stops the capacitance is only C_1. Why the sparks in the secondary were stronger with $C=aC_1$, a a whole number, is hard to say without a more exhaustive analysis.

The note at the end of the entry indicates his satisfaction with the results and that he felt it necessary to continue research in the same direction.

23 June

The two formulae are in fact identical if the thickness of the wire is neglected, because then

$$\mathcal{Q}^2 = (2\pi r N)^2 = 4\pi S N^2$$

24 June

Rarefied gases had long interested Tesla, and his work on their conducting properties, especially at high frequencies, is well known, e.g. the patents on an electric lighting system and an incandescent lamp[37]. He presented detailed analyses of the same problems in his famous lectures[4,5,6]. Rarefied gas as a conductor is also referred to in his patent application "System of transmission of electrical energy"[13].

---

* Drawings reproduced on p. 403 are taken from Tesla's original slide, now in the Nikola Tesla Museum, Belgrade, show four modifications of the transmitter.

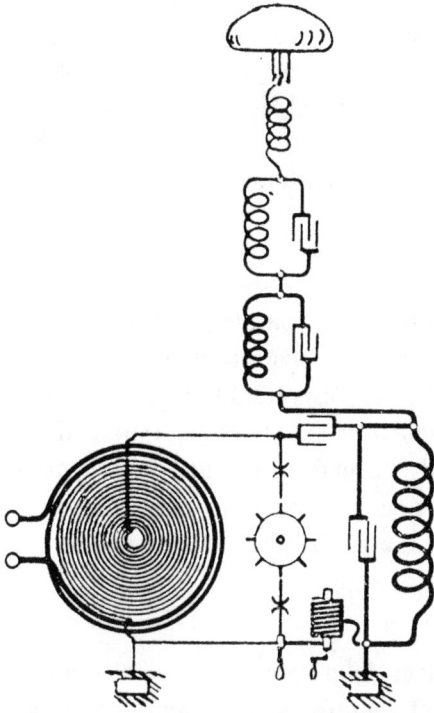

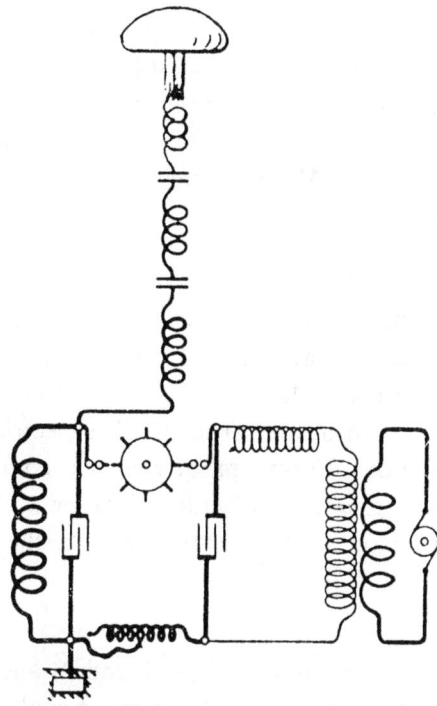

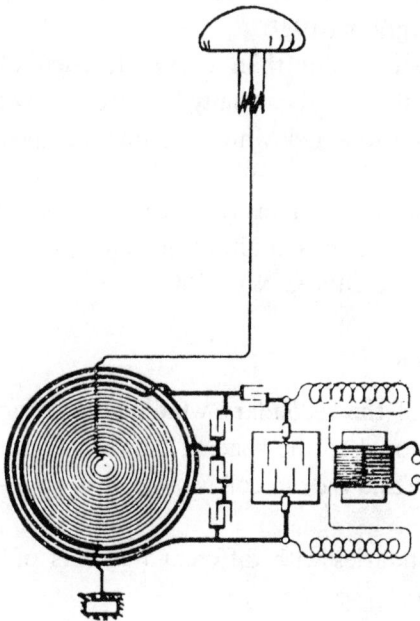

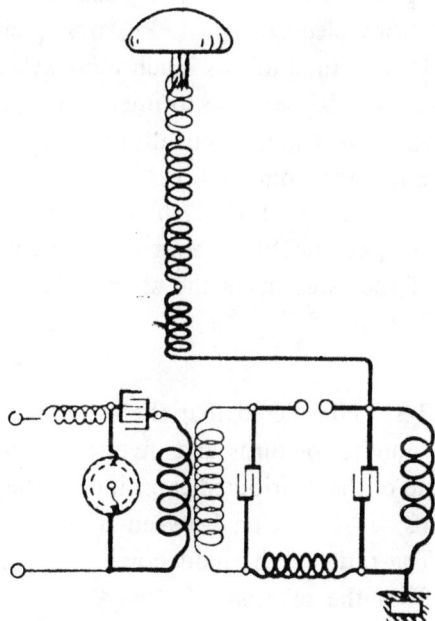

WITNESSES

INVENTOR.

BY

ATTORNEY.

The device shown in the drawing was intended to amplify the vibrations, in the following way: some of the power driving cylinder A is converted by friction of brush b against A into vibrations of b. Since the friction is a function of the current in the electromagnet, the vibrations of b have a time variation similar to the time variations of the current. If the circuit of the electromagnet includes a microphone and a battery, then the device should amplify the speech signal, brush b vibrating in synchronization with the speech pressure but with much more energy. This amplified signal could be used in a modulator (see June 13th and 14th). The drawing shown on p. 405 (from Tesla's slide in Nikola Tesla Museum, Belgrade) illustrates how Tesla thought of implementing some of these ideas*.

26 *June*

The principle of this device using high voltages to separate gases would be that the molecules (in fact ions) of the different gases would behave differently because of their different mass:charge ratios. It is not known whether Tesla tried to verify this idea experimentally. In a later article[28] on electrical oscillators he mentions among the possible applications "formation of chemical compounds through fusion and combination; synthesis of gases; manufacture of ozone . . ." but does not mention separation of gases, so that it may be he never went any further than the initial idea.

27 *June*

The transmitter (Figs. 1 and 2) and receiver (Fig. 3) having several tuned circuits, the transmitter generating several signals at different frequencies and the receiver responding only when all these signals act at the same time, were the subject of two patent applications filed 16 July 1900 (subsequently granted)[38].

This method allows much more selective reception than a single-frequency channel, and is much less sensitive to interference, and the signal can only be decoded by a special receiver. In his patent applications Tesla likens it to a lock which can only be opened when one knows the combination.

The entry of June 27th was subsequently brought in evidence in a dispute before the U.S. Patent Office about priority to the idea of a multi-frequency system[68]. The back of the page bears the stamp "U.S. Patent Office, Nov. 1902".

28 *June*

Tesla considered that the self-capacity of the secondary winding was proportional to the number of turns and inversely proportional to the spacing between turns, so that the ratio of the distributed capacities of the new and the old coil is $N_1 d / N d_1$ (N — number of turns, d — spacing between turns).

The ratio of the inductance of the secondaries with different numbers of turns he finds from the relation

$$\left(\frac{N_1}{N}\right)^2 \frac{l}{l_1} = \left(\frac{N_1}{N}\right)^2 \frac{Nd}{N_1 d_1} = \frac{N_1 d}{N d_1}$$

---

* It has not so far been established whether Tesla patented or tried to patent this modulator. It appears that the slide is a copy of a drawing intended for a patent application.

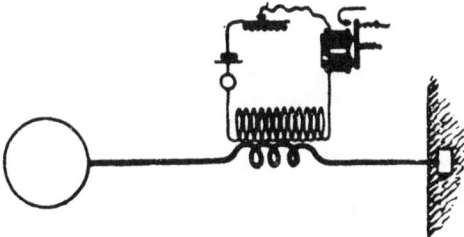

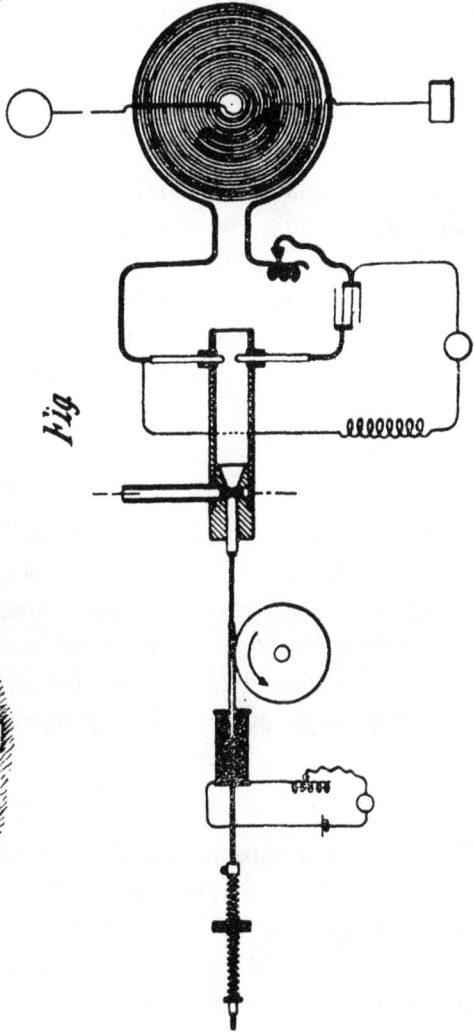

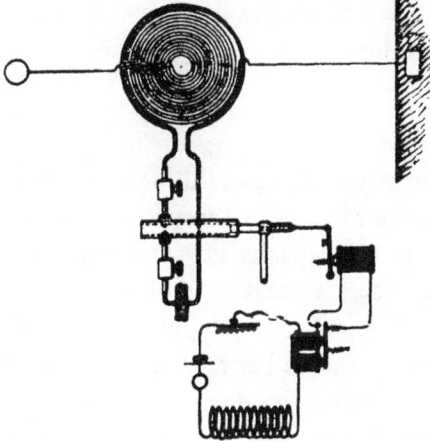

Fig.

Fig.

Fig.

Fig.

Witnesses:

Nikola Tesla, Inventor

by

Attys

derived from the expression for an infinitely long coil, and yielding the same ratio as in the case of capacitance.

The numerical value for the capacitance of the old coil appears here for the first time, without explanation.

The receiver experiments were probably done in preparation for a patent application. Leonard E. Curtis appears a number of times as a witness to Tesla's patents (see for example refs. 8, 10), or as one of the attorneys (on many patents from 1896 on).

30 *June*

Description of electric circuits in terms of mechanical analogies was at one time very popular. The resonance of an electrical circuit was likened to the swinging of a pendulum, and coupled resonant circuits to two pendulums linked together[39]. Maxwell and his followers even tried for a long time to describe the electromagnetic field in terms of a mechanical model[40]. Tesla's comparison of his "additional coil" to a pendulum is not precisely formulated but rather intuitive. He correctly discriminates between the excitation (initial conditions) and the Q-factor. He does not fully explain how he imagined that the vibrations of the three systems, the primary, the secondary and the "combined system", would be the same. By "freeing" the additional coil he means a weakening of the coupling between it and the secondary exciting it. He obviously had a clear understanding that a circuit can oscillate at its own resonant frequency if the coupling with an excitation circuit is loose.

2 *July*

Here Tesla gives the calculation of values for the spark gap oscillator in the fullest detail so far. However, the analysis does not include all the magnitudes relevant to the functioning of the oscillator, e.g. the primary/secondary coupling of the transformer and the distributed capacitance of the secondary. The power equation is also not fully explained and justified. However, by means of this approximate calculation Tesla did get a valuable rough guide relatively quickly and easily.

3 *July*

The distributed capacitance of the secondary windings is difficult to determine. It depends on the coil diameter, the dimensions of the wire and the insulation and the winding pattern. In a single-layer coil it is due mostly to the capacity between neighboring turns, and this is the way Tesla calculated it. He considers a greatly simplified model in which it is taken that the parasitic capacity per turn is equal to $A/4\pi d$, where $A = r\pi l$, half the surface area of the wire in one turn, and d is the distance between turns. The capacitance is calculated as that of a plate condenser of area A and gap d with air between the plates. This model is open to a good many criticisms, but it must not be forgotten that Tesla had to find some solution, whatever its shortcomings. It is also not correct that the total inductance and capacitance of the secondary circuit with the "additional coil" are additive, but Tesla was himself aware that this was guesswork, and often mentions the words "roughly", "estimate", etc.

In an earlier calculation (see June 20th) he had started from the primary circuit and worked out the values for the secondary, whereas here he attacks it from the other end:

from the resonant frequency of the secondary circuit and the known primary inductance (one turn) he finds the required capacity of the primary circuit. He then checks whether this capacity can be used with an LF transformer of the given power. The formula is approximate, but gives a good rough guide for the power in the mains transformer. The peak power rating of the transformer must be even greater than the value found because the condenser is not charging all the time but only in short pulses.

5 July

It is possible that Tesla was planning to construct a balloon to take an antenna to great height[13, 14], and was therefore interested in the generation of hydrogen. He does not give any indication, however, of whether he actually carried out any experiments in this direction, or of the grounds he had for expecting the desired decomposition to take place.

7 July

For the "resonance method" Tesla envisaged two possible types of resonant transformer: one with loose coupling between the primary and secondary, and the other with tight coupling but only with part of the secondary inductance*. This latter type he protected under the patent "Apparatus for transmitting electrical energy", for which he applied on 18 January 1902[44]; a good deal of his time at Colorado Springs was spent in developing it.

His conclusions about various parameters of the oscillator indicate that he had by then gained sufficient experience to be able to design such devices with improved performance in the parameters he wanted. As the experiments proceeded he gradually increased the voltage of the LF power supply. On June 20th he had calculated with an excitation voltage of 20 kV, but he had assumed a much higher rate of charging of the condenser, so that he obtained then a greater power than now with 40 kV. The difference in the number of chargings per second is nowhere explained, nor had he ever previously described how it was calculated. The first time he had probably taken it as being equal to the number of breaks on the rotary disharger, and the second time as double the mains frequency. In this light the accuracy of "the capacity of condenser which the transformer will be able to charge" is dubious. However, Tesla did not take the value he calculated as limiting the capacitance in the primary, noting that it did not take into account resonance and other factors which might enable the transformer to charge a much larger condenser.

8 July

From observing the behavior of his oscillator Tesla came to an interesting conclusion concerning the shape of the conductor of the primary winding, i.e. that a strip conductor was better than a wire of circular cross section because all other conditions being the same it did not get so hot. He believed that there was a special reason for this "not yet satisfactorily explained". Since the dimensions of the strip conductor are not known we cannot work out the reduction in resistance relative to a circular section conductor due to the

---

* It is easily demonstrated that these two methods are similar. If in the second case a part L'_2 of the secondary capacitance is coupled to the primary with a coupling coefficient of k_2, while in the first case the entire secondary inductance L_2 is coupled with a coefficient of k_1, then the response of the secondary to the primary will be the same if $k_1 = k_2 \sqrt{L'_2/L_2} < k_2$.

skin effect. The surface area of a strip will always be greater than that of a round conductor, the more so the flatter the strip: for a width to thickness ratio of 10 : 1 a strip will have about 1.8 times more surface area; this could effect a considerable reduction in resistance, which would explain, at least in part, the phenomenon which Tesla discovered.

In connection with coils, a problem to which Tesla often returned was that of the velocity of propagation of phenomena through the circuit. In order to achieve the maximum voltage across the secondary terminals without the addition of capacitance Tesla considered that the length of the windings should be equal to a quarter of the wavelength. This would be perfectly correct in the case of a straight conductor with one end grounded. Such a system, when excited, would certainly have the maximum voltage at the free end, but its magnitude would depend greatly on whether the conductor were horizontal (when radiation is small, so that the Q-factor of the resonant system is high) or vertical (when radiation is efficient so that the damping is high). With a helical conductor as in Tesla's oscillator, radiation is low as with a horizontal conductor, so that high resonant voltages are possible unless they are reduced by parasitic capacity. In fact, helical winding increases the distributed inductance and capacitance so that the velocity of propagation of current through the coil is reduced, which means that the wire must be made shorter to achieve maximum voltage across the terminals. If the secondary is terminated with a capacitive load (e.g. a metal sphere) the winding length must be still further reduced in order to maintain the same resonance conditions. Tesla took both these effects into account in designing the secondary.

Figures 1—8 illustrate several ways of reducing the distributed capacitance of the secondary. The solution of placing the turns far apart (Fig. 6) is still used today when it is necessary to reduce parasitic capacitance.

9 July

In calculating D (the ratio of the turn spacing of the old and new secondary) Tesla accidentally took the frequency instead of the period, so that he got $D=83$ instead of $D=2.45$. A second numerical error occurred in the formula relating D and C (38 omitted from under the square root) so that C came out to be 10 000 cm instead of 227 cm. Since he never made use of these results, Tesla naturally never discovered his mistakes.

Tesla's method of measuring the oscillator frequency by means of an auxiliary coil is interesting. This coil, with its own distributed capacity, in fact constituted an absorptive resonator. The size of the spark across its terminals provided an indication of the amount of power it absorbed. (In some respects it resembled Hertz's resonator). Tesla adjusted its resonance by varying the number of turns for the biggest spark. He then calculated the wavelength on the assumption that at resonance the length of the coil winding was one quarter of a wavelength. The wire length he determined by measuring the coil resistance, the resistivity per unit length of the wire being known. This method embodies a systematic error due to neglecting the reduction in speed of propagation through the coil[45], and it is applicable for oscillators of high power. However, it was the most reliable method Tesla had used to determine oscillation frequency up to that time.

For theoretical calculation of the oscillation period Tesla used two formulae: one which neglects the influence of the secondary (as for example at the beginning of this entry), the other taking this influence into account. In the latter case it is taken that the primary inductance is reduced by a factor $(1-M^2/NL)$, which would be the case were the secondary short-circuited. How far this is justified it is difficult to say because an oscillator which discharges heavily does not satisfy the simple theory of the resonant transformer oscillator: the secondary is then heavily damped and free oscillations in it decay rapidly, so one would have to apply a theoretical treatment for heavily damped oscillators.

10—11 *July*

In order to try and increase the secondary voltage of the HF transformer by keeping down the distributed capacity of the secondary Tesla added a third oscillatory circuit, thus obtaining an oscillator with three resonant circuits of which two are tightly coupled*. The third circuit will not necessarily be most strongly excited when its resonant frequency coincides with that of the primary and secondary (assuming these are the same) and the primary and secondary are tightly coupled. If the spark in the primary circuit lasts long, then the tightly coupled primary-secondary system will produce two distinct oscillations, and the third circuit will be most strongly excited if it is tuned to one (strictly speaking to near one) of these two frequencies. On the other hand, if the spark is of short duration the tightly coupled system may oscillate strongest at the resonant frequency of the secondary, and then the third circuit will be excited the strongest when all three have the same resonant frequency. Tesla believed that his system of coupled circuits was producing a single vibration, which under certain conditions is in fact feasible.

12 *July*

Early on in the diary Tesla mentioned a method using a condenser to store energy from weak impulses arriving at a receiver. In the circuit drawn here, the condenser is charged by a battery via a self-inductance coil and a coherer shunted by the secondary of an oscillation transformer. In the absence of an external signal the resistance of the coherer is large so that the charging current is small. The circuit breaker periodically discharges the condenser through the primary of the transformer generating alternating current in the secondary which biases the coherer. When an external signal is received the resistance of the coherer is reduced and the charging current rises rapidly, which in turn increases the AC bias on the coherer which therefore soon gets to full conductivity (in fact there is a feedback loop).

14 *July*

He had tried out the devices shown in these drawings earlier on, some of them for wireless remote control of a boat. Patent No. 613809, "Method of and apparatus for controlling mechanism of moving vessels or vehicles" of 8 November 1898 (application field 1 July 1898) mentions the possibility of using electromagnetic resonance but does not give the circuit diagram of the transmitter referred to here.

15 *July*

Earlier on (see the entry for June 3rd) Tesla presents a general scheme in which the 'dynamo principle" is referred too as one of the ways of accumulating energy from weak

---

* Similar systems were analyzed in 1906 and 1907 by M. Wien, in 1907 by C. Fischer, and in 1909 by J. Kaiser[46]. From their papers it may be seen that the effective value of the current in the loosely coupled circuit will be a maximum if its resonant frequency is the same as that of the other two coupled circuits but if they were loosely coupled.

signals. The circuits given here illustrate how he implemented this principle. The "sensitive device" has a resistance which varies as a function of the antenna signal, and is connected so as to alter the excitation of a DC (Figs. 1, 2, 3) or AC (Fig. 4) dynamo.

Although he says that apparatus using this principle had already worked well in New York, none of these receivers, nor the principle they embody, appeared in any of his patents.

17 *and* 18 *July*

This is a continuation of the work described in the entry of June 12th, with different combinations of the same components plus relay R for registering the signals received. In all the circuits the sensitive device has an accumulating function. He experimented with different modifications trying to optimize sensitivity and reliability. The circuit in Fig. 1 of July 18th has two batteries, and that Fig. 5 an autotransformer instead of the usual transformer with a primary and secondary.

19 *July*

This is the first mention of a device which functions either as a transmitter or, with certain modifications of the power supply and antenna circuits, as a receiver. The transmitter is powered from the mains, the receiver from two batteries, B_1 biasing the sensitive device a with AC pulses obtained by discharge of condenser C through the primary of an HF transformer when the mercury switch closes.

The modification in Fig. 2, in which the relay is the secondary of the oscillator transformer, is simpler, but cannot be used as a transmitter.

21 *July*

In this setup a small excitation of one sensitive device is rapidly amplified by a feedback loop which acts via a transformer on the other sensitive device. Figure 10 shows how the receiver was excited by aerial (elevated metal ball C or C_1) — earth system.

22 *July*

Figure 8 shows the circuit of a receiver obtained by modification of the transmitter Tesla was then experimenting with. When functioning as a transmitter it is powered from the mains and is in fact a standard Tesla oscillator with a mercury interrupter between the condenser C and the primary P. The relay, sensitive device a_1 and battery B_1 are omitted and the secondary is connected to the antenna and ground. It may be noted that Tesla did not use the best receiver modification (as in Fig. 6), probably to simplify reconnection as a transmitter.

23 *July*

The "sensitive device" Tesla used for detecting electrical waves is usually known as a coherer[47]. It consists of a tube of some insulator with contacts at either end and metal powder (chips) inside. Its resistance is normally high, but drops rapidly when a large EMF is applied. Munk of Rosenschoeld described the permanent increase of conductivity of a mixture of metal chips and carbon after a Layden jar was discharged through it in 1835. In 1856 Varley noted that the resistance of metal powder was reduced during natural electrical discharges. A major advance was Branly's observation in 1890 that a spark changed the conductivity of a metal powder at a distance. He carried out many experiments with various metal powders, determining their change in resistance by connecting them in series with a galvanometer and battery. In 1894 Lodge* showed that the conductivity of a metal powder could be altered by an electromagnetic wave; this was the final step which preceded the widespread introduction of coherers for the detection of radio waves. From the period 1895—1896 the coherers used by Popov and Marconi are well known.[43, 47]

Once activated, a coherer remains in the conducting state. To reestablish the high-resistance state it has to be shaken. The strength and timing of the shaking have to be properly adjusted. A novel method of decoherence of powders was invented by Popov, and used by him in his receiver and later by others[43]. In 1898 Rupp[48] found that constant slow rotation of the coherer keeps it sensitive. The decohering effect of rotation had been discovered earlier, in 1884, by Calzecchi-Onesti[49].

Tesla mentions that he had worked with a rotating coherer in the New York laboratory, so it is possible that he used decoherence by rotation before Rupp. He finds it superior to other methods of decoherence, because then the sensitive device behaves like a selenium sell, conducting only when radiation acts upon it. Also its sensitivity can be controlled by changing the rate of rotation.

24 *July*

From the pagination of the manuscript it may be seen that the entry for this day was divided into three parts (the previous day two parts). The first part, three pages, refers to experiments with a 35-turn secondary on the oscillator, the second part, five pages, to a resumption of these experiments, and the third, three pages, to the determination of the capacity of the 35-turn secondary.

Tesla adjusted the regulating coil in the primary to obtain the maximum secondary voltage, judged by the size of streamers. He connected an "extra coil" to the free terminal of the secondary. He investigated the operation of the transformer at harmonic frequencies by doubling the primary capacity** and making fine adjustments of the primary frequency by varying the inductance in order to get maximum response of the secondary to the harmonic of the primary.

On resuming the experiments Tesla sought an explanation for the occurrence of the largest streamers from the secondary when the regulating inductance was practically cut out. He found it confusing that the highest voltage at the free terminal of the extra coil (connected to the secondary like in Fig. 2 of July 11th) was not obtained when the frequency of the excitation was equal to the natural resonant frequency of the coil. After an extensive analysis he came to the correct conclusion (unlike that of June 30th, which was valid only for a special case), that when free oscillation of the secondary becomes influential, the parameters of the primary have to be adjusted to get maximum voltage across

---

* The term "coherer" is due to Lodge, and denotes a device containing particles of metal such that its resistance is normally high, but is reduced under the influence of electromagnetic radiation.

** For the primary to oscillate at half the frequency the capacity would have to be quadrupled. It is possible that instead of connecting the banks of 8—9 jars in series, equivalent to the capacitance of 4—4 1/2 jars, Tesla connected the previously series connected jars in parallel, achieving an equivalent of 16—18 jars, i.e. four times the capacitance of the series configuration.

the secondary, and resonant frequency of the extra coil has to be equal to the resonant frequency of the coupled primary-secondary system. It seems that it did not occur to Tesla that when the coupling was tight the combined system produced different spectra during and after the spark. It would seem therefore, all the more significant that he was able to reach this correct conclusion, through a combination of empirical results, simple theory and intuition.

Tesla notes that during discharge in the secondary sparks went across the lightning arresters. Since the arresters were connected to the power line, Tesla thought that the HF voltage came from a wave propagating through the earth and getting into the line somewhere else. We have no evidence which would support this statement or establish whether it was not due to coupling between the oscillator and the mains via the power transformer.

The third part of this entry refers to measurement of the capacity to ground of the secondary coil as a whole. Tesla does not explain how he performed the comparison with a standard capacitor nor at what frequency.

26 July

This entry is concerned with much the same topics as that of June 30th. He investigated the influence of the HF transformer primary-secondary coupling on the 8th of July.

27 July

In his first condenser discharge oscillation transformer for generating high frequencies in 1891[4, 15] Tesla used a simple air gap for regulating the charging and discharging of the condenser. However, a year later he had already described several improvements on simple spark gaps using a magnetic field or an air current for rapid extinction of the arc thereby reducing the period of the charge-discharge cycle. He also described the advantages of a splitted arc across several smaller air gaps: with the same total gap length the breakdown voltage is higher, so that smaller gaps can be used and the losses are less*. A fourth form of improvement which he invented was the use of various rotary interrupters[5].

In the period from 1893 through 1898 Tesla patented several types of interrupter, or "electric circuit controllers". It is interesting that all these patents refer to various types of rotary interrupter, with or without an air gap. Some rotary interrupters were protected within patents for high-frequency generators, including the following:

the combination with discharge points immersed in oil. The turbine whose blades make and break the condenser circuit is driven by oil under pressure[50]

mechanical make-break controllers for DC[51]

synchronous controllers with and without regulation of the interrupt timing, for use with AC sources[52]

commutators for alternate switching between two condensers in the primary circuit of a Tesla oscillator[53].

In 1897 and 1898 Tesla was granted a number of patents for "electric circuit controllers". The principle requirement was that they should make and break a circuit at the highest possible rate, i.e. that they should perform a large number of operations per unit

---

* The total resistance of series air gaps is less than the resistance of a single air gap with the same breakdown voltage.

time. In eight patents[27] Tesla gives designs for rotary interrupters with conducting or conducting and insulating fluids, usually mercury and oil, respectively. In some designs the interruption takes place in an inert gas under pressure. He gives ingenious designs for using a mercury jet playing on a toothed metal rotor, and for producing two mercury jets (fluid contact).

The rotary interrupter with two auxiliary air gaps shown in the figure was a new idea. One of the reasons Tesla added these air gaps was probably the high voltages with which he was working, since they allowed him to regulate the excitation. That this could be done may be seen from the statement that by adjusting these gaps the period of charging from the secondary of the mains transformer could be shortened. At the end of the entry he records that the best results were obtained with two rotary interrupters (with toothed disks) rotating in opposite directions. He does not explain how he chose the tooth ratio so that the number of interruptions was equal to the product of the number of teeth.

28 *July*

This entry provides one of the most detailed descriptions of the receiver with two rotating coherers and a condenser for accumulating the energy from weak signals. At point b the circuit $C—P$ is periodically made and broken and the resulting AC pulses bias sensitive device A' in the secondary. Sensitive device A is still poorly conducting so the charging current of C via damping coil L is small. When an arriving electromagnetic wave reduces the resistance of A, C charges much faster and the voltage induced in secondary S also rises rapidly. The resistance of A' drops rapidly and current from battery B' activates relay R. Judging by Tesla's report, the receiver was very sensitive to distant electrical discharges.

29 *July*

To check out his theoretical conclusions about the free oscillation of the "extra coil" (see 30 June and 26 July) Tesla made a new coil with a higher inductance. As this was his first experiment with the new coil, he had to adjust the circuit parameters by trial and error.

Tesla's ingenuity found full expression in the way in which he developed condensers for high voltages. He filed a patent application on his design for a fluid electrolyte condenser on June 17th 1896[67].

30 *July*

To try and verify his hypothesis about the rejection of harmonics with appropriate coils, Tesla changed the connection of his "extra coil" as shown in Figs. 4 and 5. To understand his way of proceeding one must take into account his ideas from 1893[6] concerning the induction of earth currents via an aerial-earth system. However, the standing waves in terms of which he tried to explain the arcing over the lightning arresters cannot be significant at these frequencies.

31 *July*

Tesla made the condensers for the primary circuit out of mineral water bottles filled with a saturated solution of rock salt, and standing them in a metal tank of the same solution, thus creating a condenser bank with one common plate. The other plates (the electrolyte in the bottles) could be connected in parallel as desired. The smallest capacity adjustment possible was equal to the capacity of one bottle.

After various tests of what voltage the glass dielectric of the bottles could stand, Tesla returned his attention to the secondary of the oscillator, in which rightly way the limiting factor for obtaining higher voltages. His analysis of the distributed capacity of the secondary is a good illustration of his inventiveness in a little known field and how he sought to reduce problems to a simple but mathematically and physically sufficiently accurate model. It must not be forgotten that these are Tesla's working notes, which is sufficient justification in itself for some of the hypotheses which the reader might otherwise rightly object to.

2 *August*

A receiver of this type is mentioned in the entries of July 12th (the principle), July 28th (circuit diagram with two sensitive devices and relay), July 30th (in connection with earth waves). The transformer here has a frame similar to that of July 28th but with somewhat more turns. The sensitive device was described on July 21st.

Tesla often worked on several problems in parallel. Here for example we have entries concerning the receivers, the development of condensers for the primary of the big oscillator, and the power equation for a new configuration of the oscillator primary circuit. The condenser C_1 in Fig. 2 protects the mains transformer against overload but has the drawback that it reduces the initial voltage on C_2. Tesla's analysis refers to the case of two condensers in series, neglecting all transient phenomena. It may be that he was induced to think about protecting the mains transformer because of his doubts about the ability of the dielectric to stand the voltages which he intended to use.

3—14 *August*

These experiments are a continuation of some earlier research. Here Tesla investigates various modifications of his "condenser method of magnifying effects". All the circuit diagrams of receivers, over 50, include at least one battery, sensitive device, condenser, rotary interrupter and HF transformer. Some of them show a relay for registering the signal received, while in others its presence is understood. Likewise, in all except one case (5 August, Fig. 1) the plates which brings the excitation to the sensitive device are not shown. Tesla says that these plates can be in one or two media, meaning that both can be in the air, both in the ground, or one in the air and the other in the ground, preferably elevated. In the patent[8], referring to these plates, he also says: ". . . they may be connected to conductors extending to some distance or to the terminals of any kind of apparatus supplying electrical energy which is obtained from the energy of impulses or disturbance through the natural media."*

As regards mode of operation, the various receivers have in common that the sensitive device is biased by a battery. They also include a Tesla oscillator (clockwork rotary interrupter) which creates an added bias on the sensitive device (or devices). This AC pulse bias acts as positive feedback, avalanching the sensitive device into conduction as soon as an arriving signal starts to cause some change. In the receivers with two sensitive devices

---

* It is interesting to note a similarity of such receiving system and the contemporary ELF grounded wire radiator. In the Nikola Tesla Museum in Belgrade few drawings, showing something that resembles a single grounded wire radiator and a parallel array ELF antenna[72], are found.

the one which receives the external signal is usually in the primary side and the other, which activates the relay, on the secondary side. When there is only sensitive device it usually shunts the transformer secondary (which has a high impedance so as not to reduce the performance of the device), thus creating an efficient feedback loop.

A general feature of all Tesla's receivers is their delicacy. Very careful adjustment was necessary to get the sensitive device at the threshold of avalanching. Most of the sensitive devices were rotated (see June 23rd) so that they were only good conductors during the action of a signal. In some cases, however, this did not achieve satisfactory deactivation of the coherer. Then he used an electromagnetic buzzer to periodically interrupt the excitation of the sensitive device (see Fig. 2 of August 8th). Probably the circuit in Fig. 2 gave him the idea for that in Fig. 3, where the rotary interrupter is replaced by a buzzer as an electromagnetic interrupter. He then used a buzzer in various other configurations (Figs. 5 and 6 of August 8th), with the aim of reliably biasing the condenser, and hence also the sensitive device, to threshold.

Tesla did not measure the sensitivity of his receivers by any definite method, but there is no doubt that he did compare them in some way. From his notes very little can be deduced about their sensitivity, i.e. the power required to activate them. A rough idea is given by data from July 4th, when he used similar receivers to register electrical discharges. He estimated that he registered waves produced by lightning at least 200 miles away, and continued to receive signals (at periodic intervals) later when the weather had already cleared. He records that with the receiver shown in the figure of July 28th he was in one instance able to register lightning discharges at a distance of 500 miles. He estimated the distance from the periodicity of the signals as the storm moved away.

13 August

The last experiments with the oscillator were described July 31st, with numerous comments and the remark "this to follow up". Probably he had prepared a new condenser bank in the meantime for work with higher voltages (he measured the capacitance of the new bottles on August 11th, and tried them out with the highest voltage so far from the power supply transformer.

15—21 August

With the new condenser bank the secondary had to be modified, and on August 15th he worked out the length of wire required. He calculated the period of the primary from the capacity of the new bottles and the inductance per turn of the primary found earlier (mentioned on June 20th as 7×10^4 cm, probably one quarter of the value measured for two turns on June 17th). It was also his intention to adjust the oscillator to the "extra coil".

The entries for 16, 17, 20 and 21 August give some new circuit diagrams for the oscillator which he thought would be more suitable for working at high excitation voltages. They bear witness to Tesla's constant search for improvements involving only limited changes in the apparatus which he used for lower voltages. The chief problem was overloading of the power supply. It is recorded elsewhere that Tesla's experiments with his spark oscillator (probably on some other occasion) burnt out the generator of a power station five miles away[36].

In this entry he returns once again to the receivers. He tried out two receiver circuits using one battery and one sensitive device. He changed the capacity in the primary circuit over a wide range, but it is not clear why 1 μF proved best. It remains unexplained what was the relationship between the frequency of the incoming signal and that generated by the receiver itself. Could it perhaps be, if the rotating coherer behaved as a nonlinear element, that the signal was amplified as in a heterodyne receiver[55]?

23 *August*

He now put the extra coil in the center of the primary, retaining this configuration from then on. After the usual adjustment of the oscillator he got sparks 2 m, and later 4 m long, indicating a voltage of around 2 million volts.

26 *August*

Tesla experimented with twice the interruption rate. The oscillator worked better and there was heavy sparking across the lightning arresters (Fig. 4). Investigating the cause of this sparking he inserted a coil in the lead of the metal sphere (Fig. 1) to reject high frequencies. In an earlier experiment (see July 30th) inserting such a choke coil in the ground line had stopped sparking across the arresters. This time it did not, so Tesla tried the circuit in Fig. 2. Still there was no marked change, the sparking across the arresters was only slightly reduced. After this experiment he began to wonder whether the grounding point of the secondary was not perhaps a peak rather than a node of the standing wave. It must be understood that Tesla thought that standing waves were set up around the transmitter (like waves on an open transmission line. With shorter waves the rate of change of amplitude with distance would be faster (i.e. maxima and minima would occur at shorter distance intervals), so he thought that a large potential difference could be obtained with a short distance between the grounding of the secondary and that of the lightning arrester.

In order to explain what happened when the sphere was not grounded (which would mean that there were no short waves) but the sparking across the arrester did not stop, Tesla found it necessary to formulate a new hypothesis: "Could the sparks be produced by static induction upon wire through the air and not chiefly by conduction through earth?" The experiment with which he tried to verify this hypothesis did not yield any definite answer.

27 *August*

Although he has noted several times already that good results were obtained with various decoherence techniques (rotation, interruption of the excitation current), this reexamination of his old ideas shows that Tesla is still seeking a more reliable solution. One of the ideas he was gathering together for further investigation is illustrated by the diagram in Fig. 4, in which a rotary interrupter, condenser, choke and battery provide bias for the sensitive device. When interrupter d breaks, the voltage on C can be higher than the battery voltage. With proper choice of the values the coherer can be biased to threshold, making it very sensitive.

Tesla's idea of the Earth as a perfectly conducting sphere lead him to a mistaken hypothesis about the general behavior of the electromagnetic field around the grounding of the transmitter. What he expected at frequencies of the order of 10 kHz in fact occurs at much lower frequencies[72], at which, as far as can be seen from his notes, he did not work in Colorado Springs. He correctly observed that the decisive factor determining whether predominantly waves of the "Hertzian type" or the waves which he thought to be propagated through the earth (in fact waves in the spherical condenser constituted by the Earth and the ionosphere) would be excited was the excitation of the "Earth". Tesla was also certainly in error when he tried to make generalizations concerning the wave frequency, and in his conviction that he needed extremely high voltages to "create" the second conductor for a system of wireless power transmission. He could not know that this conductor already existed permitting transmission at very low loss of very low frequency waves, and that it would not matter whether the energy transmitted was high or low.

29 *August*

Although the circuit looks simple enough, an analysis of Tesla's receiver with a "magnifying effect" is rather complicated, because transient phenomena have to be taken into account and the resistance law of the sensitive devices as a function of voltage has to be known. It was not easy to adjust a receiver like this to work properly.

Apparently there was an earphone T in the secondary circuit of the transformer, but it is not mentioned in the notes. The sensitivity of an earphone would normally be much greater than that of a relay, so it would be interesting to find out how this apparatus performed. Unfortunately, earphones are practically not mentioned anywhere in the diary.

Tesla here at last makes a few remarks about how the sensitivity of receivers was estimated. To test its response he put a "small capacity" across sensitive device a, but of what value, and whether it was charged or not he does not say.

3 *and* 4 *September*

The aim of these experiments is not explained, but it was probably associated with the "experimental" coil with which he examined currents in the water pipe. This was a resonant coil which in the receiver played a part analogous to that of the "extra" coil in the transmitter. Its purpose was to maximize the received signal. Since Tesla connected one terminal to ground, it appears that he wanted to pick up electrical vibrations from the earth. In this case too he found that it was not sufficient just to increase the Q-factor $\frac{pL}{R}$, but also that it was necessary to keep the coil's distributed capacity as low as possible. This conclusion was consistent with what he had earlier found about the influence of distributed capacity of the coil on the length of wire needed to achieve resonance. Conclusion (5) is interesting in that it shows Tesla was aware that the secondary and the extra coil, although excited by the same primary, would each oscillate at its own resonant frequency, and if these were not the same, they would beat.

5 September

After a number of experiments, including a few outside the laboratory, Tesla once more concludes that parasitic capacity is very harmful, so he decides to try winding a coil to have minimum capacitance. Unfortunately he does not describe how this was done. In his desire to get the maximum possible voltage from the coil he went as far as thinking that it was best to have no capacity at the free terminal. From one aspect he was right (theoretically a coil gives the highest Q-factor with the least capacity in the resonant circuit), but without the "elevated" metal sphere the received signal was much weaker because the free terminal of the coil no longer had a monopole antenna. In the circuit which he in fact used he did not, however, go to such extremes. He added the "experimental" coil but left the metal sphere (aerial capacity) connected to one end of the sensitive device.

6 and 7 September

In calculating the wavelength for the cable and ball Tesla made an arithmetical error. For the calculated T, the wavelength ought to be about ten times less, so that his assumption that on September 7th, he got vibrations of the system consisting of a ball of capacity 38 cm and 120 feet of cable is probably false. It is more likely that the experimental coil was excited by the coupled system of primary, secondary and extra coil.

11 September

Tesla probably thought that he would more easily detect standing waves in the vicinity of the laboratory if the wavelength was shorter. He assumed that the ball-cable system would produce waves which could be registered by the receiver. However, although he measured the electromagnetic field up to a mile away, he probably did not find the expected variation, and could only conclude that electrical disturbances were registered.

13 September

From a document found in the archives of the Nikola Tesla Museum in Belgrade it may be seen that Westinghouse Comp. sent Tesla a 50 kW power transformer for a primary voltage of 200/220 V and secondary tappings of 40, 50 and 60 kV. This is probably the Westinghouse transformer which he often mentions.

15—17 September

The receivers described on September 5th, Fig. 3, and September 11th, Fig. 2, include "tuned" coils whose function is similar to that of the "synchronized" coils shown in the diagrams of September 15th. Tesla did not make a detailed analysis of these receivers, nor do any of his patents on receivers refer to similar circuits. It therefore seems that we do not have sufficient information to draw any reliable conclusions about their sensitivity or their ultimate purpose (for example it is not clear whether they are just for registering signals or for receiving intelligence).

18 and 19 September

As already remarked, oscillators like those he was working with here are not the classical Tesla oscillator with a resonant transformer. The "extra coil" essentially changes the loading of the secondary circuit, and this alters the mode of oscillation. Also, shunting the secondary with capacitance (as in the diagrams of 18 September and 19 September Figs. 2, 3 and 4) alters the spectrum of the oscillation in comparison with that yielded by an oscillator with two oscillatory circuits. Configurations such as those shown in Figs. 5 and 6 of September 19th can be considered as typical Tesla oscillators with a loosely coupled third circuit consisting of the extra coil and capacitive load. Then the greatest voltage at the free terminal of the extra coil is obtained when the natural resonant frequency of this circuit (together with the ball antenna) is the same as that of the strongest component in the spectrum of the oscillator.

22 and 23 September

Having investigated the tapering secondary Tesla started making a new, 15 m diameter cylindrical secondary. The criterion that the weight of copper in the primary and secondary should be the same follows from the requirement of equal losses in the two windings (losses in the copper). This way of calculating the gauge of the primary and secondary conductors is applied in designing LF transformers, but for HF transformers it only provides a rough guide, for a number of reasons, e.g.: the current ratio may differ considerably from the turns ratio, skin effect is not taken into account, etc.

25 September

As he often did earlier, before finalizing a set up Tesla measured the inductance of the primary and the regulation coil which he usually used as an added, adjustable primary inductance. The value he obtained for L_{2p} differs from that obtained earlier (see July 17th) by the same method.

26 September

By this method the frequency of an oscillator is found with a help of a resonant circuit of known parameters. When its resonant frequency is adjusted to coincide with the frequency of the oscillator, the voltage across its terminals, estimated by the strength of the spark across an "analyzing gap", is a maximum. Tesla says that the excitation must be "convenient". Since he introduced regulation of the excitation by means of the small gap b, it is clear that "convenient" excitation was obtained with loose coupling. Loose coupling between the primary and secondary circuits of a spark oscillator ensures that the two frequencies which such an oscillator normally produces are very close. Up to a certain degree of coupling, Tesla's oscillator produces a single frequency. According to Fleming and Dyke[31], with an ordinary spark gap the maximum coupling coefficient for monochromatic oscillation is around 0.05 (certainly less than 0.1), while with a rotary break producing pulse excitation a coefficient of up to 0.2 gives good results. With higher coupling coefficients three components are obtained, even if the primary and secondary circuits by themselves have the same resonant frequency.

27 September

True to the principle that measurements should be checked by calculation, Tesla calculates the inductance of the same coil using the formula for a coil of infinite length, but does not obtain agreement. Since distributed capacitance increases the effective inductance at frequencies below the natural resonance of the coil, the second possible reason which he mentions (inexactness of the coil dimensions) could have some influence, but the main reason is the poor approximation provided by the formula when applied to a coil with this length: diameter ratio.

28 September

The circuit diagrams are of great interest because they illustrate a new approach to feeding the antenna (now known as shunt feed[73]) which obviates the problem of insulating the aerial pole. Unfortunately the explanations Tesla gives are too cryptic to be fully comprehensible. The figures do not clearly show whether the lower terminal of the antenna is grounded or insulated. Tesla's conclusions that a standing wave is set up along the antenna and that the distance between points of equal potential is half a wavelength are correct.

The frequencies he was using were not high enough for his antennas to work in the manner shown by the figures (in which case they would be much more efficient radiators than he usually had), so that this contribution to the theory of wire antennas was never properly formulated.

29 September

Tesla says that he experimented with the antennas shown in the drawings, but he does not compare them with a grounded antenna.

The shortness of the antennas relative to the wavelength made them inefficient radiators. The configuration shown in Fig. 3 was best probably because it had the greatest terminal capacity, providing the most favorable current distribution on the antenna. Lack of coil and ball dimensions makes it impossible to go into any more detailed analysis of these antennas.

3 October

The drawing of several of the coils which Tesla often used offers some interesting information about the laboratory which cannot be seen from the numerous photographs. One sees that there was a wooden floor raised 30 cm above ground level, and the drawing shows the dimensions of the coils and how the HF transformer of the oscillator was wound.

4 October

Tesla was primarily interested in the change of capacity of a ball with height, so he measured the primary capacity for two elevations of the ball. In both cases he tuned for resonance of a "special coil". In the first measurement he had $L_p C_{p1} = L_{sc} C_{b1} = 1/\omega_1^2$ and in the second $L_p C_{p2} = L_{sc} C_{b2} = 1/\omega_2^2$, where p refers to the primary circuit and b to the ball. These equations neglect the effect of interaction between the primary and secondary. They readily yield Tesla's equation

$$\frac{C_{p1}}{C_{p2}} = \frac{C_{b1}}{C_{b2}}$$

The inductance of the primary L_p and of the "special coil" L_{sc} do not figure in the capacity ratio equation. In deriving this equation all distributed capacitance in the secondary and the secondary coil itself are neglected.

From the measured inductance of the new secondary and mutual inductance of the primary and secondary, and the primary inductance measured earlier (see September 25th), it follows that the coupling coefficient was 0.58, i.e. tight coupling*. An oscillator with this much coupling will probably produce three pronounced components, even with very rapid interruption of the spark in the primary; this is indicated by the results Tesla obtained with spark oscillators with looser coupling (see the commentary on 26 September).

As before, Tesla determined the wavelength of the oscillator from the period of the primary circuit (see, e.g., June 20th). He compares one quarter wavelength with the total length of wire in the secondary, special coil and extra coil (when no other coils were used, he considered that the length of the secondary windings should be one quarter wavelength).

9 *October*

He made the last measurements of the change of capacity of a sphere with height on October 5th, but did not give the calculation results. He subsequently improved the apparatus as a whole and in the present entry describes a different way of connecting the "special coil", the chief effect of which was to loosen the coupling, which immediately proved its advantages. With weaker excitation it was easier to adjust the "special coil" to resonance because there were no streamers. Parasitic capacities were reduced, mainly to the distributed capacity of the "special coil".

Tesla first determined the distributed capacity of the "special coil". He assumed that the ball circuit resonated at ω_0, determined by the primary circuit, so that one can write

$$L_{p1} C_p = L_{sc}(c + C)$$

where L_{p1} and C_p are the total inductance (including the regulating coil) and capacity of the primary circuit, L_{sc} is the inductance of the "special coil" (including connecting wires), C is the distributed or parasitic capacity of the "special coil", and c the capacity of the ball.

Subsequent changes in the height of the ball changed the capacity in the circuit of the "special coil". To bring the oscillator into resonance with this circuit again, Tesla changed the inductance in the primary circuit. When resonance is achieved, according to Tesla, one can write

$$L_{p2} C_p = L_{sc}(c' + C)$$

Dividing this by the preceding equation yields

$$c' = \frac{L_{p2}}{L_{p1}}(c + C) - C$$

which is in fact the equation Tesla uses to find c'. Because of an arithmetical error in calculating C, Tesla's numerical results for the ball capacity are about 10% higher than they should be, but this does not essentially affect the conclusions. To calculate the distributed capacity of the coil** he uses the relation $L_p C_p = L_{sc}(C+c)$ for the ball at

---

* The regulating coil in series with the primary reduced the coupling. The new coupling coefficient is found to be $k' = k\sqrt{L_p/(L_p+L_{rc})} < k$.

** By distributed capacity Tesla used to mean the total capacity between turns of the coil. Here he uses a different definition of "internal capacity" similar to that normally used today.

a height such that he could consider its capacity close to the theoretical capacity of an isolated sphere.

11 *October*

Tesla obviously did not sleep much the previous night since he was photographing the oscillator in operation both late at night and early in the morning. The Nikola Tesla Museum in Belgrade possesses several photographs which date from this period, but they are too faded to be worth showing. One of the better preserved photos is shown on p. 221.

12 *and* 13 *October*

To calculate the inductance of cylindrical coils Tesla used the formula for a coil of infinite length, which always gave values too large, especially when the diameter: length ratio of the coil was not much less than unity. However, when the proper corrections are made (Russel[57]), the inductances obtained differ from Tesla's values by less than one percent.

15 *October*

Because of the arithmetical error made on October 9th (see commentary), he mistakenly concludes that the capacities of the ball are now somewhat less than before. Had he used the correct values, his conclusion would have been just the opposite.

17 *October*

The 122 ft metal pole bearing the 30″ ball is the antenna to be seen in the middle of the laboratory on many photographs. The bottom end of the antenna is insulated by a wooden pole. This is a single-pole antenna of small electrical length. At around the highest frequencies which Tesla used the h/λ ratio was about 0.015. The terminal capacity made the effective height somewhat greater than h, but it still remained an electrically short antenna.

20 *October*

Tesla was measuring the capacity of the coil which he had used for determining the change of capacity of a sphere with height (up till October 9th he had called it a "special coil"). Considering the dimensions of the primary (coil diameter 15 m) and the coil being tested (diameter 64 cm, length 145 cm), the coupling between them was obviously loose, so that the frequency found from the parameters of the primary circuit (provided that the main secondary of the oscillator did not influence the oscillation of the primary) can now be accepted as accurate. It is not stated how resonance was determined, but it was probably from the sparks at the terminals of the test coil. Similar resonance methods are given in recent textbooks on electrical measurements[56]. It must be noted, however, that determination of the distributed capacity of a coil from the resonance of the coil alone is not reliable, it depends on the mode of excitation and always gives lower values. It is therefore recommended to measure it with an added lumped capacity in the circuit.

In a thorough analysis of all details of his measuring apparatus, Tesla did not omit a determination of the parasitic inductance of the connections, by an interesting method which he says he used often in the New York laboratory. Varying the primary inductance and capacitance but keeping a constant frequency of the oscillator (as determined with an auxiliary resonant circuit), one has

$$C_{p1}(L_{p1} + L_{con}) = C_{p2}(L_{p2} + L_{con})$$

where C_{p1}, L_{p1} are the first and C_{p2}, L_{p2} the second capacitance-inductance pair in the primary giving the same frequency. From this equation one can find the parasitic inductance of the connections L_{con}, which Tesla denotes by x.

23 *October*

In further experiments to determine change of capacity with height Tesla uses an apparatus similar to that of the previous day. As far as can be judged, the coupling between the oscillator and the measuring circuit (coil with elevated ball) was loose. The lower terminal of the latter was connected with a condenser of the oscillator circuit. Loose coupling is evidenced by the relatively weak sparks obtained across the air gap of coil L (see figure) in comparison with the sparks obtained when a similar coil was excited by the secondary of the oscillator, tightly coupled to the primary (as for example on October 4th and 5th). Under these conditions the spark oscillator would generate a single frequency, determined by the parameters of the oscillatory circuit with the spark gap.

26 *October*

Tesla had already been using the 689-turn coil for several days in experiments to determine change of capacity with height of a ball. On October 18th he calculated its inductance using the formula for an infinitely long coil. Now he determines it by measuring the current and voltage at a frequency of about 140 Hz, knowing the resistance. He gives the results of two sets of measurements. He is convinced that the second set, for which he used a small dynamometer, gave low values, and this was probably so. The first set gave an inductance slightly less than calculated, but a correction of the theoretical value for the finite D/l ratio* gives a value about 6% less than that measured. Thus the calculated value ought to have been 0.023 H, while the experimental result was 0.024 H. The accuracy of the measurement method cannot now be verified but in view of the small difference between reactance and resistance it is doubtful whether it could be of the order of a few percent.

27 *October*

Tesla does not explain how he made the comparison with a standard 0.5 μF condenser. The number of bottles used in the condenser bank is indeed impressive. He did not

---

* Russell[57] gives the inductance of a coil at very low frequencies as

$$L = (\pi\, Dn)^2\, l\left[1 - 0.424\,\frac{D}{l} + 0.125\left(\frac{D}{l}\right)^2 - 0.0156\left(\frac{D}{l}\right)^4\right]$$

Substituting $\pi D^2 = 4S$ (D is the mean diameter of the coil), and $n = N/l$ (number of turns per cm), the first term in the above equation yields the expression Tesla used. l is coil length. When all quantities are expressed in units of cm, L is also obtained in cm.

carry out measurements on individual bottles to determine what kind of tolerance they had. Since he only measured complete banks, i.e. rather large capacities, and mentions "readings with 7 cells battery", he probably made the comparison in terms of stored charges.

28 October

After several days gathering data and making further measurements of inductance and capacitance, he finally proceeds to the calculation of the unkown capacity of the sphere, using the readings of October 23rd and 26th. Consistent with his general principle, he checked the measured values by (usually approximate) calculations.

He calculates the capacitance of the vertical wire by the formula for an isolated ellipsoid of high eccentricity. It is not known whether Tesla first had the idea of using this formula, but it was used later for a similar purpose[58]. All things considered, the agreement between the calculated and measured values is very good.

Tesla then calculated the ratio of the capacities in the lowest and highest positions. In the lowest position the sphere makes little difference to the total capacity. In the highest position it increases the capacity in the coil circuit by 18.7 cm. This is less than the theoretical value for an isolated sphere of 18″ diameter, which does not agree with some of his earlier measurements (see the results of October 21st for a 30″ ball). However, if a comparison is to be made, it must be noted that the new results are probably better because the apparatus had been modified and the parameters checked.

29 October

His remark about eddy currents in the sphere is interesting. To prevent their formation he slit the tinfoil with a knife. Did he assume that in the vicinity of the coil the sphere would behave like a short-circuited turn? It is readily shown that if this effect is pronounced (and not taken into account) the measured capacity of the sphere will be too low. This might be an explanation for the reduction of the effective capacity of the sphere in the lowest position (see the calculation of October 28th for an 18″ sphere in the lowest position).

1 and 2 November

The new extra coil was larger in diameter but shorter than the previous one (see August 23rd). The formula for an infinite coil introduces a rather large error, but with the correction referred to in the commentary to 26 October the agreement with the experimental results is good. The correction terms are significant because the ratio D/l is even greater than unity, and the correction is more than 30%. The corrected value is 0.0198 H, 2—3% less than the measured values.

3 November

A repeated measurement of the capacity of the vertical wire and the 30″ sphere by the method of October 29th but with a new coil L (see October 31st). A new feature is Tesla's attempt to eliminate the wires of the spark gap, estimating the excitation solely from the streamers.

5 November

Photographs of the Colorado Springs laboratory always show the pole rising from the center of the building. Its dimensions are given in the entry of 17 October. Now Tesla calculates the capacity of the pole as the sum of the capacities of its parts of different thickness, using the formula first cited on October 28th. His final remark indicates that he had thoroughly understood the physical essence of the phenomenon.

6 November

Tesla carefully measured the capacitance of the aerial pole by the resonance method, from the known inductance of the 550-turn coil (see September 8th) and known frequency of the oscillator, with two measurements, one with and the other without the capacity to be measured. He did not make use of his earlier results for the inductance of the regulating coil and connections from October 30th, although he could have done. A calculation check shows that the results of October 30th were rather high (by as much as 10% for $2\,^1/_8$ turns of the regulating coil), but Tesla probably thought that the new procedure was better and so did not use the old results. An analysis shows that if the old values had been used the final result would not have been essentially affected, so that Tesla's conclusion that the measured capacity of the pole was less than the theoretical value of November 5th remains valid.

From the relatively lengthy discussion following the measurements it may be seen that Tesla expected just the opposite. As usual when his expectations were not fulfilled, he considers ways for getting more reliable results.

7 November

Measurement of the capacity of the structure at two frequencies was intended to demonstrate the reduction of effective capacity with increasing frequency. Tesla did in fact obtain a small difference, but it is dubious proof considering the accuracy of the measurements. The frequency difference was quite large, from 50 kHz to nearly 250 kHz (using "extra" and "experimental" coils).

8 November

The primary inductance values cited are from November 5th. The other values given in the table do not agree with those derived from the measurements of October 30th. Also, earlier data do not include values for half a turn of the regulating coil. It must therefore be concluded that the measurements from which the tabulated values were calculated are not described in the diary.

It seems that in measuring inductance from voltage, current, frequency and resistance Tesla had difficulty because of unreliability of the frequency determination. He therefore used the voltage ratio, when it is only necessary for the frequency to be constant. By this method he measured the inductance of the regulating coil plus connections, for various numbers of turns.

9 November

The measurements of mutual inductance in terms of the inductance of the primary when the secondary is open and short circuited are noteworthy. They were made at constant current and frequency, simplifying the calculation.

To reduce the oscillator frequency, in some cases Tesla used two special coils which he refers to only by wire gauge number. He compares the calculated and measured values for these coils. The values measured by the voltage ratio method are about 2% less than those found from voltage, current and frequency. The calculated values are lower than either. Correction of the measured values as described in the commentary to 26 October does not make much difference (about -5%) because the D/l ratio is relatively small.

10 *November*

Had Tesla published the measuring methods he developed in New York and Colorado Springs, his name would probably be frequently encountered in earlier textbooks and handbooks on electrical measurements at high frequencies. As it is, we can only remark his exceptional ingenuity in designing measuring devices and the accuracy with which he determined the resonance of oscillatory circuits. An especially interesting feature is his method using a lamp already heated up by a supplementary power source, greatly increasing its sensitivity to small amplitude changes around the resonance peak of the oscillatory circuit.

11 *November*

In measuring the capacity of a sphere at different heights Tesla here uses a loosely coupled circuit containing a lamp to determine resonance. The results for a 50 ft wire differ somewhat from those of October 28th, but are within the limits of error of the method. The values for the capacity of the sphere are somewhat higher than before, but not in proportion to the diameter of the sphere.

12 *November*

Measurements of the pole capacity, like those of November 7th, but now using a coil with 1314 turns. Resonance was determined by means of a small lamp in series with a coil loosely coupled to the measuring circuit. The value obtained was again similar, so Tesla concludes that it is near the true value of the effective capacity.

13 *November*

Tesla uses an improved method for determining the resonance point, with the light bulb in a dark box for more precise detection of luminance, and determines the capacity of the iron piping once more, obtaining a value about 10% less than in earlier measurements (see November 7th and 12th).

15 *November*

Tesla again measures the capacity of the sphere on top of the metal pole as on November 7th and 12th, but with the secondary coil of the oscillator instead of the earlier "supplementary" coils. The results did not agree with those obtained earlier. Tesla puts this down to the large distributed capacity of this coil, but it would seem that other factors influenced the accuracy as well. Because of the tight coupling between the primary and secondary of the oscillator, it was probably producing a compound spectrum.

Main entrance to Colorado Springs Laboratory in the early phase of development. Tesla is looking through the door (Tesla's own photograph now at the Nikola Tesla Museum, Belgrade)

Capacity measurements made during the period 16—22 November agree on the whole with those made earlier. Tesla does not explain why he repeated similar measurements, e.g. those of November 16th and 18th when he determined the distributed capacity of the supplementary coil and the vertical wire. Nor does he explain why he repeated the measurements of the change of capacity of the sphere with elevation (see November 18th and 20th). He may only have wanted to confirm the earlier results.

On November 17th and 19th he measured the capacity of a vertical wire of various lengths and gauges. From his comments on November 17th it may be seen that at greater lengths he expected some inductive effect. A check of the wavelength, however, reveals that all Tesla's antennas were short in comparison (h/λ of the order of 0.01), so that divergence between the theoretical and measured values cannot be ascribed to an inductive effect.

21 *November*

For some reason which he does not explain, Tesla was interested in the capacity of the same wire when vertical and horizontal, which he measured by the usual resonance method, repeating it with a different capacitance in the primary as a check. Although the results from the two sets of measurements differ appreciably, the value obtained with the wire horizontal was somewhat higher in both cases. The formulae which Tesla used July 24th here yields 54.37 cm for the vertical wire and 58.43 cm for the horizontal. These values agree well with his measurements, especially the first set.

24—26 *November*

To check the values for the inductance in the primary circuit (of the oscillator) which he had earlier measured by the voltage, current and frequency method (see October 30th), Tesla repeats these measurements using the resonance method. He described the procedure on October 21st and made some measurements but did not follow them up with calculations. This time he made both measurements and calculations, but only for one L_pC_p combination. He compares them with values derived from the table given November 8th using linear interpolation. He was probably satisfied with the agreement, and did not make further checks. He had measured the capacity of the same structure, but without the protective cap and using the "extra" and "experimental" coils, on November 7th. On November 12th he had made similar measurements using the 1314-turn coil. In the 26 November entry he refers to the result of 7 November with a new "extra" coil. There is also one more result obtained with an "extra coil", using the best method he had developed for detecting resonance (see 13 November). This result, which differs appreciably from the others, is not mentioned November 26th.

The remark closing this entry suggests the possibility of a systematic error in the determination of resonance, and Tesla emphasizes that it has to be checked.

5 *December*

In this, as in earlier measurements, he found a "reduced inductance of the primary because of the reaction of the secondary". This interpretation of the functioning of the oscillator diverges from Oberbeck's theory[29]. If the spark duration is relatively long the oscillator starts to produce oscillations of two frequencies, and when the spark is broken it gives a third frequency which is determined by the secondary oscillatory circuit. With a third circuit ("extra coil") the oscillation of the system becomes even more complicated, the oscillations during break being determined by the secondary circuit and the "extra coil". Neglecting for the moment the "extra coil", the three frequencies which a Tesla oscillator with tight inductive coupling[31] and equal natural resonant frequencies of the coupled circuits can be expected to produce are

$$\omega_0 = \frac{1}{\sqrt{LC}} \qquad \omega_1 = \frac{1}{\sqrt{LC(1-k)}} \qquad \omega_2 = \frac{1}{\sqrt{LC(1+k)}}$$

where k is the coupling coefficient. Thus ω_1 can be interpreted as the natural frequency of a circuit with capacity C and inductance $L\,(1-k)$. For the primary inductance of Tesla's oscillator (see 9 November) one obtains the "reduced" L, i.e. $L\,(1-k)=23{,}094$ cm; Tesla measured $L=24{,}063$ cm.

6 December

The photographs of the inside of the laboratory show the 100-turn "extra coil" raised above the floor in the center. With this coil Tesla again got similar results for the "reduced" inductance of the primary. However, aware of the indeterminacy of this "reduction", and hence also of the oscillation frequency, he notes that the secondary should be broken at more points when the primary is used as a measuring inductance. This would ensure monochromatic oscillation of the oscillator by reducing the coupling of the primary and secondary (i.e. the circuit of the "extra coil"). The measurements with the secondary eliminated are more reliable, and the accuracy with which the values sought are determined depends mainly on the accuracy to which the inductance and capacitance in the exciting circuit of the oscillator are known.

1 January

Photograph XVII shows lamps connected into a resonant circuit consisting of one square turn. According to the data Tesla gives, one side of the square was about 1.3 m from the secondary coil of the oscillator. The capacity of the oscillatory circuit consisted of two condensers in parallel. The lamps are paralleled.

Tesla calculates the inductance of the square turn from the formula for the inductance of two parallel conductors, as if there were two such pairs connected in series. The formula for a square coil (Fleming, p. 155),

$$L = 8\,l\left(\ln \frac{d}{r} - 0.774\right)$$

yields a value 12.6% less than Tesla found. The calculated resonant frequency is therefore somewhat higher than it should be, so that the inductance of the oscillator primary, as Tesla calculates it, is still less. In fact, because of the tight coupling of the secondary the oscillator must have been producing a complex spectrum, probably with its strongest component at the resonant frequency of the oscillatory circuit of the square coil.

In connection with photographs XVIII—XXI showing the secondary producing intense discharges, Tesla makes an interesting remark about signalling over great distances. Comparing this with other induction apparatuses he had constructed, he concludes that

Interior of Colorado Springs Laboratory

one could expect signals to be picked up at distances of a thousand miles or more, even on the Earth's surface. The diary does not mention any measurements at great distances, but in an article[41] he published soon after finishing work at Colorado Springs he states that he observed effects at a distance of about 600 miles.

2 January

In this entry of 21 pages (the longest in the Notes) Tesla describes 11 photographs.

The explanation to Photograph XXII concerning the transmission of power from the excited primary circuit to the "extra coil" via the earth is similar to that he gave in 1893[6]. The experiment to which the photograph refers was made with the aim of estimating the power of the oscillator from the thermal effect of the HF current. What Tesla calls the "total energy set in movement" would correspond to the total energy transferred to condenser in the secondary (i.e. the power) if an energy of $\frac{1}{2}CV^2$ is transferred in each half-cycle. It can be shown that the active power dissipated in the circuit is much less than this and is inversely proportional to the Q-factor of the oscillating circuit.

The next few photographs show a movable coil which powers light bulbs by means of the high-frequency power which it picks up. One end of the coil is grounded, the other free or just connected to a short piece of wire. The bulbs are inductively coupled to the resonant coil via the auxiliary secondary. Tesla gives no data about the distance of the resonant coil from the oscillator coil.

Tesla's commentary on photograph XXVIII illustrates that he still retained a lively interest in the problem of electric lighting, even after a period of over ten years. His earlier discovery of the luminescence of the gas and not only the filament with HF currents was here again confirmed[5].

In photograph XXVIII the bulb is connected in series with the terminal capacitive load. In the calculation Tesla does not use the "total energy set in movement" but assumes that $\frac{1}{2}CV^2$ of electrostatic energy is consumed in the bulb in each half-cycle. A similar comment applies to photograph XXIV.

Several times Tesla remarks that the principle energy transfer from the oscillating to the receiving coil takes place via the earth. He finds confirmation for this in the experiment described on p. 363 (photograph XXX). He found that the voltage induced in the receiving coil was greatly reduced if the ground connection was broken. It may be that such experiments led him to the conclusion that "transmission" through the earth was a more efficient method of wireless transmission of power than the "inductive method".

Photograph XXXI is an X-ray picture of a finger. Tesla's comments on this experiment illustrate his interest in this type of radiation, already referred to (see the commentary to 6 June 1899).

3 January

After describing some photographs of the laboratory, in the commentary to photograph XLI Tesla explains some transformations of the streamers. He mentions the splitting of streamers near the floor, splitting and reuniting, the phenomenon of luminous parts on the streamers (which he then refers to as sparks), and the breaking up of sparks into streamers and fireballs. His remarks concerning the genesis of fireballs are particularly noteworthy. This phenomenon has been a source of interest since ancient times. Some references to it can be found on Etrurian monuments, in the works of Aristotle, Lucretius and other old sources[63]. Fireballs are considered to be a form of electrical discharge generated during thunderstorms. They are rare in nature, but a fair-sized body of observations has nevertheless been assembled upon which several theories of their origin have been founded. Some hypotheses maintain that fireballs are an optical illusion (an opinion shared by Tesla until he produced them himself), others that they are the traces of meteors. The first genuine scientific approach to the problem was Arago's analysis of some twenty reports of fireballs in 1838. After the publication of his work they became a legitimate subject of scientific interest, but to this day have remained something of an enigma.

A fireball is a luminous sphere occurring during a thunderstorm. Fireballs are usually red, but other colors have also been observed: yellow, green, white and blue. Their dimensions vary, a mean diameter being about 25 cm. Unlike ordinary lightning, fireballs move slowly, almost parallel to the ground. They sometimes stop and change their direction of motion. They can last for up to 5 seconds. Their properties vary greatly from case to case, so that it is believed that there are various types. According to Singer[63] it can be stated that as yet no single theory can explain the occurrence of fireballs in nature.

Despite numerous attempts, only a few types of fireball have been created, and not entirely successfully, in the laboratory. These include the weakly luminescent fireballs generated when ordinary lightning strikes some object. Tesla mentions phenomena of this type several times as the result of sparks or streamers striking wooden objects (see e.g. photograph XL). According to recent theories, fireballs consist of a plasma zone created by electrical discharge. The latest research and calculations by Kapitsa[64] show that the lifetime of a fireball cannot be explained by the energy it receives at the time of genesis, but that it must receive energy from its surroundings. Kapitsa theorizes that this external energy is produced by a naturally created electromagnetic field. The small zone of ionized gas created by the initial lightning or other electrical phenomenon during the storm subsequently expands at the expense of the external electromagnetic field. The diameter of the plasma sphere is determined by the frequency of the external field, so that a resonance occurs. The usual dimensions of fireballs would require that the electromagnetic field have a wavelength of between 35 and 100 cm. According to this theory standing waves created by the reflection of natural electromagnetic waves from the earth would play a certain role. The theory has obtained partial experimental confirmation, but there are still many points on which it is unable to give a satisfactory explanation. It has been found that to maintain a lump of plasma in air requires a power of the electromagnetic field of about 500 W, which is much less than power which can be produced by an electrical discharge. However, too little is known about natural electromagnetic waves to allow any reliable conclusions to be drawn.

Tesla's hypothesis on the origin and maintenance of fireballs includes some points which are also to be found in the most recent theories, but it also bears the stamp of the time. For instance, like Kapitsa, Tesla considers that the initial energy of the nucleus is not sufficient to maintain the fireball, but that there must be an external source of energy. According to Tesla this energy comes from other lightnings passing through the nucleus, and the concentration of energy occurs because of the resistance of the nucleus, i.e. the greater energy-absorbing capacity of the rarefied gas than the surrounding gas through which the discharge passes. In nature the probability of other discharges passing through the nucleus of a fireball is small, so Kapitsa's hypothesis that act via electromagnetic standing waves is more logical. It is possible that in Tesla's experiments the "passage" of a number of later discharges through the same nucleus was more frequent.

7 *January*

This is the last entry in the diary. Apart from the usual description of photographs, Tesla writes about experiments he intends to carry out on his return (where?). He qualifies the experiments to date as satisfactory, considering that his aim was "to perfect the apparatus and make general observations". The apparatus which he was then envisaging for future experiments was to be an improved oscillator which would enable better results than any he had so far obtained.

REFERENCES

(1) N. Tesla, LECTURES, PATENTS, ARTICLES, published by Nikola Tesla Museum, Belgrade, 1956 (from now on: Tesla), "The transmission of electric energy without wires", *Electr. World and Eng.* March 5, 1904, A-153.

(2) Tesla: "Alternating electric current generator", U. S. Patent 447 921, March 10, 1891, Appl. Nov. 15, 1890, P-129.

   "Method of operating arc lamps", U.S. Patent 447 920, March 10, 1891. Appl. Oct. 1, 1890, P-205.

(3) Tesla: "Phenomena of alternating currents of very high frequency", *The El. World*, Febr. 21, 1891, A-3.
   "Electric discharge in vacuum tubes", *The El. Engineer*, July 1, 1891, A-16.

(4) Tesla: "Experiments with alternate currents of very high frequency and their application to methods of artificial illumination", a lecture delivered before the AIEE, May 20, 1891, L-15.

(5) Tesla: "Experiments with alternate currents of high potential and high frequency", a lecture delivered before the IEE, London, Febr. 1892, L-48.

(6) Tesla: "On light and other high frequency phenomena", a lecture delivered before the Franklin Ins. Philadelphia, Febr. 1893, L-107.

(7) Tesla: "On Roentgen rays", *El. Rev.* March 11, 1896, A-27, A-32
   "On reflected Roentgen rays, *El. Rev.* April 1, 1896, A-34
   "On Roentgen radiations", *El. Rev.* April 8, 1896, A-39
   "Roentgen ray investigations", *El. Rev.* April 22, 1896, A-43
   "An interesting feature of X-ray radiations", *El. Rev.* July 8, 1896, A-49
   "Roentgen rays or streams", *El. Rev.* August 12, 1896, A-51
   "On the Roentgen stream", *El. Rev.* Dec. 1, 1896, A-56
   "On the hurtful actions of Lenard and Roentgen tubes", *El. Rev.* May 5, 1897, A-62
   "On the source of Roentgen rays and the practical construction and safe operation of Lenard tubes", *El. Rev.* Aug. 11, 1897, A-69.

(8) Tesla: "Method of intensifying and utilizing effects transmitted through natural media", U.S. Patent, 685 953, Nov. 5, 1901, Appl. June 24, 1899, P-297.

(9) Tesla: "Method of utilizing effects transmitted through natural media", U.S. Patent 685 954, Nov. 5, 1901, Appl. Aug. 1, 1899, P-303.

(10) Tesla: "Apparatus for utilizing effects transmitted from a distance to a receiving device through natural media", U.S. Patent 685 955, Nov. 5, 1901, Appl. Sep. 8, 1899, P-312.

(11) Tesla: "Aparatus for utilizing effects transmitted through natural media", U.S. Patent 685 956, Nov. 5, 1901, Appl. Nov. 2, 1899, P-319.

(12) Eccles, W. H. WIRELESS, Thornton Butterworth Ltd, London, 1933.

(13) Tesla N. "System of transmission of electrical energy", U.S. Patent 645 576, March 20, 1900, Appl. Sept. 2, 1897.

(14) Tesla: "Apparatus for transmission of electrical energy", U.S. Patent 649 621, May 15, 1900, Appl. Sept. 2, 1897, P-293.

(15) Tesla: "System of electric lighting", U.S. Patent 454 622, June 23, 1891, Appl. Apr. 25, 1891, P-208.
   "Electric incandescent lamp", U.S. Patent 455 069, June 30, 1891, Appl. May 14, 1891, P-213.

(16) Tesla: "On electricity", *El. Rev.* Jan. 27, 1897, A-101.

(17) Tesla N. "The stream of Lenard and Roentgen and novel apparatus for their production", Lecture before New York Academy of Science, Apr. 6, 1897 (Nikola Tesla Museum, Belgrade)

(18) Tesla: "High frequency oscillators for electro-therapeutic and other purposes", a lecture delivered before the American Electro-Therapeutic Association, Buffalo, Sept. 13, 1898, L-156.

(19) Fleming, J.A. THE PRINCIPLE OF ELECTRIC WAVE TELEGRAPHY AND TELEPHONY, Third ed. 1916, Longmans Green & Co, London (from now on: Fleming), p. 877.

(20) TRIBUTE TO NIKOLA TESLA, Nikola Tesla Museum, Belgrade, 1961 (from now on: Tribute), Fleming, A. P. M "Nikola Tesla", Jour. of Instit. of Electr. Eng., London, vol. 91, February 1944, A-215.

(21) Wait, J.R. "Historical background and introduction to the special issue on extremely low frequency (ELF) propagation", *IEEE Trans. on Communications*, Vol. **COM-22**, No. 4, April 1974.

(22) Fleming: p. 22.

(23) Hertz, H.R. UNTERSUCHUNGEN ÜBER DIE AUSBREITUNG DER ELEKTRISCHEN KRAFT, dritte auflage, Leipzig, 1914, Johann Ambrosius Barth.

(24) Цверава Г.К. НИКОЛА ТЕСЛА, изд. Наука, Ленинград, 1974.

(25) Tesla: "Means for generating electric currents", U.S. Patent 514 168, Febr. 6, 1894, Appl. Aug. 2, 1893, P-225.
"Method of regulating apparatus for producing currents of high frequency", U.S. Patent 568 178, Sep. 22, 1896. Appl. June 20, 1896, P-228.
"Apparatus for producing electric currents of high frequency and potential", U.S. Patent 568 176, Sep. 22, 1896, Appl. April 22, 1896, P-233.
"Method and aparatus for producing currents of high frequency", U.S. Patent 568 179, Sep. 22, 1896, Appl. July 6, 1896, P-237.
"Apparatus for producing electrical currents of high frequency", U.S. Patent, 568 180, Sep. 22, 1896, Appl. July 9, 1896, P-241.
"Apparatus for producing electric currents of high frequency", U.S. Patent, 577 670, Feb. 23, 1897, Appl. Sep. 3, 1896, P-245.
"Apparatus for producing currents of high frequency", U.S. Patent, 583 953, June 8, 1897, Appl. Oct. 19, 1896, P-249.

(26) Tesla: "Electrical transformer", U.S. Patent, 593 138, Nov. 2, 1897, Appl. March 20, 1897, P-252.

(27) Tesla: "Electric circuit controller", U.S. Patents:

609 251, Aug. 16, 1898, Appl. June 3, 1897, P-256.
609 246, Aug. 16, 1898, Appl. Febr. 28, 1898, P-272.
609 247, Aug. 16, 1898, Appl. Mar. 12, 1898, P-276.
609 248, Aug. 16, 1898, Appl. Mar. 12, 1898, P-279.
609 249, Aug. 16, 1898, Appl. Mar. 12, 1898, P-282.
613 735, Nov. 8, 1898, Appl. Apr. 19, 1898, P-285.
"Electrical circuit controller", U.S. Patents:
609 245, Aug. 16, 1898, Appl. Dec. 2, 1897, P-262.
611 719, Oct. 4, 1898, Appl. Dec. 10, 1897, P-267.

(28) Tesla: "Electrical oscillators", *El. experimenter*, July, 1919, A-78.

(29) Oberbeck A. "Ueber den Verlauf der Elektrischen Schwingungen bei den Tesla'schen Versuchen", *Wied. Ann. der Physik*, 1895, vol. **55,** s. 623.

(30) Fleming: p. 792.

(31) Fleming J.A. and Dyke G.B. "Some resonance curves taken with impact and spark-ball dischargers", *Proc. Phys. Soc. London*, vol. **23**, Feb. 1911, p. 136.

(32) Попов А. С. "Прибор для обнаружения и регистрирования электрических колебаний", *Журн. русского физ.—химич. об-ва*, 1896, т. **22**, *4. физич.*, отд. 1, вып. 1.

(33) Fleming: p. 513.

(34) Galeys J. TERRESTRIAL PROPAGATION OF LONG ELECTROMAGNETIC WAVES, New York, Pergamon Press, 1972.

(35) TESTIMONY U.S. District Court, New York, Oct. 3, 1916. Samuel, M. Kintner and Halsey M. Barrett vs. Atlantic Communication Comp.

(36) Hawks E. PIONEERS OF WIRELESS, London, Methuen Co. Ltd, 1927. p. 205.

(37) Tesla: "Incandescent electric light", U.S. Patent 514 170, Febr. 6, 1894, Appl. Jan. 2, 1892, P-216.
Also see ref. (15).

(38) Tesla: "System of signaling", U.S. Patent 725 605, Apr. 14, 1903, Appl. July 16, 1900, P-337.
"Method of signaling", U.S. Patent 723 188, Mar. 17, 1903, Appl. June 14, 1901, P-352.

(39) Fleming: p. 287.

(40) Born. M. and Wolf E. PRINCIPLES OF OPTICS, Third ed. Pergamon Press, 1965, Oxford, p. XXVI.

(41) Tesla: "The problem of increasing human energy", *The Cent. Illustr. Mon. Magazine*, June 1900, A-109.

(42) Tesla: "Art of transmitting electrical energy through the natural mediums", U.S. Patent 787 412, Apr. 18, 1905, Appl. May 16, 1900, P-331.

(43) ОЧЕРКИ ИСТОРИИ РАДИОТЕХНИКИ, изд. Академия Наук СССР, Москва, 1960.

(44) Tesla: "Apparatus for transmitting electrical energy", U.S. Patent 1 119 732, Dec. 1, 1914, Appl. Jan. 18, 1902, P-357.

(45) Pocklington H.C. "Electric oscillations in wires", *Proc. Camb. Phyl. Soc.* Oct. 25, 1897, vol. **ix**, p. 324.

(46) Zenneck J. LEHRBUCH DER DRAHTLOSEN TELEGRAPHIE, Verlag, Stuttgard, 1915.

(47) Fleming: p. 467.

(48) Fleming: p. 483.

(49) Calzecchi — Onesti T. "Sulla conduttivita elettrica delle limature metalliche", *Nuovo cimento*, 1884, v. **16**, p. 58, 1885, v. **17**, p. 38 (Pisa).

(50) Tesla: see ref. (25), P-225.

(51) Tesla: see ref. (25), P-228, P-233.

(52) Tesla: see ref. (25), P-237, P-241.

(53) Tesla: see ref. (25), P-245, P-249.

(54) Jackson J.D. CLASSICAL ELECTRODYNAMICS, John Wiley, 1975, New York.

(55) Fleming: p. 706.

(56) Terman F. E. and Pettit J. M. ELECTRONIC MEASUREMENTS, McGraw Hill, New York, 1952.

(57) Russell A. "On the magnetic field and inductance coefficients of circular cylindrical, and helical currents", *Phyl. Mag.* April, 1907.

(58) Fleming: p. 637.

(59) Tesla: "Method of and apparatus for controlling mechanism of moving vessels or vehicles", U.S. Patent 613 809, Nov. 8, 1898, Appl. July 1, 1898, P-363.

(60) Maxwell J.C. "A dynamical theory of the electromagnetic field", *Phyl. Trans. Roy. Soc.*, 1865, vol. **155**, p. 419.

(61) Erskine-Murray J. A HANDBOOK OF WIRELESS TELEGRAPHY, Crosby Lockwood, London, 1913, chap. XVII.

(62) Маринчић А. "Тесла — один из основоположников современой электротехники", *Вопросы истории естествознания и техники*, Акад. Наук, СССР, Москва, 1976, Выпуск 2 (**55**)

(63) Singer S. THE NATURE OF BALL LIGTNING, Plenum Press, 1971, New York.

(64) Капица, П.Л. "Шаровая молния и радиоизлучение линейных молний". *Жур. тех. физики*, **88**, 1829, (1968)

(65) United States Reports, vol. 320, Oct. 1942, Oct. 1943, Washington, MARCONI v. s. U.S.

(66) Bowers B. X-RAYS, Science Museum, London, 1970.

(67) Tesla N. "Electrical condenser", U.S. Patent 567 818, Sep. 15, 1896, Appl. June 17, 1896.

(68) Testimony in behalf of Tesla, Interference No. 21,701, United States Patent Office, New York, 1902.

(69) Letter of Tesla to Morgan J.P. Jan. 9, 1902. (Nikola Tesla Museum in Belgrade).

(70) Tesla: "Apparatus for the utilization of radiant energy", U.S. Patent 685 957 Nov. 5, 1901, Appl. Mar. 21, 1901, P-343.
"Method of utilizing radiant energy", U.S. Patent 685 958 Nov. 5, 1901, Appl. Mar. 21, 1901. P-348.

(71) Wheeler L.P. "Tesla's contribution to high frequency", *Electr. Engineering*, New York, August 1943, p. 355, also Tribute: A-211.

(72) Wait J.R. "Propagation of ELF electromagnetic waves and project Sanguine/Seafarer", *IEEE Jour, of Ocean. Eng.*, Vol **OE-2**, No. 2, April 1977.

(73) Morrison J.F. and Smith P.H. "The shunt-excited antenna", *Proc. IRE*, 1937, v. **25**, No. 26.

www.ingramcontent.com/pod-product-compliance
Lightning Source LLC
Chambersburg PA
CBHW082001190326
41458CB00010B/3040